The Revenge of Athena

Science, Exploitation and the Third World

The Revenge of Athena

Science, Exploitation and the Third World

Edited by *Ziauddin Sardar*

Mansell Publishing Limited
London and New York

First published 1988 by
Mansell Publishing Limited, *A Cassell Imprint*
Artillery House, Artillery Row, London SW1P 1RT, England
125 East 23rd Street, Suite 300, New York 10010, U.S.A.

British Library Cataloguing in Publication Data

The Revenge of Athena: science, exploitation and the Third World.
1. Developing countries. Society. Effects of technological development.
I. Sardar, Ziauddin
303.4'83
ISBN 0-7201-1891-3

Library of Congress Cataloging in Publication Data

The Revenge of Athena: science, exploitation, and the Third World/edited by
Ziauddin Sardar.
        p.   cm.
Partly based on a Consumer Association of Penang (CAP)
seminar, held 21-26 Nov. 1986 in Penang, Malaysia.
Includes bibliographies and index.
ISBN 0-7201-1891-3: $60.00 (U.S. : est.)
1. Science—Developing countries.   2. Science—Developing countries—
International cooperation.   3. Science—Philosophy.   4. Science—Social aspects.
I. Sardar, Ziauddin.
Q127.2.R48   1988
303.4'83'091724—dc19   88-19216 CIP

This book has been printed and bound in
Great Britain. Typeset in English Times by
Colset (Private) Ltd., Singapore, and printed
and bound by the Camelot Press, Southampton,
on Solent Wove paper.

for Uncle Idris

Contents

Contributors

Claude Alvares, an independent journalist and radical philosopher, is a frequent contributor to *Indian Express* and the *Illustrated Weekly of India* and is the author of the highly acclaimed *Homo Faber: Technology and Culture in India, China and the West, 1500 to the Present Day*, 1979.

Munawar Ahmad Anees is a biologist and Director of Research and Development at East-West University, Chicago. He is the author of *Guide to Sira and Hadith Literature in Western Languages*, 1986, and the forthcoming *Islam and Biological Futures*.

J.K. Bajaj is Assistant Editor of *Indian Express*, New Delhi.

J. Bandyopadhyay is attached to the Research Foundation for Science, Technology and National Resource Policy, India.

David Burch is Lecturer at the School of Science, Griffith University, Australia.

Merryl Wyn Davies, anthropologist and journalist is the author of *Knowing One Another: Shaping an Islamic Anthropology*, 1988.

Glyn Ford, formerly with the Department of Liberal Studies on Science, University of Manchester, now represents Greater Manchester in the European Parliament.

Susantha Goonatilake is President of the Sri Lanka Association for the Advancement of Science. His books include *Aborted Discovery: Science and Creativity in the Third World*, 1984, and *Crippled Minds: An Exploration into Colonial Culture*, 1982.

D.L.O. Mendis is President of the Institution of Engineers, Colombo, Sri Lanka.

Seyyed Hossein Nasr is Professor of Islamic Studies at the George Washington University, Washington, D.C., and is the author of *Science and Civilization in Islam*, 1986, *Islamic Science: An Illustrated Study*, 1976, and many other books.

Khor Kok Peng is Research Director of the Consumers' Association of Penang, Malaysia.

Alejandro Gustavo Piscitelli is Assistant Professor in the Sociology Department, Universidad Nacional de Buenos Aires, Argentina.

Jerome R. Ravetz, formerly Reader in the Department of History and Philosophy of Science, University of Leeds, is the author of the classic study, *Scientific Knowledge and its Social Problems*, 1972. He now works as an independent consultant on risks.

Amulya Kumar N. Reddy is Director of the Indian Institute of Science, Bangalore, India. He edited *Rural Technology*, 1980, and authored many papers on alternative technology.

Ziauddin Sardar is Director of the Center for Policy and Future Studies, East-West University, Chicago, and is the author of *Science, Technology and Development in the Muslim World*, 1977, *Science and Technology in the Middle East*, 1982, and several other books.

Dhirendra Sharma, India's foremost expert on nuclear risks, teaches at the Jawaharlal Nehru University, New Delhi. Author of numerous books on the Indian nuclear industry, he edits the quarterly journal, *Philosophy and Social Action*.

V. Shiva is attached to the Research Foundation for Science, Technology and National Resource Policy, India.

Rakesh Kumar Sinha is at the Institute of Technology, Benares Hindu University, India.

M.D. Srinivas is Director of PPST Foundation, Madras, India.

Lawrence Surendra is Co-ordinator of the Asian Regional Exchange for New Alternatives in Manila, Philippines.

G.K. Upawansa is a consultant at the National Engineering Research and Development Centre of Sri Lanka.

Introduction
The Revenge of Athena

Ziauddin Sardar

When the ancient Greek philosophers were laying the foundations of 'civilization' as we know it, the people of Athens turned to a particular goddess for
intellectual and social guidance. She was Athena, the daughter of Zeus — master of gods and men throughout the whole of the Greek world — from whose
head she sprang fully armed. Athena personified the Hellenic ideal, being the
goddess of both war and reason. At a very early date Greek artists endowed her
with attributes which made her easily recognizable at first sight: a helmet, lance,
and, in particular, a shield of goat-skin on which the petrifying head of the
Gorgon was attached. So revered was Athena that in the fifth century BC the
Parthenon was built in her honour. Inside the temple, people worshipped a
forty-foot ivory statue of the goddess dressed in gold. Her right hand held a
statue of Nike, goddess of victory, her left hand rested on a twenty-foot shield.

As much of contemporary western civilization draws its inspiration from the
Greeks, so Athena continues to represent the intellectual and social ideal of our
time. She is best personified by modern science where reason and war fuse to
produce a violent enterprise. Modern science — by which I mean science as it is
practised today with its origins in the seventeenth-century European Enlightenment — is based on the extreme use of reason directed towards the extreme use
of violence.

Modern, western science incorporates a fundamentalist attitude to reason: it
is a tool of reduction with an essentially exclusionist methodology, and its use is
limited strictly within an ontological and epistemological framework. Reason is
exclusive in the sense that there is no place in science for issues of morality or
values for it is pure, clinical and neutral; only those aspects of a phenomenon
which are amenable to pure reason are really worthy of investigation. It is
exclusive as only those who have been specially trained in the use of scientific

1

reasoning have the right of access to knowledge and are the true judges of what constitutes scientific knowledge. And finally, it is exclusive in that reason constitutes the only legitimate way of knowing and is the only arbitrator of truth. As a tool of reduction, the use of reason in modern science is based on the theological belief that all phenomena can be reduced ad infinitum, that all systems can be broken into smaller and smaller components, and components of a system consist of discrete and atomized parts, that all systems operate on the same mechanical processes, and that it is possible to know the whole system by studying the components. Material objects are reducible to sense-data; mental events and processes are reducible to physiological, physical or chemical events and processes in the brain; social structures and social processes are reducible to relationships between actions of individuals; biological systems are reducible to physical systems; philosophy is reducible to analysis; mathematics is reducible to logic; and what is not reducible is irrelevant. The ediface of modern science is built upon this exclusivist and reductive use of reason. By raising reason to the level of a god, by exclusively limiting its use to a particular methodology by denying the existence of all other forms of knowing, western science has taken a fundamentalist position which can be defended only by declaring war on everything and everyone else.

That western science is a theology of violence — with its own belief system, priesthood and temples — was announced at its inception; Francis Bacon's dictum that nature gives up her secrets under torture has been its motto. But the violence intrinsic in western science is not limited to nature. It is directed towards people and their built and natural environments, towards the lifestyles, culture and the modes of knowing, doing and being of those who live outside the borders of the Hellenic ideal — the contemporary 'barbarians'. Moreover, it manifests itself in extreme forms; consider, for example, the fact that over 80 per cent of all scientific research is devoted to the war industry aimed at a scale of violence that could destroy the earth several times! The war-like nature of modern science, as Vandana Shiva has pointed out, manifests itself in four distinctive ways:

1. Violence against the subject of knowledge. It is perpetuated socially through the sharp divide between the expert and the non-expert — a divide which converts the vast majority of non-experts into non-knowers even in those areas of life in which the responsibility of practice and action rests with them. But even the expert is not spared: the fragmentation of knowledge converts the expert into a non-knower in fields of knowledge other than that of his or her specialization.

2. Violence against the object of knowledge. This becomes evident when modern science, in a mindless effort to transform nature without a thought for the consequences, destroys the innate integrity of nature and thereby robs it of its regenerative capacity. The multidimensional ecological crises all the world over are eloquent testimonies of the violence that reductionist science perpetrates on nature.

3. Violence against the beneficiary of knowledge. Contrary to the claim of modern science that people are ultimately the beneficiaries of scientific knowledge, the people — particularly the poor — are its worst victims: they are deprived of their life-support systems in the reckless pillage of nature. Violence against nature recoils on man, the supposed beneficiary of all science.

4. Violence against knowledge. In order to prove itself superior to alternative modes of knowledge and be the only legitimate mode of knowing, reductionist science resorts to suppression and falsification of facts and thus commits violence against science itself which ought to be a search for truth.[1]

The Revenge of Athena explores just how science perpetuates violence against the people, societies, economies, environments, traditions, cultures, ontologies and epistemologies of the Third World; and what possibilities the Third World can itself develop to meet the challenge of western science. The book grew out of a Consumer Association of Penang (CAP) seminar entitled 'The Crisis in Modern Science' held during 21–26 November 1986 in Penang, Malaysia. While not all the papers presented have been included in this volume, a number have been added to give a focus to the book.

CAP is one of the most noted and active non-governmental organizations of the Third World.[2] Its seminars, an annual occurrence, are attended by scholars, scientists, journalists, activists and intellectuals of all shades of opinions and background and are renowned for their radical stands. Whatever the subject under discussion, whether the environment, development, the media or science and technology, the emphasis is always on what the Third World itself can do to improve its situation, how developing countries can free themselves from the spiral of underdevelopment. But CAP seminars are not simply occasions for debate and discussion; they often have a far reaching impact. The proceedings of the seminars are frequently used for teaching purposes in Malaysian universities, and CAP often uses the resolutions and recommendations of its seminars as a springboard to launch consciousness-raising or reformist campaigns directed towards communities, industries and local and national governments. Many of these campaigns have been successful and have resulted in reforms that have benefited the consumer, rural communities and the poor in Malaysia.

'The Crisis in Science' seminar for 1986 focused on the impact of science and technology on Third World societies and possible alternatives that the Third World can provide to meet the crisis. It is not surprising then, that the contributors to this volume start with the assumption that there is a crisis both in and of science: Part One, 'What is Wrong with Science?', is devoted to an analysis of this crisis.

That the crisis of science may be a component of the ideological crisis of western civilization has long been argued by Marxist philosophers, historians and critics of science. Indeed, much of our contemporary awareness of the political and ideological dimensions of science is due to the work of Marxist scholars in this field. Glyn Ford's essay, therefore, provides an ideal starting point for the discussion by surveying the radical science movement and its critique of the ideological dimension of science. He traces the inception of the Marxist critique of science to 1931 when the Russian scholars attending the Second International Congress of the History of Science in London startled the gathering by arguing that science is intrinsically linked to ideology: 'It is done in a particular social order and reflects the norms and ideology of that order.' The Congress provided the spark for the writings of J.D. Bernal, J.B.S. Haldane and Joseph Needham who influenced a whole generation of Marxist scholars right down to the Radical Science Journal Collective. The Marxist perspective on science, as Ford points out, is of particular relevance to the Third World; after all, it is concerned 'to ensure that all the people in society receive the full benefits of science and technology'.

But how realistic is it to assume that modern science and technology, even when stripped of all their ideological layers, will prove to be beneficial? Is the production of destructive side-effects inherent in science? Jerome R. Ravetz argues that along with knowledge science is also increasing our ignorance. Ravetz starts by pointing out that the carefree days when science produced facts, either in its own pursuit or in response to social problems, are over. The theme of choice has been appreciated as being vital to the direction of science and technology. Science is now big business, and technology cannot depend on an automatic mechanism of a market to turn inventions into successful innovations. In each case there must be policy enabling direction to be given, and choices to be made, in accordance with general strategic objectives. And it is institutions necessarily and closely aligned with the general political/economic structures of the society which shape this strategic objective. Where does that leave the objectivity of science? Ravetz points out that objectivity is not guaranteed by the materials or the techniques of science, but emerges partly as a result of the integrity of individuals and partly from open debate on scientific results.

Ravetz's thesis is that not just scientific knowledge but also our ignorance is socially constructed. In addition, there are a growing number of scientific problems, solutions to which are crucial for our survival, that cannot be solved, either now or in any planned future. We cannot predict, for example, when, or even whether, the Earth's mean temperature will rise by two degrees centigrade due to an increasing carbon dioxide content in the atmosphere. Yet this prediction can be cast as a scientific problem for which there are both empirical data and theoretical models. Such insoluble problems are increasing. Indeed, the problem that faces us now is that the sum of knowledge and power is insufficient for the preservation of industrial civilization. 'Scientific ignorance' is paradoxical in itself and directly contradictory to the image and sensibility of our inherited style of science and its associated technology. Scientists, urged on by politicians, now exist in the world of pure fantasy as in the case with 'Star Wars'.

In 'Radical Sociology of Science: From Critique to Reconstruction', Alejandro Gustavo Piscitelli takes the arguments further. He insists that the fusion between science and politics comes about 'neither from a perversion of the post-Renaissance scientific ideal, nor from a corruption of the scientific ethos lured by the temptations of industry and consumerism. Politics instantiates possibilities in modern science which were previously hidden.' In other words, there is something inherent in modern science that will always make it amenable to political and ideological manipulation; given its present status science cannot be isolated from political and ideological influences. Piscitelli notes that the Popper–Kuhn debates are identical to those which have gone on for some two hundred years in the realms of political, social, economic, ethical and legal theory. (And, before that, for over eight hundred years in Muslim civilization in the form of Asharite and Mutazalite clashes.) The clash between Popper and Kuhn, between the Mertonian and the anti-Mertonians, is almost a pure case of the opposition between the Romantic and the Enlightenment ideologies.

The way forward, Piscitelli argues, is the notion of self-organizing paradigm (SOP) presented by the Chilean biologist Francisco Varela. SOP seeks to escape the 'polarized concepts that make reference to a privilege direction in the flow of time. The concepts strike different emotional chords whether we are inclined to espouse casual (events "pushed" by the past and headed towards disorganization) or teleological (events "pulled" by future explanations and adding order to an otherwise inert universe) explanatory models'. The most important epistemological consequences of SOP is that 'the complexity of systems is much more a function of the observing systems than that of the observed system. It is impossible to understand the social behaviour of man without taking into account the fact that knowledge and behaviour have both a biological and social basis and that observation must be inscribed in a system of alternative perspectives'.

When it comes to tackling the problems of traditional societies, modern science certainly casts its observations in a perspective that is quite alien to the people, environment and needs of the Third World. Part Two assesses the impact of science and technology on the Third World. As J. Bandyopadhyay and V. Shiva point out in their essay on exploitation of the natural resources, when science arrives in a traditional society it takes control over resources out of the hands of indigenous peoples and local communities and puts them into the hands of a minority. The 'experts' play a critical ideological role in this transfer by creating a knowledge system which produces epistemological conditions for transfer of control; by devising technological systems which divert or destroy resources for commodity production; and by constructing a legitimization system when the transfer of control is challenged. Yet, traditional people, as those in the Chipko movement, have consistently shown themselves to be the more ecologically sound users of natural resources.

Within the science community itself control rests in the hands of a chosen few. As Dhirendra Sharma illustrates in his article on 'How Indian Atomic Energy Policy Thwarted Indigenous Scientific Development', before partition Indian science was on a satisfactory course. But after independence, it was largely controlled by a single person, Homi J. Bhabha, who developed the Indian atomic energy plan at the expense of all other areas of science. Throughout the developing world science is controlled by a few individuals who approximate to the role models of western gurus of science; it is western scientists, therefore, who directly or by proxy, control science in the Third World.

But control in science is not limited to external mechanisms. There are internal movements to ensure that control remains in white, male, elitist hands. In his survey of 'Sex, Race and the Nèw Biology', Munawar Ahmad Anees demonstrates how reductionism in biology has taken this most fundamental of sciences 'straight into the black-hole of determinism'. Biological determinism is presented as biological inevitability. Under this blanket, almost any ideological, political and racist claims can be presented as pure science. Thus, the emergence of sociobiology — 'the systematic study of the biological basis of all social behaviour' — and new reproductive technologies now threaten to transform radically the fundamental attributes of human life. But to whose benefit would the new transfer of human attributes accrue? Anees has no doubts. Indeed, he seems to argue that the methodology of biology itself has an inbuilt mechanism which leads it towards 'proving' that white males are the best product of evolution and hence the natural masters of the world.

Just as western medicine has taken control from women over their own bodies, including the natural process of birth, so also has contemporary technology destroyed indigenous methods of solving problems and replaced tried and trusted solutions with high-technology modern techniques. As Rakesh Kumar Sinha argues in 'Science and Efficiency: Exploding a Myth', modern technology operates on the 'logic' of centralization of production and concentration of economic and political power. But is this the most efficient way of doing things? Sinha demonstrates that modern technology, contrary to popular belief, is a rather inefficient and wasteful enterprise. Indeed, when compared with traditional technologies, for example agricultural techniques and steel making, they turn out to be rather inferior.

We can say the same about medicine. As I argue in 'Medicine and Metaphysics', the western medical establishment systematically destroyed Islamic medicine in India, Egypt and Tunisia throughout the colonial period. The idea was not that a superior system of medicine would serve the colonies, but that a particular world-view and life-style should be established. Both diseases and illnesses, as well as systems of medicine, are products of world-views. Modern medicine is completely true to the world-view of its origins: reduction is its methodology, capitalism is the dominant mode of production, power and control is its prime goal, violence is its eventual outcome and an endless quest for meaninglessness is its ultimate direction. Such a system of medicine cannot serve the health needs of a traditional people. Even after over two centuries of suppression, it is the traditional systems of medicine which are serving the rural populations of India and Pakistan, Bangladesh and Malaysia, Egypt and the Sudan. These systems of medicine may not be able to deal completely with the diseases and illness of modern lifestyles — for they have their being in different systems of thought — but when it comes to meeting the health needs of rural communities and the urban poor, they have again and again shown themselves to be far superior to their modern counterpart.

The suppression of traditional medicine and import of modern medicine in the Third World is intimately connected to poverty. As Claude Alvares shows in 'The Redundancy of Drugs', in a typical developing country some 5,000 patented drugs may be imported. But only about 200 of these can be regarded as essential drugs making a positive contribution to the health of the society. The rest are either purely ineffectual or harmful — all are a major drain on foreign exchange and a source of fat profits for the pharmaceutical multinationals. Moreover, traditional medicine can provide cheap and effective alternatives for most essential drugs. The Third World therefore has little use for the western medical system which has epistemologically removed society from the domains of medicine. Western medicine is not concerned with promoting health in developing countries, but with underdeveloping them, with waging war on disease which it sees as a commodity, and with safeguarding the interests of the privileged and the powerful. The replacement of health-orientated traditional medicine in the developing countries with profit-motivated, high-technology western medicine has played havoc with health care systems in the Third World.

But even where science has specifically sought to be benevolent it has managed only to destroy the traditional strengths of developing countries. The Green Revolution was specifically designed to increase agricultural productivity and thereby reduce malnourishment and hunger in the Third World. As J.K. Bajaj demonstrates so brilliantly in his detailed analysis of the impact of the Green Revolution on Indian agriculture, it has managed to devastate Indian agriculture, reduce agricultural productivity and increase hunger. It produced seeds which benefited only a selected group of people: 'no revolutionary improvement in the production and productivity of Indian agriculture as a whole occurred with the so-called Green Revolution. If anything happened, it was that the rates of growth of Indian agriculture declined. What looked like a revolution was merely a spurt in the growth of a few commercially important foodgrains in a few areas which were already surplus. This growth too was achieved at a very high cost of resources, and at a cost of an enormously enhanced dependence of agriculture on external, often imported, inputs. The increased costs pushed up prices all round and made the subsistence farmers — who were not protected by inputs subsidies and were not helped by higher output prices, since in any case they had no surpluses to sell — even more improvished. The yields in those subsistence farms consequently seem to have declined below the pre-Green Revolution levels.' The scientific solution to the food problem in the Third World has thus contributed to aggravating the hunger of the poor.

The Green Revolution and the new seed technologies also have a much more sinister side. The end product of the Green Revolution is not just the destruction of Third World agriculture, decimation of traditional agricultural practices, and transfer of arable land into vast wastelands, it also has serious consequences for the Third World's future supply of food. Lawrence Surendra, in 'Plant Genetic

Resources and the Impact of New Seed Technologies', points out that biotechnology and agricultural research networks — centred around such institutions as The International Rice Research Institute (IRRI) in the Philippines — are being used to transfer invaluable germplasm from the developing countries to the rich North. Surendra communicates a chilling warning: 'Biotechnology as yet does not create new genes, it mutates existing ones. This means that seed germplasm has to be found wherever it is located. This of necessity has involved gene drain from the South to the North, and affects the world's pool of PGRs [plant genetic resources] in two distinct ways. First, successful mutations of genes and the large-scale use of new varieties adversely affect the existing plant varieties in nature. Sometimes the effects are devastating in their reach and plant varieties can simply disappear . . . The second consequence of the South–North gene drain demands very serious attention and action. This is the heavy germ plasm losses caused by commercial plant breeders and seed multinationals who plunder the germplasm of the South but do not use it at all or preserve it. Private firms exercise ''life and death'' powers over germplasm under their collection and storage . . . As we look to new plants to feed humanity in the future, control over major crop germplasm also could become a form of political control. About 55 per cent of collected germplasm is with the North.'

It seems then that biotechnology, often described as a boon for mankind, is set to increase further the dependency of the Third World. Can the Third World rely on the industrialized countries, on the transfer of technology, to get itself out of the ever increasing spiral of dependency? Not likely, says David Burch. His examination of the 'Trends and Outcomes of the Transfer of Technology in the 1980s', which takes a detailed look at the British aid policies during the eighties, leads him to the conclusion that aid programmes will repeat what has happened in the past: they will destroy local capabilities to the benefit of local and foreign interest groups. Is there anything in western science and technology that can remotely benefit the Third World, help it out of its present impasse? With the aid of economic models, Khor Kok Peng illustrates in 'Underdeveloping the Third World' that, given the present structure, process and system of science, western science can never meet the basic needs of the Third World or solve any of its pressing problems. Indeed, present trends and priorities indicate that the situation can only continue to get worse.

All this amounts to a savage and devastating indictment of western science. But these charges, despite what some critics may argue, are not based on any anti-science feeling or sentiment; they are the result of decades of experience, observations, analysis and reflections. If they do have a base in sentiments it stems from the fact that there is something intrinsic in western science that reduces a traditional people into a state of helplessness, ridicules their world-view and way of knowing, attacks everything they hold sacred. As Indian philosopher and social worker, S.N. Nagarajan pointed out from the floor during the CAP seminar, 'So far as modern science is concerned, there is nothing sacred. The idea of sacredness is nothing more than weak sentimentalism. If you declare something sacred, how can you dissect it? And if you cannot dissect it and destroy a thing how can you know it? So if you seek genuine objective knowledge you have to forget and reject the very notion of sacredness. So nothing is holy, nothing is sacred. It is said that these sentiments are the biggest obstacles for true knowledge. But are they not the idols that hide the truth?'

Nagarajan went on to ask: 'What does western science tell us, the traditional people of the world? It tells us that:

1. Your crafts are useless.
2. Your crops and plants are useless.
3. Your food is useless.
4. Your houses are useless.
5. Your cropping patterns and agricultural patterns are useless.
6. Your education is useless.
7. Your knowledge is useless.
8. Your religion and ethics are absolutely useless.
9. Your culture is useless.
10. Your soil is useless.
11. Your medical system is useless.
12. Your forests are useless.
13. Your irrigation system is useless.
14. Your administration is useless.
15. You are finally a useless fellow.

Accept what science and scientists tell you, obey their dictates. That is what God Almighty has ordained. Modern biology will finally generate human bulls to produce at least some half-breeds which may be better.'

Nagarajan's accusation may be enveloped in sentimental terms, but its contents cannot be disputed. Western science has tried to write off the entire corpus of traditional thought, downgraded traditional lifestyles, and devastated traditional modes of existence. Just as it has declared war on nature and environment, it has performed unforgivable violence on traditional world-views and those who move within them. In the name of reason, Athena has exacted an horrendous revenge on the people of the Third World for adhering to non-Hellenic worldviews.

Accusations apart, can the Third World offer positive substitutes to western science? In Part Three we explore the possibilities of indigenous science in the Third World.

It is worth emphasizing that we are not looking for 'alternatives' to western science. The notion of alternative assumes that there is a norm which, by definition, is superior by virtue of the fact that it is the norm. Alternative acknowledges the existence of an external frame of reference, an external yardstick, by which the new possibilities are judged and measured. We are looking for other non-western systems of science: these modes of doing science may be just as objective, rational and universal as the western mode of doing science, however they would draw their legitimacy not from the criteria of objectivity defined by western civilization but from their own world-views.

There is another reason for not labelling non-western sciences as alternative. Conventionally, alternative movements in science and technology have focused on the end results. The search for alternatives has been based on the argument that it is not science but the use to which science has been put that is perverted. Scientific knowledge can be employed to achieve quite different goals, it can be put to use for military or peaceful purposes, inhuman or human ends. The processes of nature are blind, scientific laws describe well-defined, constant relationships between certain variables. What is needed is to use the knowledge of the laws of science and processes of nature to promote ecologically sound practices, conservation of resources and bring the benefits of science to all of mankind. Such arguments have firmly set the alternative debate within political boundaries. What is being sought is an alternative political use of technology; this is why the alternative movement has concentrated solely on the creation of alternative technology.

But this argument overlooks a fundamental observation. Everywhere on the planet, in the industrialized democracies of the West or in the peoples' dictatorships of the eastern bloc, despite ideological variations or differences in political systems and institutions of decision making, science has generated the same basic problems: alienation, wasteful consumption, pollution, suppression of traditional practices, domination and control of what are seen as 'non-scientific' cultures and people. Alternative use would no doubt focus on alternative political and social goals, but it would leave scientific knowledge essentially the same, the reductive and violent nature of science would continue unabated.[3]

What many Third World scholars are seeking is not an alternative within the world-view of western civilization, but a way of knowing and perceiving the external world and solving problems which has its bearing in non-western epistemologies, in traditional world-views. We are thus looking for sciences which are different in nature, style, characteristics and contents. This is not a political but an epistemological goal.

Contributors to Part Three all challenge the basic premises on which modern science and technology are based. Every scientific theory is an attempt at answering a definite set of questions. These questions make sense if certain assumptions about nature and reality, time and creation, are taken *a priori* and accepted unconditionally. These assumptions are the epistemological starting point for research. Western science for example is based a mechanistic world-view, on the supposition that the physical universe is the prime reality, reduction is the sole analytical tool, and that man is superior to and removed from nature. The approach of western science makes sense only when one accepts these assumptions; but scholars from non-western traditions do not and consequently think that modern science is not asking the right questions.

The basic assumptions of modern science are also rejected by many western scholars developing new paradigms of thought. Fritjof Capra, for example, rejects the mechanistic framework formulated by Descartes, Newton and Bacon and the associated methodology of reduction that goes with it. In *The Turning Point*[4] he shows that the old paradigm belief that in complex systems the dynamic of the whole can be understood from the properties of the parts is now untenable. In the new paradigm, the relationship between the parts and the whole is reversed. The properties of the part can be understood only from the dynamics of the whole. Ultimately, there are no parts at all. What we call a part is merely a pattern in an inseparable web of relationships. In the mechanistic world-view, it is thought that there are fundamental structures, and then there are forces and mechanisms through which these interact, thus giving rise to process. In the new paradigm, every structure is seen as the manifestation of an underlying process. The entire web of relationship is intrinsically dynamic. While in the Newtonian outlook scientific descriptions were believed to be objective, independent of human observer and the process of knowledge, in the new paradigm it is believed that epistemology — the understanding of the process of knowledge — has to be included explicitly in the description of natural phenomena. The old metaphor of knowledge as a building is being replaced by that of a network. As we perceive reality as a network of relationships, our descriptions, too, form an interconnected network representing the observed phenomena. In such a network there won't be hierarchies or foundations. While the Cartesian paradigm is based on the belief in the certainty of scientific knowledge, the new paradigm recognizes that all scientific concepts and theories are limited and approximate. Science can never provide any complete and definite understanding. Scientists do not deal with truth; they deal with limited and approximate descriptions of reality. The new paradigm thinking also rejects the patriarchal idea of 'man dominating nature' formulated by Francis Bacon, and which has dominated science and technology ever since with disastrous consequences. New paradigm thinking in science, Capra argues, will have to be based on different methods and values if we are to survive.

And where does Capra look for new methods and values? Towards eastern wisdom and particularly towards Hinduism and Zen Buddhism. Why Capra and other new paradigm seekers are attracted towards Zen Buddhism we shall return to shortly; first, what do traditional Indian thinkers make of western scholars' overtures to Hindu thought?

In his second contribution to this volume, Claude Alvares appears rather unimpressed with the new paradigm thought. In 'We Have Been Here Before' he agrees with Capra's analysis of reductionist science; but it is his attempt to cast eastern wisdom in the mould of western physics that disturbs Alvares. He is discovering features of the East that the East did not even know it possessed. Alvares argues that two methods of knowing with a bearing in different epistemologies cannot be integrated: 'It is just not proper to make Indian metaphysics squat with a seventeenth century, ethnocentric methodology. The values of both are directly opposed: they do not cancel out, but stand as two fuming bulls in the ring.'

Despite his own analysis which shows western science to be destructive and inherently violent, Capra, alleges Alvres, does not reject western science but continues to hold modern physics as a reasonably reliable theory of knowledge. Like the Indian scientist before him, Aurobindo, he is trying to relate the dominant obsession of his time, sub-atomic physics, to Indian thought. It is a blatant attempt at co-option. 'A theory of knowledge that can suit different empirical facts, relating to different periods of man's history, has zero truth value.' In the end, Capra is guilty of the very crime he has accused western science of: reductionism. For, in trying to capture mysticism in the bottle of elementary particle physics, he brings mysticism down to a reductive level, 'to an understanding articulated by an analysing mind'.

Alvares's conclusion: 'What Capra is proposing then in his "complementary" solution to the crisis in modern science is a totalitarian hypothesis. On the one hand, what he thinks is a reasonably reliable interpretation of reality, fabricated by analysis, by scientific method. On the other is this other view that has always issued from eastern traditions, which seems to be in agreement with the scientific picture today. Capra is not providing merely a new view, but a final picture of the world.' Capra's attempt to implant Indian metaphysics into the parochial, idiosyncratic perception of the era, is an attempt to improve science; the metaphysical bleakness of science encourages constant foraging in other traditions. Thus, to Capra, science is in a fresh phase of colonization.

The one particular tradition from where new paradigm thought — in the form of Capra, Gary Zukav, William Irving Thompson, Ken Wilber and many others — has sought to find values and with which it has tried to forge some synthesis is Zen Buddhism.[5] The affinity 'New Age' thinkers feel for Zen Buddhism is not altogether surprising. Modern science offers a secular, highly structured, totalitarian system of thought that permits no diversity. Those seeking to enrich the banality and meaninglessness of western scientific thought would naturally be attracted to a secular, highly structured, totalitarian system of metaphysics that permits no diversity. The 'Pacific shift' that William Irving Thompson talks about is nothing more than a marriage of two secular and structured systems: since it is secularism in a number of different manifestations, including its manifestation as modern science, which is the root cause of the contemporary predicament of mankind, a synthesis of Zen Buddhism and western thought is not likely to move us forward to the goal of producing the 'new paradigm'.[6]

However, Third World traditionalists are not in the game of creating the alternative paradigm to western science. Shifts in paradigms involve changes in beliefs and values; if Third World traditionalists were to create a new universal paradigm, they would have to admit changes in their beliefs and values — a daft enterprise, since these beliefs and values have stood the test of ecology for thousands of years, acknowledge diversity in thought, cultures, lifestyles and ways of knowing, and have proved to be clearly superior to the secular paradigm of domination and control. What the traditional thinkers and scholars of the Third World are looking for is a contemporary expression and understanding of their values and belief systems. An altogether different undertaking from searching for a new paradigm.

In 'A Project for Our Times', Susantha Goonatilake articulates these ideas in his characteristic style. Before the emergence of modern science, he point outs, the topography of world knowledge consisted of several hills of knowledge reflecting the civilizations of China, India, Islam, Europe as well as other regional civilizations. Since the European Renaissance, other hills have been levelled and a single hill with its base in Europe has emerged. But this is not a world hill; it is only a regional hill masquerading as a universal phenomenon. The goal now should not be to create yet another regional hill — even were it to combine two traditions such as western thought and Zen Buddhism — with its base in Europe, North America and Japan, and declare it to be a universal mountain. The project of our time is to recreate the topography of several new hills, 'in our own back yard'. Each great civilization must create a knowledge structure based on its own unique world-view, on its own way of knowing. 'The search for a truly universal hill and of a truly "universal" global science can begin only after this re-emergence.'

One of Goonatilake's hills would be Islamic science. In 'Islamic Science, Western Science: Common Heritage, Diverse Destinies', Seyyed Hossein Nasr, one of the most noted exponents of Sufi metaphysics, argues that western science and Islamic science share the same historic roots: 'Both were heirs to the sciences of the same world and their knowledge of the natural order, concept of law, causality and general cosmology drew from the same sources although each developed these inherited concepts differently.' As a result Islamic science and western science enjoyed a much closer relationship than medieval Latin science and Chinese science or even Indian science and Chinese science. Thus, even as late as 'the thirteenth century medieval European science was developing along lines parallel to and usually based upon Islamic science'.

So why the radical divergence, such diverse destinies? Nasr identifies two main factors. The first is the disappearance of the 'sapiential' (by which Nasr means mystical or gnostic) aspects of Christianity. 'Every science of nature relies upon a world-view concerning the nature of reality. Medieval Christianity shared with Islam a world-view based at once upon revelation and a metaphysical knowledge drawn from the sapiential dimension of the tradition in question, although, as far as the metaphysical significance of nature was concerned, this knowledge was not fully integrated into the mainstream of Christian thought. Once this knowledge was eclipsed and for all practical purposes lost, there was no means whereby a science based on metaphysical principles could be cultivated or even understood.' The second factor is the rise of nominalism in the fourteenth century. This denied the very meaning of universals (an act of reduction) and based religious truth upon faith rather than upon both faith and knowledge, diminishing both theology and philosophy; the next step, the rise of Cartesianism was a natural outcome. Western thought became blind to the language of symbolism. Thus followed a rapid process of desacralization of the cosmos and the quest for absolute power over nature.

So what differentiates western science from Islamic science? The absence of the sapiential tradition in the western world-view and the 'presence of metaphysical and cosmological doctrines of Suhrawardi, ibn 'Arabi and Sadr al-Din Shirazi at the heart of Islamic intellectual tradition'.

Nasr's basic assertions — that there is an absolute metaphysical vacuum at the core of western intellectual tradition, that faith is completely divorced from knowledge, that western thought is totally blind to symbolism — cannot be challenged. However, while they differentiate the two world-views and their intellectual traditions, they add little to our picture, historic or contemporary, of Islamic science. But there is another aspect of Nasr's thought that is rather disturbing; his presentation of the doctrines of Suharwardi, ibn 'Arabi and Sadr al-Din Shirazi as the central core of Islamic thought is both partisan and a gross misrepresentation of the rich diversity of thought that flourished in the civilization of Islam. True, throughout the history of Muslim civilization, Sufi scholars and metaphysicians were always present and made their presence felt; but so were scholars who exemplified other schools of thought such as the Asharis, the Mutazilahs, the Zahiris and so on. Scholars from other schools of thought almost always rejected the metaphysics of the Sufis. Islamic intellectual tradition, threfore, has never been a monolithic one; its essential strength lies in its diversity of thought which stemmed from a single world-view and a single ontology but encapsulated a whole array of opinions, views, methodologies and ways of knowing. Like Capra, Nasr is engaged in a totalitarian exercise, although this time it is a traditionalist who is guilty of unnecessary violence: he rejects (or consistently ignores which amounts to the same thing) non-Sufi traditions of Islam and offers his variety of Sufism as the only complete solution to all problems.

Moreover, one cannot develop a gnostic tradition into a practical methodology for solving problems. Intuition, the basic mode of knowing in gnostic traditions, does have a place in science: as the history of science shows many theoretical insights have emerged by accident, by intuition, as unintended by-products. Thus one must acknowledge the existence and importance of intuition in creative work; but one must also accept that intuition cannot be systematically formulated and made a cornerstone of a scientific methodology. Hossein Nasr's notion of Islamic science based on the methodology of the gnostic tradition, more particularly Sufism, does not work. While it fulfils certain criteria for being a science, a system for solving practical problems and puzzles, it fails to meet certain other, equally important criteria. To be classified as science, scientific theories must not only have the characteristic of an axiomatic system, they must be consistent, repeatable, able to be corroborated independently and should be potentially accessible to all segments of society. Gnosis is neither amenable to empirical analysis, nor can one have access to it independently and at will, neither can its results be repeated or openly distributed to mankind at large. While the world-view of Islam recognizes its existence and acknowledges its importance, it certainly did not build its science on the methodology of gnosis. The exponents of Islamic science must go beyond mere gnosis to produce something that is clearly distinguishable as science.

Enter Munwar Ahmad Anees and Merryl Wyn Davies. They are concerned with the rediscovery of an Islamic science which can clearly be recognized as a science. In presenting an overview of the current literature they place the emphasis on the epistemology derived from the immutable values and concepts of Islam. The gateway to Islamic science is revitalization of *ilm*, the Islamic concept of knowledge. *Ilm*, they point out is a multi-dimensional, integrative concept that regards knowledge as an organic unity that can be pursued only within the framework of values. This has major implications for Islamic science. The Islamic world-view takes a much more encompassing view of the possibilities of human cognition, the balanced interaction of revelation and reason, it opens a whole range of methodological approaches as relevant and necessary for science. It also firmly establishes goals and objectives for science founded upon accountability and social responsibility for attaining human betterment within a social and cultural milieu. Islamic science, argue Anees and Davies, is not to be equated with re-inventing the wheel, a subtle undermining of the cumulative human labour of amassing wisdom. What Islamic science does mean is the development of a whole system of knowledge that questions and evaluates what constitutes wisdom based upon its own holistic definition of human betterment.

Having made an Islamic theory of knowledge central to Islamic science, they castigate both the 'Islamization of knowledge' movement — currently the dominant preoccupation of most Muslim scholars — and 'Bucaillism' — looking for science in the Qur'an and justifying belief according to the dictates of modern science, as diversionary follies. Both Islamization and Bucaillism accept the integrity of western science and do violence to the integrity of the Islamic world-view. Islamization of knowledge, they argue, unwittingly amounts to a westernization of Islamic knowledge and by basing itself upon western defined knowledge groups perpetuates a fragmented approach to the organic unity of knowledge, the limitations of which are being belatedly recognized even in western science. Bucaillism is simply a logical fallacy as well as being reductive in its assumptions and implications. (Bucaillism is a Muslim parallel to Capra's equation of modern physics with eastern mysticism.) Both approaches encourage mental inertia amongst Muslim scientists by suggesting there can be synthesis, an emendation of western science by addition of certain tempering values, without what Anees and Davies regard as the essential characteristic of the Islamic outlook, a critical attitude.

The critical outlook of Islamic science makes it *subjectively objective*, an open-ended system of knowledge operating within a framework of values. It is a science that thrives on values to perpetuate values with value clarification as an essential procedure. The matrix of Islamic science they see as composed of ten essential Islamic concepts: *tawhid* (unity), *khilafah* (trusteeship), *ibadah* (worship), *ilm* (knowledge), *halal* (praiseworthy), *haram* (blameworthy), *adl* (social justice), *zulm* (tyranny), *istislah* (public interest), and *dhiya* (waste).[7] When translated into values these concepts make the parameters of Islamic science. However, to operate within these parameters, science would have to adopt a different role and social institutionalization, one geared to social relevancy and social responsibility for problem solving within a particular social and cultural milieu based upon universal norms and values. Islamic science must answer the needs of today, but it cannot be parochial in conception or operation. Thus, Islamic science aims at global change and is unashamedly universal in its character.

The full-fledged emergence of non-western sciences, like Islamic science, depends to some extent on discovering non-linear systems of logic. M.D. Srinivas's exploration of 'Logical and Methodological Foundations of Indian Science' shows that non-western logic is not just possible but exists and can be regarded as superior in some respects. Indian logic does not follow the rules of content-dependent, purely symbolic or formal language; it is not a study of propositions and their logical forms, but of cognition, and awareness. It is a logic of *jnana* (cognition, awareness). While it is just as rigorous as western logic, it is not a reductive but a constructive logic. It does not divorce itself from ordinary language into 'pure' symbols, but has its foundations in a natural language which it tries to free from inaccurate reasoning and ambiguous statements. While *jnana* has a concrete occurrence in Indian philosophy it does not have a logical structure of its own but a structure that becomes evident after reflective analysis. There are rules which clarify the modes under which ontological entities become evident in *jnana*.

The object of Indian logic is to make the logical structure of cognition clear and unambiguous by reformulating it in a technical language. Indian logic insists that formulation of universal statements, apart from being unambiguous, should be phrased in accordance with the way such cognition actually arises. Such a formulation involves the use of two negatives. In contrast to the simple notion of negation in western logic, Indian logic conceives of absence as a property by a hypothesis of denial. *Abhava* (negation) is thus conceived as the object of negative cognition and hence as a separate entity. In such a system, the relationship of the absence of an object to its locus of being is naturally and automatically emphasized: 'whenever we assert that an absence of an object "a" (say a pot) occurs in some locus (say, the ground), it implied that "a" could have occurred in, or, more generally, could have been related to, that locus by some definite relation. Thus, in speaking of absence of "a" we should always be prepared to specify this such-and-such relation, that is, we should be able to state by which relation, "a" is said to be absent from the locus.'

Indian logic also has its own way of constructing theories. Srinivas points out that just as the western axiomatized formal theories find their paradigm example in the exposition of geometry in Euclid's *Elements*, the Indian method of theory construction finds its paradigm example in the Sanskrit grammar of

Panini, the *Astadhyayi*. 'The technical terms of a theory (*samjna*), the metarules (*paribhasha*) which circumscribe how the rules (*sutras*) have to be used, the limitation of the general (*utsarga*) rules by special (*apavada*) rules, use of headings (*adhikarasutra*), the convention of recurrence (*anuvrtti*) whereby parts of rules are considered to recur in subsequent rules, the various conventions on rule-ordering and other decision procedures as also the various so called 'metalinguistic' devices such as use of markers (*anybandhas*) and the use of different cases to indicate the context, input and change — all these and many other technical devices employed in *Astadhyayi*, are now coming to be more and more recognized as the technical components of an intricate but tightly-knit logical system, as sophisticated as any conceivable formal system of modern logic, linguistics, mathematics or any other theoretical science.'

Indian logic certainly presents a very powerful tool for the formulation of scientific theories. It demonstrates simultaneously that perception can be part of analysis and that methodological tools radically different from the modern mathematical logic or the attendant formal systems exist and need to be studied and researched further. Clearly this is an area of vast potential for those who seek to develop sciences based on non-western metaphysical, philosophical and sociological assumptions.

But exploration of non-western sciences should not be limited to theoretical realms; the strengths and limitations of traditional technologies have to be explored thoroughly so that they can be enhanced and given a contemporary image. Here, the appropriate technology (AT) movement offers a useful starting point. After more than fifteen years of involvement with the appropriate technology movement, A.K.N. Reddy — one of the most articulate Third World defenders of AT — finds himself redefining the whole endeavour. In 'Appropriate Technology: A Reassessment', he finds himself redefining AT and sees it as much more than low-level, low-cost alternatives. AT is technology which promotes the satisfaction of basic human needs, social participation and control and ecological soundness. The test for appropriateness of technology is whether it reduces inequalities, strengthens self-reliance and is in harmony with the environment. This redefinition now implies that AT is not against industrialization or modern technology and while it is not against traditional technology, AT does not constitute a return to traditional technology. Moreover, many of the old features of AT have now been shed: it is not limited production or low or intermediate or small-scale technology, or even a strategy for rural areas only. And AT, he states bravely, is not a western concept, despite its origins in the work of Schumacher and other western gurus; neither is it a task for western institutions even though they may be set up by kind-hearted individuals. AT is a continuum of technology, people and institutions. It is not a substitute for social change, and it cannot be achieved without popular participation.

While AT may not constitute a return to traditional technology, it must be based on the principles of traditional techniques and methods for two basic reasons. Firstly, ecologically healthy practices which relate directly to the people of the Third World are traditional practices. If the objective is popular participation and meeting the basic needs of society, then incorporation of traditional principles in any contemporary appropriate technology is essential. Secondly, we have had too many technological experiments at the expense of Third World people; it is time tried and trusted methods and techniques which actually work and meet the needs of the Third World people were implemented. Only traditional technological systems meet this demand: as D.L.O. Mendis illustrates, the ancient irrigation system of Sri Lanka is far superior and more ecologically sound than any modern expensive irrigation scheme undertaken in Sri Lanka, such as the Walawe or Lungugamvehera irrigation schemes, which have massively increased the debt burden of the country. Many traditional techniques use the simplest and cheapest available resources to considerable effect. G.K. Upawansa describes how traditional farming methods in Sri Lanka manage to prevent crop damage simply by timely planning in conjunction with lunar cycles; minimal tillage, using buffaloes and cattle, not only saved energy but ensured a better crop; mixed cropping was used to promote better photo-synthesis and reduced competition for plant nutrients; pest damage was kept to a minimum by the use of certain pest-controlling plants, by allowing areas for birds to feed, by inviting birds to certain fields to consume crop-eating caterpillars. The challenge facing Third World scientists, technologists and policy-makers, is to use such traditional practices to evolve new techniques and methods for solving the problems of their societies.

As a first essential step, Third World policy-makers need to integrate schemes for revitalizing traditional techniques and technologies in their science and technology policies. The Penang Declaration on Science and Technology, which constitutes the recommendations and resolutions of the CAP seminar on 'The Crisis in Modern Science', spells out in some detail the steps we need to take to safeguard traditional world-views, lifestyles, cultures, technologies and modes of knowing and the kind of policies that have to be implemented to ensure the survival of a vast majority of the people in the Third World.

The Revenge of Athena demonstrates that non-western sciences which differ in fundamental ways from the dominant mode are possible, even if being 'scientific' is defined in terms of logical (mathematical) description, prediction, empirical evidence and reproducible experiments. Full-fledged systems of science have existed in the civilizations of the Third World — the challenge before us is to rediscover and contemporize them. It is a formidable challenge; but it is a challenge that has to be met if the onslaught of western science and technology, and its associated world-view which combines the use of reason with violence, is to be checked.

To be fair, one must note that Athena too had her better sides. She protected all heroes who fought for the good of mankind, such as Heracles and Theseus, and aided all who represented the ideals of Hellenism such as Ulysses, the whole race of the Achaeans during the Trojan war, and most of all, the people of Athens whose social patron she was. She presided over many peaceful activities and was frequently invoked by women weaving and spinning, by workers and craftsmen. But during the entire span of such engagements she remained an innocent virgin and guarded her virginity with prudish sensitivity. It was only when she was raped — rape is a central metaphor in modern science — that her character actually changed and she started on the path that led to the havoc and revenge that is the legacy, in our time, for Third World societies.

Notes and References

1. Vandana Shiva, 'The Violence of Reductionist Science', *Alternatives*, **12** (2) 243–61 (April 1987).
2. For a general introduction to CAP's work, see Ziauddin Sardar, 'The Fight to Save Malaysia', *New Scientist*, **87**, 700–703 (4 September 1980).
3. For an interesting discussion on the possibility of alternative science, see Friedrich Rapp, 'The Chances of Alternative Science and Technology', *Research in Philosophy and Technology*, Vol. 7, 159–76, JAI Press, 1984.
4. Fritjof Capra, *The Turning Point*, Flamingo, London, 1983.
5. Fritjof Capra, *The Tao of Physics*, Flamingo, London, 1976; Gary Zakav, *The Dancing Wu Li Masters*, Flamingo, London, 1979; William Irving Thompson, *Pacific Shift*, Sierra Club, San Francisco, 1986; and Ken Wilber, *Eye to Eye: The Quest for the New Paradigm*, Anchor, New York, 1983.
6. For various positions on the new paradigm thinking, see the fascinating presentations and discussions of the Symposium on 'Critical Questions About New Paradigm Thinking', *Re-Vision*, **9** (1) 5–98 (Summer/Fall 1986).
7. For a more detailed discussion of the conceptual matrix, see Ziauddin Sardar (ed.), *The Touch of Midas: Science, Values and the Environment in Islam and the West*, Manchester University Press, Manchester, 1984.

Part One

What's Wrong with Science?

1

Science and Ideology
The Marxist Perspective

Glyn Ford

For anyone brought up in an advanced capitalist country, the importance of science would seem difficult to deny. People learn about science in schools and places of higher education, they rely upon its manifestations in their everyday lives, checking the time on digital watches, cooking in microwave ovens, watching videos — and they may sometimes worry about its getting 'out of control', as may be witnessed in the ongoing debate in Britain and America over the freezing of human embryos. It may therefore seem surprising that a large proportion of the political left in the West, perhaps even a majority, have shirked the task of analysing the relationship between science and society. In Britain this has been largely undertaken by a small group of Marxist scientists and academics working in the fields of the history and sociology of science for whom the analysis is seen as politically fundamental. Believing, in many cases, that science in the West is somehow unresponsive to the needs of people, they have been motivated to understand its specific form of operation under capitalism and to formulate proposals for its transformation.

The 1930s witnessed the first real interest in a radical analysis of science in Britain. Influenced to a degree by Soviet views — the 1931 Second International Congress of the History of Science and Technology, held in London, representing a significant landmark — a group of socialist scientists at Cambridge University are widely acknowledged to have initiated the radical science movement in this country. Their writings, particularly those of J.D. Bernal, were arguably of profound importance in shaping the 1960s Labour Party policy of the 'white hot technological revolution'. Others such as Christopher Caudwell, G.D.H. Cole and Michael Oakeshott from the Communist, Labour and Conservative political parties respectively, also considered the role of science in society but their ideas were not as influential as those of the Cambridge scientists.

Subsequently the belief grew in radical intellectual circles that the view of science espoused by the Cambridge writers was providing the left with a grossly distorted picture, involving dangerous prescriptions for the transformation of science which the new radicals believed had found their way into the Labour Party.

There are several issues involved in this continuing debate. For example, is science autonomous, developing independently of the socio-economic context? Is it disinterested and progressive, furnishing an increasingly accurate and objective picture of the world about us? Or might it not be impregnated and even saturated by (capitalist) ideology? Should a 'scientific attitude' be elevated to a uniquely privileged position? And how can science be used for the benefit of society as a whole?

This debate may appear fairly negative, but it by no means exhausts Marx's thoughts on science. Time after time he alludes to the 'social advances which scientific work permits', a point most strikingly made when he writes of 'the historical significance of capitalist production in its specific form . . . the development of the social forces of the production of labour'. The point would seem to be that it is not science itself which is an 'evil' force — quite the contrary, for Marx's respect of 'science' is luminously apparent. Nor is it really the application of science to production that is the problem. It is rather the capitalist mode of production that fetters the benefits of science to humanity, rendering it a mere tool for the extraction and appropriation of profit. Ironically, though, even the exploitation of scientific work under capitalism ultimately benefits society as a whole, Marx's well-known thesis propounds the idea that the development of the productive forces in capitalism is an essential step along the road to socialism. Finally, he believed, scientific advances and their application to production would be enjoyed by all.

Soviet interpretation and development of Marx's and Engel's perspective on science and technology were conveyed for the first time to British scientists by a delegation to the 1931 Congress. Scientists were regaled with glowing reports of scientific achievements in the Soviet Union together with assertions that these were possible only in a socialist society. Specific papers challenged the cherished beliefs of those, perhaps even including Marx himself, who viewed science as a disinterested pursuit and the history of science as a study of the successes of great men. The 'Social and Economic Roots of Newton's Principia' presented by Boris Hessen was especially critical, displaying Newton's theories not as disinterested knowledge but as a response to the technical needs of the seventeenth-century bourgeoisie. This in effect introduced British scientists to the idea that capitalist ideology permeates the content of science and suggested that there were two kinds of science — a bourgeois and a socialist variety.

Some participants at the conference dimissed these views as incorrect and even dangerous; others, however, were profoundly affected. This was especially true for the group of Cambridge scientists, notably J.D. Bernal, J.B.S. Haldane, Hyman Levy, Lancelot Hogben and Joseph Needham, who went on to develop some of the most influential views on the relationship between science and socialism. They had immense respect for science and believed that its application in production would be fully enjoyed by all people only in a socialist society. But more in the spirit of Marx rather than Soviet writers, they adopted a use/abuse model of science, that is, scientific knowledge is intrinsically neither good nor evil and external pressures account for its destructive aspect. This view allowed them a degree of reformism. They believed that there was no need to wait for the advent of socialism to change the role of science within capitalism. Hence they all attempted to formulate and popularize proposals for transforming science in the existing society. Where they differed among themselves was in their vision of a socialist society and in their attitude to science. This can be clearly seen by comparing J.D. Bernal's views with those of Joseph Needham.

Needham took great pleasure in illustrating how some Chinese scientific achievements preceded those in the West. The 'soul of the mechanical clock', for example, was not, Needham tells us, the invention of an unknown artist in Europe around 1280 but rather that of a Tantric monk and mathematician, I-Hsing and his collaborator Lsiang Ling Tsan in China in 725. By exposing similar myths in the history of science Needham was able to demonstrate the racist attitude of Europeans to non-westerners. Western historians, he argues, continually refer to 'our' science and 'our' modern culture assuming that all great scientific advances are European in origin.

Needham's message was an important one. There have been further illustrations that scientific achievements in other non-western countries have been similarly neglected. George Sarton, for example, has provided solid details of Islamic achievements in *An Introduction to the History of Science* (three volumes, 1927–1948). Seyyed Hossein Nasr in *Islamic Science: An Illustrated Study* (London 1976) shows how Muslim scientists contributed to scientific knowledge within the *dar al-Islam*. In Professor Nasr's work the link between science and Muslim culture is apparent. The Muslim quest for knowledge was guided by the Qur'an. This influenced both the problems which the Muslim scientist tried to solve and the types of solutions that were offered.

The Cambridge scientists continued to popularize their views on science and socialism. In the 1960s, some helped to found the British Society for Social Responsibility in Science at a time when science was coming increasingly under attack. At this time science was popularly believed in the West to be 'out of control'. Many scientists in Britain, the USA and also in France, Italy, Belgium and West Germany became concerned that their findings were being used for destructive purposes. The war in Vietnam and the proliferation of nuclear, chemical and biological weapons, as well as the degradation of the environment, confirmed their fears.

Although the protests were different in every country, there were two main kinds of responses to this seemingly worsening situation. Many scientists concerned to protect the name of science argued that it was being misused (the use/abuse model) and that scientists had a responsibility to agitate outside their laboratories to ensure that the results of their experiments were not applied to the destruction of people or the environment. The other response was not to see external pressures as the problem but science itself. There were also developing among radical intellectual circles at this time new critical analyses of the role of science in society. In Britain left wing scientists centering in and around the British Society for the Social Responsibility of Science and The Radical Science Collective played a prominent role in formulating these critiques. Their analyses drew upon contemporary studies in the philosophy and sociology of science from the English speaking world and were very different to those offered by their predecessors.

The Roses have extended this analysis to other areas in science. In a critique of neurobiological sciences they argue that many theories and associated technologies of neurobiology, from drug therapy through to IQ testing, are fundamentally biologistic. Biologism, in their opinion, takes one part of the explanation of the human condition, excludes others and then claims that it provides the explanation for aggression, war, love and hate. This implies that it is absurd to attempt to change ourselves or the world. Biologism, however, for all its claims to be scientific, is ideological in the sense that it helps to legitimate the status quo. Moreover, the reductionism inherent in a biologistic approach, originally a tool for examining specific problems under rigorously defined conditions, becomes saturated with ideological connotations and at the same time obscures the ideological bias within science.

Others have taken up the themes of ideology *within* and *of* science and revealed that science is sexist and racist. One of the themes, for example, in Brian Easlea's book on *Science and Sexual Oppression* (London, 1981) is that physicians during the nineteenth century promulgated the belief that women are unpassionate, an idea legitimized and reinforced by subsequent medical ideas and practice. Hilary Rose and Jalna Hanmer (in H. Rose and S. Rose (ed.), *The Political Economy of Science*, Macmillan, 1976), have also argued that sexism is ingrained in the current developments of reproductive technology from genetic engineering to hormone time capsules. This characterization of science is opposed to that of radical feminists who see technology as essentially neutral.

While many socialists would now concede that there is capitalist, racist or sexist ideology in science, they would claim that some science, even under capitalism, is objective. This view, however, is rejected by some sections of the Radical Science Journal Collective, in particular by Robert Young who argued in 'Science is Social Relations' (*Radical Science Journal*, no. 5, 1977) that *all* science is ideological. In so doing, Young effectively denied any objectivity to science — in other words, science does not provide a view of the world which corresponds to reality. This also implies that objects do not exist but are merely manifestations of social relations, an implication that is thus the antithesis of the traditional Marxist position where science was seen as possessing at least a degree of relative autonomy. It is possible that Young's main contribution was to stimulate interest in the question of the objectivity of science among radical scientists. A whole series of articles appeared in various socialist papers on the subject, some directly addressing Young.

The Roses were especially critical, arguing that he mistakenly believes that the social determinants of science completely dissolve the phenomena itself. Restating Marx's position, they argue that science is relatively autonomous and that materials have properties which are open to analysis, despite the intrusion of ideology into scientific discourse.

Allen Callincos, writing in 1979 in the January issue of *Socialist Review* put forward a related view. He stated that science is not neutral but is able to provide us with a knowledge of reality. Moreover, science cannot simply be reduced to an ideology that reflects what happens in class struggles because, as the history of science tells us, science is continually transforming conceptually. It is this process, he argues, that gives science its relative autonomy. Yet another view on the subject was provided by Dave Albury in the same issue of *Socialist Review*. He argued that there is an objective reality but our conceptual tools for trying to comprehend it are determined by 'socio-economic imperatives' that are operating in society. Science therefore provides us with partial truths but not the whole truth.

While there may be disagreements about the extent to which science is imbued with ideology, most of those participating in the debate would agree that it is politically necessary to expose the myth of scientific neutrality. By this they mean the ideology in and of science, or more generally, that science is not divorced from society. Showing that science is not neutral is an essential step in their attempts to make the discipline serve the needs of the people.

For Marx, the term science referred to the natural sciences (such as physics, chemistry, astronomy, biology and geology) and mathematics. He described it as 'the general intellectual product of the social process' which, while presupposing 'a certain level of material development' in order to flourish, seems to have been regarded as possessing at least a relative degree of autonomy in its evolution. The crucial distinction was between science *per se* and its application to the production process in the form of technology. Science, Marx tells us, is pressed into the service of capital. This is most glaringly apparent in the case of machinery, which necessitates 'the replacement of the rule of thumb by the conscious application of natural science', although the utilization of scientific knowledge is much more pervasive than this, revolutionizing the materials of production, facilitating greater economy in the use of physical productive resources and, in the sphere of agriculture, permitting enhanced crop yields via the use of fertilizers.

Of course, in Marx's analysis of the situation capitalists do not make use of science in these ways for socially philanthropic reasons. They do it for profit and in so doing the workers suffer. Take the case of machinery. With its intro-duction, the innovating capitalist enjoys a competitive advantage over rivals in the same sphere of production, the point being that unless this was so, there would be no incentive to be innovative in the first place. The capitalist knows, however, that this situation will not persist indefinitely. His or her competitors will sooner or later be compelled to introduce the new technique themselves — the 'coercive wind of competition' ensures that this is so. To get the most of the temporary advantage over rivals, and guard against what Marx calls 'moral depreciation' — the possibility that competitors will introduce even more advanced techniques — production is speeded up and a shift system enforced. Moreover, the effect of introducing machinery may well render obsolete the skills of particular workers, thus depressing wages within the factory, as well as making work dull, repetitive and dehumanizing. As Marx puts it, the introduction of machinery — the application of science — transforms the worker into 'part of the detail machine'; mental and physical labour are disunited and the worker becomes a 'crippled monstrosity'.

The application of science in the service of capital need not be confined to the development of machinery. For example, Marx discusses advances in chemistry which enable waste to be recycled. Again, this benefits the capitalist in the form of increased profitability but, with equal predictability, workers may suffer financial hardship if the process of innovation renders partially obsolete existing techniques of production, for then the capitalists will try and compensate by forcing them to work longer hours, preferably with lower wages.

There is also a general cost to those working in the field of scientific applications, for applications of their work take place 'in isolation from the knowledge and abilities of the individual worker'. This creates further mystification of the social relations of production and, indeed, of the nature of science itself, which takes on the appearance of being 'the direct offshoot of capital' rather than something which is used *by* capital.

Bernal's main analysis of science and socialism is to be found in his 'Social Function of Science' published in 1939. He argued that science was inherently progressive but would reach its full potential benefit only in a socialist society. Looking to the Soviet Union he agreed with Marx that science required a certain level of material support and also some degree of planning. He then offered a dismal vision of socialism in which a rationally planned and organized science would be the chief means of technological transformation. Most of his writings reflect an overwhelming respect for the authority of science. He elevates the role of scientists in society by arguing that only a trained scientific elite could ensure that the application of science meet peoples' needs, and at times he tends to see science as the agent of change rather than the people.

Bernal's views were certainly taken seriously. In 1953 he was awarded the Stalin Peace Prize and in the same decade joined a group of scientists bent on changing the Labour Party's perspective on science and technology. These meetings were regularly attended by Hugh Gaitskell, James Callaghan, Harold Wilson and Richard Crossman and out of the discussions arose a new commitment on the part of the Labour Party to modernize its image and fight the 1964 general election on the basis of a 'white hot technological revolution'.

Bernalism, some argue, still exists today in the Labour Party in the belief that science will become a force for liberation providing there are more funds and better management.

The embryologist Joseph Needham completely altered his view of the history of science after the Soviet visit. In 1931 he had written an internalist account of the history of embryology where he recorded the intellectual advances of 'great men' in the field. By 1938 he had rejected this way of representing history, arguing that scientific thought should always be studied in relation to the social and economic context of the time. He was, more than Bernal, worried about scientific development in the Soviet Union. After the Lysenko affair, when whole areas of genetics in the Soviet Union were obliterated because they were perceived to be ideologically unsound, he argued that there were dangers in the ruthless application of central planning in science. He also had reservations about Bernal's rationally planned society and respect for the authority of science, arguing that too many Marxists were dangerously intoxicated by the 'scientific opium'.

Needham's rejection of excessive scientific authority was related to his continuing interest in religion. As a Christian Socialist who had passed through Anglo-Catholicism, a variety of modern western religions and Quakerism, his ideal society was one where human beings possessed both scientific pride and religious humility. Rather than looking to the Soviet Union for his ideal social-ist society, he turned instead to China where he believed people had not fallen prey to the idea that science represents the only valid way of understanding the universe and where science satisfies all the peoples' needs. His studies on Chinese scientific traditions have been published in several volumes of *Science and Civilisation in China* (Cambridge, 1956–1965).

Before discussing some of their perspectives on scientific objectivity, trans-formation and ideology, it is necessary to mention briefly three studies in the philosophy and sociology of science which have arguably informed the radical scientists' positions. In the 1960s, Thomas Kuhn, a sociologist of science, pub-lished the *Structure of Scientific Revolutions* (Chicago, 1962). This book ques-tioned the objective nature of scientific knowledge and its development as a continuous, linear process unlocking the secrets of the universe. In his account, scientific progress closely parallels the Marxian dialectic, alternating between 'normal' and 'revolutionary' phases in which scientists (respectively) make piecemeal advances, or choose between rival grand systems. Hence the genuine 'progress' of science becomes impossible to account for or guarantee in 'revolu-tionary' or 'normal' science. This aspect of Kuhn's work has been used to rebut the idea of an objective universal basis of scientific knowledge. M. Anis Alam, for example, argues in 'Science and Imperialism' (*Race and Class*, 1978) that as subjective, personal and partisan considerations play a decisive role in the acceptance of a new paradigm science can hardly be said to be objective.

Paul Feyerabend, an American philosopher of science, writing in the 1960s and 1970s, denied any special status to science as a branch of knowledge. One of the themes of his book, *Against Method* (London, 1975) is that for any rule or method enunciated by philosophers of science there has been an important occasion where it was broken by some great scientist, thereby allowing the authority of science to be questioned. In *Science in Free Society* (London, 1978) he claimed that (scientific) 'rationality is one tradition among many rather than a standard to which all traditions must conform'. Moreover, in a free society, science would have no claim to status at all. He called for the removal of science from its dominant position. 'This volume has one aim; to remove obstacles intellectuals and specialists create for traditions different from their own and to prepare the removal for specialists [scientists] themselves from the life centres of society.' His views therefore had the declared liberatory intent of making knowledge and power more accessible.

If we examine the debates among the radicals of the 1960s and 1970s, it is clear that one concern has been to demonstrate that science is impregnated with capitalist ideology or, as some would argue, it is ideological. The idea that capitalist values penetrate scientific knowledge was, as noted earlier, introduced to British scientists by the Soviets in the early 1930s. The Cambridge scientists, with the exception of Hogben, never went on to explore the relationship but the issue was taken up by Stephen and Hilary Rose in the early 1970s and subsequently by others.

For the Roses, science is an ideologically laden activity. It is done in a particular social order and reflects the norms and ideology of that order. Moreover, it is part of an interacting system in which internalized ideological assumptions help to determine the very experimental designs and theories of scientists themselves. They are able to demonstrate this by comparing how the language and central metaphors changed in the biology of the 1930s and 1940s and in the biology after the mid-50s. In the 'Myth of the Neutrality of Science' (Fuller ed., *The Political Economy of Science*, Routledge & Kegan Paul, 1971), they argue that in the later period the central metaphors of cybernetics were used — control, feedback, regulation etc. — whereas in the earlier period pre-Keynesian metaphors such as currency, energy bank, deposit account were common. In their opinion, this clearly shows that when scientists are puzzling over problems in biology, they use values which they have internalized from the wider society.

For the Roses, the socialist strategy of making science and information available to the working classes and other oppressed groups is meaningless because what is being offered is 'bourgeois science'. But they do nevertheless believe that this strategy would gradually transform science for the better. They argue, with Robert Young, that socialists should not separate theory from practice and suggest that more collective rather than elitist ways of organizing work should be the norm in universities and scientific laboratories. Above all, they argue in the 'Radicalisation of Science' (*Socialist Register*, 1972) that to radicalize scientists, the critique of science must be integrated in a general political analysis and to achieve this, scientists must bring their work into the area of activity of Marxist groups. Of course, scientists are a long way from being 'radicalized' and while the Roses are disillusioned with this state of affairs, they remain optimistic that with the increasing industrialization of science, more scientists will become politicized as they become alienated.

For others, a truly socialist society would need to transform science and seek alternative technologies. David Dickson argues in 'Technology and the Construction of Social Reality' (*Radical Science Journal*, 1976) that socialism involves not just the adaptation of new relations of production to a productive base developed under capitalism, but rather a fundamental transformation of the techniques of production and the products themselves. To create alternative technologies would, in itself, be a political struggle.

What criteria should be used in judging whether a technology is acceptable? Euro-Communists, revisionists who distance themselves from Russian and Eastern bloc experiences, offer some suggestions. Fred Steward, representing their position, suggests in *Marxism Today* (February 1981), that we should reject as the main criterion conventional economic considerations and the effect on Britain's international competitiveness. Instead, much greater emphasis should be placed on developing civilian rather than military technology as well as products that are ecologically sound, non-polluting, resource conserving and non-hazardous. At all times we should aim to fulfil important social needs rather than those of an elite and consider what impact technology may have on the labour force's skills and work satisfaction and on the civil liberties of the population.

To achieve these ends, Steward argues that the left, with the trade unions and labour movement, should intervene politically to control and direct technology. This requires some appreciation of how technology has developed in the post-war era, extensive state intervention and an enhanced role for scientific and technical expertise. Steward's vision is that there should ultimately be a co-ordinated policy for technological change which overcomes the existing fragmentation among government departments, with industrial enterprises and new democratic structures at national and local levels where people choose from alternative technical options.

Political intervention to control technology is also the theme in shop-floor worker Mike Cooley's socialist strategy. He argues that fragmentation of skills, increased work speeds and shift work began in the nineteenth century in the chemical sector but now affects computing, architecture, engineering and other areas involving scientists. He maintains, however, that the sectors are differentially affected and it is possible for workers to intervene at vulnerable points in this process so that they make sure technology is introduced on their terms rather than management's.

The strategies presented here to ensure that all the people in society receive the full benefits of science and technology are by no means exhaustive, and, like the discussion of scientific ideology and objectivity, are part of a continuing debate. Moreover, the issues are relevant for people of the Third World. As Anis Alam argues in the aforementioned article in *Race and Class*, underdeveloped countries like India have similar elitist systems of science that promote prestige areas such as nuclear, particle, solid state and space physics. Science in these countries all too often serves the needs of the ruling elite and does not relieve the misery of the population. In these countries, developing an analysis of science and imperialism and a political strategy to transform science and technology may be the first step towards ensuring that society is changed for the better.

2

Science, Ignorance and Fantasies

Jerome R. Ravetz

Our modern scientific technological culture is based on two articles of faith. The first, deriving from Bacon, is that knowledge is power, over our material environment. The second, from Descartes and his philosophical colleagues, is that material reality is atomic in structure, consisting of simple elements denuded of interconnection and of causes relating to human perceptions and values. On that basis our civilization has achieved unparalleled success in terms of theoretical knowledge and material power. But we all know that we are dangerously deficient in control. Viewed from *inside*, we may question whether we are only 'Sorcerer's Apprentices', capable of starting the magic engine but incompetent to control or stop it. Viewed from *outside*, our civilization may appear as a weed, dominating and choking out all other cultures disturbed by its material conquests.

In recent years there has been an increasing tide of criticism of this dominant world-view on all fronts. Following the explosion of consciousness among the affluent youth in the 1960s the metaphysics of this civilization has been subjected to critical scrutiny and many alternatives proposed. Some of these call for a return to world-views and religions which pre-date the rise of the present dominant civilizations. More influential (so far) are those which draw on the cultural resources of the East, particularly the Taoist style of thinking. The presence of 'complementarity' in the structure of the most advanced theories of fundamental physics, has been used strongly as evidence for the naturalness and 'scientific' character of this alternative framework of thought. The steady growth of alternative or complementary medicine, in fields where the atomistic style seems ineffectual, counter-productive or positively barbarous, gives this other world-view a firm basis in successful practice and popular experience.

In the area of technology, the focus of Francis Bacon's dream, we have been coming to see more clearly how the solution of a problem at one level, as in a technical fix, can produce more serious, perhaps insoluble, problems at other levels. The various forms of pollution, problems of the disposal of radioactive wastes, and the conversion of the former colonial world into a global slum and sweatshop for the so-called advanced nations are reminders of the inadequancy of a simplistic approach to power over nature. Here I shall develop some heuristic concepts whereby we may better comprehend such problems. I hope thereby to show how alternative styles of thought are as relevant to the control of material culture as to abstract physics or medicine.

Value-loading in Science, or the Social Construction of Ignorance

The optimistic philosophy of science of previous generations rested on a simple, linear scheme of the application of science to human benefit. Science produced facts, either in its own pursuits or in response to perceived social problems. In themselves these facts were value-free; the interests or prejudices of the individual investigator did not affect his conclusions, which were tested against the objective world of nature. But in their totality, they embodied the highest human values. For the miseries of mankind were easily seen to result from poverty, ignorance and superstition. The first two of these would be removed directly as a result of scientific enquiry; and the last would be defeated by the exposure of the real causes of human suffering, in material and intellectual culture. Those who espoused this philosophy were well aware that science would not easily succeed on its own; there had to be a struggle against the institutions that profited from exploitation and oppression; previously established religion and, more recently, an unjust social system.

The successes of this ideology, at least for the great mass of people in its homelands, must never be overlooked. Even now, when material poverty persists in the most advanced nations, there will be sharp practical contradictions between progress (realized in the relief of drudgery and the production of jobs), and an ecological awareness of the limits of growth.

However, even within those highly-developed economies, some systematic complications have been recognized. The theme of choice has been appreciated as vital to the direction of science and technology. The image of the isolated, autonomous pure scientist, following his own curiosity and accidentally producing results of social benefit, is totally obsolete. Science is now a big business requiring choices for the allocation of limited resources. And technology cannot depend on an automatic mechanism of a market to turn inventions into successful innovations. In each case there must be policy, enabling direction to be given and choices to be made in accordance with general strategic objectives.

What is the source of such a strategy? It does not come from an immediate contact with nature that is instantly and rigorously tested by results. Rather, it is found in institutions, which, since they embody power, must necessarily be closely aligned with the general political/economic structures of the society of which they are a part. The ultimate motive of such strategic planning may well be the improvement of the condition of mankind. But this aspiration will inevitably be filtered through the realities of power in any given context. Hence the science that is done (and perhaps more importantly, the science that is *not* done), reflects the values of a society as they are realized in its dominant institutions. In terms of this analysis, such slogans as 'science is not neutral' and 'science for the people' are not merely partisan rhetoric. They represent protests against the particular institutional arrangements for the productions of scientific knowledge, and also against the ideology of objectivity by which it is still reinforced.

It might be thought that in spite of these forces shaping and (by some criteria) distorting the collection of scientific materials available to society, there must still be a hard core of 'facts' independent of these forces. This is a very delicate and sensitive question; for if we abandon all belief in our commitment to objectivity in science, then there is no defence against charlatans or power-politicians deciding public policy on matters scientific and technological. Hence I argue only that objectivity is by no means guaranteed by the materials or the techniques of science, but rather emerges partly from the integrity of individuals and partly from open debate on scientific results.

I can establish this point by an example from a common element of scientific technique: statistical inference. When statisticians test an hypothesis, they cannot possibly decide its truth or falsity; at best they work to within a 'confidence limit', which (roughly speaking) gives the odds (in terms of a mathematical model of the universe to which the given data is assumed to belong) that their conclusion is correct. Different problems conventionally are investigated to different confidence limits, say 95 per cent or 99 per cent. A more rigorous confidence-limit requires a more extensive investigation. But a conclusion that states that there is no sufficient evidence is always relative to the pre-assigned confidence-limit. A more searching test might have proved a positive result. Hence the *values* defining the investigation, the costs of 'false negatives' and of 'false positives', as well as the cost of the study itself, can determine the answer. A low-cost investigation can result in an effect remaining concealed. Knowledge is costly; but the price of economy is continued ignorance.

This general point of methodology can become an issue of political struggle in the case of suspected pollutants. When one considers all the methodological problems of field investigations, ranging from the inherent imperfections of data, through the weight to be assigned to indirect evidence (as from animal studies), the assumptions of 'normal' practice, and the implicit burden of proof in any regulatory decision, it is easy to see why at the present time methodology has become overtly political, at least in those countries (as the USA) where procedures are required to be published and available for criticism. There, the typical situation is for 'the facts' provided by science to be the focus of debate in public forums, regulatory agencies, and the courts as well.

All this occurs only when a scientific issue has become salient, and there are institutions for its public debate. Until then, and generally elsewhere, the public is ignorant of environmental hazards. The ignorance is not due to an essential impenetrability of the phenomena, but to social decisions (taken in leading institutions of state and of science) to neglect certain problems in favour of others. Such problems will usually *not* be those promising prestige and rewards to a scientific elite, but rather those involving diffuse, imperceptible, chronic or delayed effect of the unintended by-products of the industrial system. In that sense, our scientific-technological establishment moulds public awareness, by negative means, as surely as did the theological establishments of earlier times by indoctrination and prohibitions. The social construction of ignorance is a phenomenon of our modern period, all the more important because it happens unnoticed and in contradiction to the received ideology of science as the bearer of truth.

Technological Blunders

Corresponding to the new uncertainties in science, we have recently discovered the possibility of massive blunders in technology. For a long time it had been recognized that the costs and benefits of technological advance are unequal in their incidence. The conquest and destruction of native peoples by those with superior means of production or destruction is no longer easily justified by the apparent progress of civilization. But we must now reckon with the deleterious as well as the supposedly beneficial aspects of technological progress. This will occur most obviously where a technology is strongly innovative and lacking the automatic controls of a competitive market. Then it can happen that ignorance in the design process and incompetence in fabrication and operation can combine to produce a resounding failure. The most notorious present case is the civil nuclear power industry in the USA. There, cost and time overruns have produced crippling burdens of debt on utilities even when plants have been completed. And when they are abandoned after the expenditure of hundreds of millions of dollars, the victims (utilities and their customers) are left with massive debts and the real possibility of bankruptcy. And incompetence in operation, resulting from the power industry's being unprepared for the sophistication of the technology presented by science, produces even more crippling burdens.

Less obvious on the ground, but equally dramatic, are those cases where chemical manufacturers proceed for years to produce hazardous substances, choosing frequently to remain in wilful ignorance of the dangers to their workforces, consumers and the general public. When this socially constructed ignorance is eventually exploded, it appears that the guilty men have been only ordinary people doing their jobs within the constraints of compartmentalized bureaucratic responsibility and generalized cost-cutting.

The question of how this happens is a real one. Engineers and plant managers of all sorts, presumably well trained and competent in their jobs, have as a group allowed major industries to cause great inconvenience and damage, for which they now face destructive popular antagonism. A part of the answer may lie in the traditional education and outlook of such persons. It has been over-whelmingly restrictive and reductionist, preparing for competence in routine operation, but providing no tools, technical or conceptual, for coping with the new problems of modern high-technology. These problems include extreme sensitivity of plant to deviations from 'normal', so that simple, unavoidable errors can have costly or catastrophic consequences, most familiar in the case of nuclear power. Further, environmental impacts, no longer the gross, obvious pollutants of nineteenth-century factories, lie totally outside the technical com-petence or experience of those who design and operate installations. Trained to solve simple problems in traditional ways, the engineers are far from being in control of the hyper-sophisticated technologies they have created.

Quality Control — The Moral Element

Such problems in technology may be viewed merely as the growing pains of some industries where progress has been a bit too rapid for comfort for a couple of decades. That may well be; only time will tell. But these phenomena do raise the problem of the maintenance of quality-control in these fields. The recent spate of publicized cases of fraud, plagiarism and the claiming of co-authorship on another's work, show that the problem is also present in research science.

The maintenance of quality-control in industrial production has become relatively straightforward. Once the quality of products is appreciated by con-sumers, quality-control is understood by management to be essential for sales and survival; and techniques for employee participation are easily transplanted between such different cultural milieux as Japan and the United States. But in science it is otherwise; there is no external set of discriminating consumers, no hierarchical management, and no simple tests of quality of unit operations. Hence research science must be self-policing; and the wide variations in quality of work between different fields and different centres, shows that the problem has no automatic solution. If we ask what motivates the individual scientist to invest the extra time and trouble to ensure the highest possible quality of his or her research, there can be several answers. The simplest is prudence; poor workmanship will be detected and rejected by his colleagues. But this presupposes a collective commitment to high quality, and so, in effect, begs the question. Other reasons lie in the personal integrity, and pride of craftsmanship of the individual scientist, operating either as a researcher or as a quality-controller in refereeing or in peer review. However, these are moral attributes; they are not automatic consequences of the research process, nor can they be instilled by simple political or administrative means.

Scientific progress is uniquely sensitive to the maintenance of quality. Innovative work is hard and risky; the minority who dare and succeed can all too easily be smothered by an entrenched mediocrity that wishes to stay comfortable in old routines of problems and techniques. Thus the maintenance of generally good quality of research is a necessary background for the emergence of excellence and originality. Governments, even industries, can survive for a long time in a state of complacency and inefficiency, even enduring corruption. When such a situation exists in a field of science, the effects are not visible to the inexpert eye: teaching, research, conferences, grant applications continue smoothly; the one thing lacking is anything worthwhile happening.

Hence the value-component of science has another essential element: the commitment by enough scientists, and particularly those in positions of political power in their scientific communities, to the production of good work, really for its own sake. Otherwise all of the world's research science would soon become like that recognizable in various backwater communities: much spurious activity, but no contribution either to knowledge or human welfare.

Similar phenomena can be observed in fields of technology where purchasers can be captured by producers, notably state (particularly military) procurement. It may seem outrageous and incredible that military authorities would endanger the lives of soldiers, and compromise the chances of victory in eventual wars, for the sake of bureaucratic convenience or advantage. But it is so; the examples are best known for the USA, but perhaps mainly because of the greater openness of government there.

Thus even in the cases of the most apparently hard and objective fields of human endeavour, we can discuss the effects of a moral environment; if not enough people *care* about quality, then it will inevitably be lost. Cyclical theories of civilizations, usually cast in terms of political and military affairs, and standards of private morality, may be seen to apply to science and technology as well.

Fantasy Hardware, the Ultimate Aberration

Before the advent of modern science, there was a well-recognized category of secrets too powerful to be revealed. Whether they were actually so, we will never know. But in any event, the optimistic faith of the seventeenth-century prophets of modern science rendered that category void. Although great material powers were promised through the new science, they were understood to be strictly limited. In the materialist world-view, effects were commensurate with causes; enhancements by spiritual or magical means were seemingly absurd. By the later nineteenth century, the technology of war was eroding that basic metaphysical assumption. Inventors were once again producing 'weapons so terrible that they would make war impossible forever'. Thus nuclear weapons were not a totally new phenomenon; they were in a continuous development ideologically as well as technically. First seen as a cheap but very dramatic extension of a means of quickly destroying a city and its inhabitants, they were indeed used, partly for their immediate effect and partly as an extension of diplomacy.

However, the second generation of nuclear weapons, involving enormously enhanced destructive power, effective means of delivery, and a sharing of the technique between the two major antagonists, did introduce a qualitatively new element into warfare. For it was universally admitted that it was highly undesirable to use such weapons, even though only a critical minority argued that a nuclear war could not be war in any meaningful sense.

The function of such weapons then shifted drastically: it became deterrence. This concept was twofold: it referred to nuclear war involving an exchange of long-range missiles, but it also extended to the discouragement of a conventional war in Europe. In the pure case of intercontinental ballistic missiles, deterrence introduced a very new sort of problem into military theory. Strategic thinking was concentrated on games of bluff and counter-bluff, with models from the theory of games and economic behaviour, and with pay-offs in megadeaths. This was very quickly exposed as an idiotic pseudo-science, by an eminent military scientist, Sir Solly Zuckerman. But he was ignored, by politicians, strategists and philosophers of science alike. Hence the gigantic machine of nuclear armament, distorting the economies and the politics of all the world's nations, and presenting an ever-increasing threat to the survival of mankind, had as its rationale a strictly nonsensical theory. What a fate for a civilization that so proudly bases itself on science!

Practical contradictions also afflicted nuclear strategy, though these took a couple of decades to mature. The defence of Europe by the threat of its obliteration through American-controlled weapons led to increasing disquiet there. Civil defence finally revealed its idiocy in American plans for evacuations, requiring (for example) the inhabitants of each of the so-called 'twin cities' of Minneapolis and St Paul to seek refuge in the other!

Independent deterrents by second-rank powers as Britain and France could be only an expensive means of maintaining fantasies of national glory. And the spread of nuclear weapons to less-responsible ruling elites poses a sinister threat that cannot now be removed. Such a situation might have seemed as bizarre as it was possible to be, until a new element was revealed in the early 1980s: the weapons themselves are unreliable. American missiles have been tested only on constant-latitude paths. Hence any talk of first-strike, counterforce attack (by missiles travelling over the pole and targeting with great accuracy and precision), is pure fantasy. Further, the coming generation of American missiles seem likely to impose a *de facto* freeze. The MX system is totally devoid of any plausible function, except to keep the air force in the nuclear arms business. The Pershing II missile is a design disaster. The Cruise missile can fly sometimes under optimal conditions, but it is so plagued by difficulties that its production-run has been seriously curtailed.

All these facts are in the public prints, and yet all sides in the nuclear debate choose to ignore them. Of course, existing weapons are in place, the spread continues, the threat to humanity is as menacing as ever. But it seems to be in no one's interest to make political use of this essential feature of nuclear weapons:

as well as being absolutely evil, they are also absolutely insane, even to the point of becoming increasingly a matter of sheer fantasy.

The metaphysics of our civilization is based on an absolute distinction between the primary quality of things, taken from mathematics, and the secondary ones, involving perceptions. Tertiary qualities, involving values, are allowed metaphysical reality only on Sundays. This world-view has been dominant for some three centuries. Now its contradictions have matured. They are most manifest in plans to base a nuclear strategy on a future missile system that will certainly never operate. This complete interpenetration of fantasy and hardware could be seen as a sort of Zen koan; and perhaps some day it will.

References

The general problem of quality-control, and the importance of morale and of moral imperatives, was discussed at length in my book *Scientific Knowledge and its Social Problems*, Oxford University Press, Oxford, 1971.

The phenomenon of the misdirection of science, to the neglect of problems of human and environmental concerns, is discussed in *Quality in Science*, M. Chotkowsky La Follette (ed.), MIT Press, Mass., 1982, particularly in the essay by Harvey Brooks, 'Needs, Leads and Indicators'.

The most recent study of the provision of low-quality or inappropriate weapons to the American military is *National Defense*, James Fallows, Random House, 1981. His most striking example is the modification of the M-15 rifle into an ineffective weapon for use in Vietnam, in the interest of the preservation of a bureaucratic monopoly on design and testing.

For the history and institutional/political theory of the development of nuclear weapons and nuclear strategy, an eye-witness account is provided by Lord Zuckerman; see his *Nuclear Illusion and Reality* (Viking, London, 1982).

3

Radical Sociology of Science
From Critique to Reconstruction

Alejandro Gustavo Piscitelli

The 'new' alliance between science and politics comes about neither from a perversion of the post-Renaissance scientific ideal, nor from a corruption of the scientific ethos lured by the temptations of industry and consumerism. Politics instantiates possibilities in modern science which were previously hidden. While the seventeenth century announced it, the eighteenth century dreamed of it and the nineteenth century set the blueprints, the hegemony of science was only realized with the industrialization of knowledge triggered by the Manhattan project (build-up of the first atomic bomb).

Science has carved so deeply into the organization of the economic blueprint that distinctions we were once happy to use to distinguish between different social systems have definitely collapsed. Whereas a century ago we could easily classify countries and regions according to the bias of religious, ethnic, ideological or even social criteria, nowadays the identity stick is measured in terms of production output and in knowledge-related investment and innovation.

In the mid 1960s an American sociologist of science attempted a detailed quantification of world scientific output for all times. He found out that during modern times 50,000 different journals had been published of which 30,000 still were in publication. The total number of scientific papers published up to the 1960s was around 10 million, with an estimated annual increase of 600,000 units. People with university degrees in the USA numbered around 1 million, compared with 1,000 in 1800; 10,000 in 1850; and 100,000 in 1900 (Price 1973).

One of Price's most significant revelations was that the exponential increase in scientific productivity was not a new trend but an entrenched tendency begun at the time of the industrial revolution (Rescher 1978). There are two extrapolations that can be based on Price's findings. Firstly, that science as a set of rule-governed activities grounded on social compromises is a massive enterprise that has greatly benefited from the increase in resources, inputs, information, purveyed mainly by the national states. Secondly, that as a productive enterprise it has been generating relentlessly both a huge literature that advertises its findings as well as a wealth of technological innovations that embed them.

Nowadays, science is a big industry. This connection to commerce and finance, much more than its increase in speed, is what separates small from big, artisanal from twentieth-century science. Industrial society is a mobilized society which concentrates its efforts in a limited number of goals among which the defence and development of scientific and technological resources have priority. The scientific estate not only endorses science as one of its main offsprings, but is characterized by the fact that scientists want their own goals to be the same as those of society. The increasing difficulty of separating the analysis of knowledge from the problems of power relies heavily on the fact that by means of its operational character modern science has transformed its theoretical nature into political discourse.

A crucial question to ask at this stage is why we know so little about science. I would suggest that it is because the analysis of knowledge production, consumption and distribution was initially in the hands of the professional philosophers and is now in those of academic sociologists of science. Both philosophers and scientists have long cherished the description of the analysis and critique of knowledge given by John Locke many centuries ago:

> . . . in an age that produces such masters as the great Huygenius and the incomparable Mr Newton, with some other of that strain, it is ambition enough to be employed as an under labourer in clearing the ground a little, and removing some of the rubbish that lies in the way to knowledge.

In spite of what philosophers claimed about their humble taks, the most perfunctory reading of their works reveals that their avowed intention got transformed into the much less humble one of setting criteria for relevance and irrelevance in the use of language, and thence changed into the more ambitious one of setting the standards by which all claims of knowledge should be assessed.

A second conception espoused by Locke was the master-scientist or metaphysical view according to which the relationship between science and philosophy consists in building up the entirety of acceptable human knowledge into a one massive, logically connected and internally consistent system of propositions. This epitomizes the Cartesianism and its twentieth century equivalent — the logical empiricist tradition.

Kant pointed out long ago that something — if not everything — must be wrong in a discipline in which answers to its central questions may be just as easily proved as their contradictions. That men cling to contradictory statements, promoting them to a self-consistent and self-sustaining system, has been restated once again by twentieth-century logical positivism in its unsuccessful attempt to marry the antithetic traditions of the under-labourer and the master-scientist conception of philosophy. In spite of their acknowledged difference both positions share an understanding of science and philosophy that is deeply anti-historical and consistently anti-localist and universalistic. This is not the place to work out the pragmatic and theoretical consequences of their far-fetched assumptions, but what is beyond any doubt is that both of them are false and mischievous (Amsterdanski 1975; Ford 1976).

As many researchers in the domains of the theory of ideology and the sociology of the science have shown (see among others Barnes 1977; Coward 1977; Thompson 1984), we are today in a position to stress that the 'pure' and uncontaminated vision of knowledge and knowledge production — as codified under the 'standard' theory of science — is more a product of a normative proposal than the description of how the scientific machinery works.

The constitution of the scientific object is deeply prejudiced by the frameworks within which we organize our experience a priori and before any knowledge has begun (Foucault 1970). The technical and practical interests of knowledge are not to be seen, therefore, as hindrances which have to be eliminated for the sake of pure knowledge: they constitute the framework which enables certain parcels of reality to be objectified and they can be made accessible to experience, although most of the time retrospectively (Habermas 1979).

Entrapped by a positivistic conception of objective knowledge most contemporary theorizing has tried to deal with social structure without taking into account social conflicts, change, and persistent social transformations in terms other than as disruptions and undesired perturbations (disorder). Cultural productions are not reducible to formal and stable systems accountable in terms of either random or purely acontextual cultural selection such as Popper or Campbell contend. Research must focus instead in the cultural practice, on the role human beings play in the system of cultural signs in which they are embedded (Bourdieu 1983). It is no longer possible to posit any human endeavour, be it as abstract as science, as being outside the cultural system.

Under different guises and for a long time now, positivism and variations thereof have directed our conception of science and knowledge. Among other important tenets, positivism has endorsed a view of an objectivistic, theory-ridden, purely descriptive and observationally neutral science. Philosophers of science from Nagel to Popper, from Lakatos to Musgrave, from Russell to Baskhar, under different guises, and using increasingly sophisticated arguments and mathematical apparatuses, have tried to sever the context of justification from the context of discovery, facts from values, ideology from science. Certainly their differences are great and some of their analysis particularly useful, but nevertheless they all share those anti-historical and anti-political traits that not only discourage any comprehensive understanding of science in its compromised relationship to society but, what amounts to the same, try to install a normative model of science production.

One of the most unfortunate situations in the sociology of science is the fact that this positivist model has been pervasive since the start. This ingrained leaning towards positivism and naturalism has been so great that not even those authors that purported to cut their ties with it (such as Kuhn for example) have been successful in their attempts.

Since some of these basic connections between the standard view of science and the naturalistic sociology of science have usually been ignored, a cursory examination of their mutual relations is due.

Against a Normal Sociology of Science

That dreaded connection between positivism and the first attempts to establish a normal science is an important factor in obscuring social scientists' analysis since it is big, ominous, polluted, entrenched with political and military interests acting as substitutes for what a model of science should be. Thus we should not be surprised that we had to wait until the mid 1930s before anything like a sociology of science saw the light. Besides, an 'external' event — the international Congress of Science and Technology held in London in 1931 with its Marxist emphasis on the social determination of knowledge — was to have a much more important influence in the setting up of the discipline than any previous academic stimulus.

The queries of Robert Merton, the founding father of the discipline, would be moulded on the aftermath of those discussions. He was influenced by the work of Karl Mannheim and Max Scheler as well, and finally published the first studies dealing with the historical development of science (Merton 1973). His important contributions focused on a definition of science as a social activity based on its own ideal norms of communalism, universalism, disinterestedness and organized scepticism. Important as the Mertonian analysis was, today we tend to appreciate it more for its philosophical and political undertones than for its renderings on science.

Merton attempted to secure for science a strong foothold where it could avoid the confusing demands of politically conflicting life. His project condoned the autonomization and absolutization of science and it cannot be severed from his positivistic allegiance and his functionalistic approach. Intimately woven to those tenets, Merton's sociology of science fostered a conception of science as a rational, cumulative and non-conflictive view of knowledge production and accumulation. Modern writers have been sensitive to the exquisite blend of a positivist epistemology with a consensualist sociology in Merton's work.

> . . . tout se passe en effet comme si la conception paisiblement cumulative, non conflictuelle et consensuelle de la science d'une part, et les exigences propres de l'analyse fonctionnelle d'autre part, se renforçaient

mutuellement pour conferer au statu quo scientifique et social de la science la justification d'une entiere rationalite (Lecuyer 1978:273)

Merton's thesis has reigned undisputed for twenty years as the main guide in the sociology of science. There were of course minor disagreements and corrections. From Hagstrom's theory of exchange to Mulkay's introduction of the psychology of the discovery, attempts were made to refine the peripheric assumptions without ever touching the core. In fact from the mid 1960s to the present day researchers have worked on many subjects in the sociology of science from a standpoint that is not strictly Mertonian in itself, but which does nevertheless pay him more lip service than is currently acknowledged. Thus current research coming from the sociology of labour and organizations, from the sociology of professions, from the science of science (mainly Solla Price's) and from the historical sociology of scientific institutions (mainly Ben David's) has again criticized Merton in order to secure his most entrenched views of science as a neutrally political activity.

For this reason the Kuhnian 'revolution' which started in the mid 1960s was to have such a wide impact in that it attempted to debunk not only a way of conceiving the internal workings of science, but all those philosophical, sociological and political presuppositions that were attached to the Mertonian model. Kuhn's thesis (1970) has been so widely discussed that to restate it may seem superfluous. Briefly, however, his view is that science is an activity governed by rules which in the last analysis happen to be external to the scientific apparatus. He postulates that:

1. Paradigms dominate normal science.
2. Scientific revolutions amount to paradigmatic changes.
3. Paradigms determine observations.
4. Paradigms set up the legitimacy criteria of scientific discourse.

In his view science is a complex activity that blends together observations, theories and models, traditions, research programmes and metaphysical presuppositions on the entities that build up the world. Here, more than in any other sociological research programme so far elaborated, one would expect to have fulfilled the old dream of a combination between the externalist and the internalist analysis of scientific production.

Almost twenty-five years later than announced this programme has been submitted to the hugest hagiographical endorsement as well as to the most devastating criticism. It is impossible to assess both the latter or the former if one does not bear in mind that the programme had as its main target the philosophical model of science growth development developed by Karl Popper in the mid thirties. Revamped on the works of Nagel, Braithwaite, Hempel *et al.*, that model was to become the core of the standard view of science. Although it is undeniable that Kuhn's model attempts to debunk Popper's, yet there is a vast area of common ground between them. The differences are often so subtle that they could be passed over. For instance, both authors agree that facts are not simple things given to us in an unproblematic, direct experience of the world. The same idea applies to truth: Popper would agree with Kuhn that no guarantee can be provided in order to ensure progress towards the goal of truth.

Nevertheless the divergences between both accounts are considerable too. Thus they give different weight to the descriptive and normative aspects; Popper stresses debate, disagreement and criticism whilst Kuhn stresses the areas of agreement that are taken for granted; Popper focuses on those aspects of science which are universal and abstract, whereas Kuhn focuses on the local and concrete aspects such as the specific pieces of work which provide exemplars for groups; Popper's vision of science sees it as a linear, homogenous process while Kuhn has a cyclical conception.

What should be remarked is that this debate in the philosophy of science is structurally identical to debates which have gone on for some two hundred years in the realms of political, social, economic, ethical and legal theory. The clash between Popper and Kuhn, between the Mertonian and the anti-Mertonians is almost a pure case of the opposition between the Romantic and the Enlightenment ideologies. The correspondence can be briefly stated:

1. the antithesis of individualistic democracy and collectivist, paternalist authoritarianism is apparent in the two theories of knowledge. Popper's theory is anti-authoritarian and atomistic; Kuhn's is holistic and authoritarian.
2. the antithesis of cosmopolitanism and nationalism is also easily detected. Popper's theory of the rational unity of mankind and the 'free-trade' of ideas contrasts with the closed intellectual state of the paradigm and with the special richness of its unique language.
3. the antithesis between the Benthamite lust for codification and clarity and Burke's claims about the rule of prejudice corresponds to the difference between Popper's methodological legislation and boundary drawing and Kuhn's stress of dogma, tradition and judgement.

(Bloor 1976)

Why does this repeated pattern of ideological conflict crop up in an esoteric area such as the philosophy of science? Why does the philosophy of science replay these themes? What do these remarks imply for the possibility of a renewed sociology of science?

The hypothesis that we want to advance, though without attempting a detailed proof, is that theories of knowledge are, in effect, transformations of social ideologies. The ideological opposition is widely diffused throughout our culture. We may have in our minds, from our experience of social life and language, the very social archetypes that appear to be having an effect on the theories of knowledge just considered. What may feel to the philosopher like a pure analysis of concepts or meanings, or the mere drawing out of logical entailments, is in reality the rehearsal of certain accumulated experiences in our epoch. What is more important for our analysis is that if knowledge is endowed with a secret character, because of the connections of the image of knowledge to images of society, then both the Popperian and the Kuhnian programmes would be equally opposed to the sociology of knowledge and to a critical sociology of science.

The fact is that they are not equally opposed. Although the Popperian accusations to Kuhnians on the grounds of irrationalism and subjectivism are strong, they do not dispel the mystifying characterization that Kuhn gives of knowledge. The difference between Popper and Kuhn is not their relative resistance to the sociology of knowledge but the strategies that each of them use in order to secure this end:

> . . . the mystifying resources of Kuhn's account are clear because of its similarities to Burke's position. The Romantic means of fending off unwelcome investigation into society, whether scientific or otherwise, is by stressing its complexity, its irrational and uncalculable aspects, its tacit, hidden and inexpressible features. The Popperian style of mystification is to endow logic and rationality with an a-social and, indeed, transcendent objectivity (Bloor 1976:67).

The 'law' at work here appears to be this: those who are defending a society or sub-section of society from a perceived threat will tend to mystify its values and standards, including its knowledge. Those who are either complacently unthreatened or those who are on the ascendency and attacking established institutions will be happy for quite different reasons to treat values and standards as more accessible, as this-wordly rather than as transcendent. The fact is that the Kuhnian sociology of science has been normalized and Kuhnians have themselves come under attack. It could be shown how the corrosive approach of still other epistemological positions — such as Prigogine's — is finally undermined by the kind of allegiances that Bloor so sternly denounces.

One point should now be clarified. The shift away from a functionalistic sociology of science that emphasized consensus and with ideals placing science in an ivory tower, was shattered by the social and economic upheavals of the mid 1960s. At that time the Kuhnian paradigm emerged only to be swiftly swept away both by its internal retreat (Kuhn's positions being progressively milder and searching for bigger compromises with its archrivals) and by an ongoing cultural conservative revolution.

If we examine the way science is practised in places other than standard academe, or normal research institutes in the developed countries, we would be able to reassess the lost critical stance of the Kuhnian programme. I should now like to refer briefly to some frontiers of research which could mould an entirely different image of science to that given either by the Mertonian or the Kuhnian approaches.

The focus will still be on fringe research since I believe that there is where we can find the sources of inspiration in order to know the exact shape science has today, and how some scientists (e.g. Varela and Maturana, 1980) have been able to build up an epistemological view of knowledge that is neither prone to the deficiencies of the positivist account nor to the naturalistic reductionism implicit in the Kuhnian approach.

The Self-organizing Paradigm

Much has been said in the last decade about the cybernetic revolution. Big research programmes have been structured under its banners and the results have been generally poor, a mere shift of names for the same things that were done before under other labels (Wilden 1980). What is less known is that here, as was to happen in the sociology of science itself, the original systemic and cybernetic movement had a critical and humanistic slant that was to be progressively suffocated by the more vociferous and pragmatical engineering mongers (Helms 1980; Dupuy 1986). With the rather eccentric exception of Gregory Bateson and the less known work of Heinz Von Foerster, there is little produced under the systemic umbrella which could resist the erosion of time and the changes in theoretical fashions. Fortunately, the work of such pioneers was resumed in the mid 1970s by two Chilean biologists (Varela and Maturana) whose reflections have initiated a new stage in the philosophy of science and from whose work extrapolations to the sociology of science have still to be drawn.

One of the most important results of the self-organizing revolution was the profound blow that it delivered to the hidden motor of the standard view of science: that is, the endorsement of reductionism and analytical thought as the main engines for the understanding of natural and social life. The general programme of reductionism has consisted historically of the attempt to account for phenomena pertaining to a certain level of organization by means of laws which explain phenomena belonging to a lower order of complexity. The degree of sophistication involved in such a programme is high. It is nonetheless ironical that whenever reductionism is espoused the historical development of science inevitably flunks it. Does not in fact the history of science teach us that the most daring attempts at reduction (the reduction of rational fractions to order pairs of natural numbers; the reduction of mechanics and chemistry to the electro-magnetic theory of matter; the reduction of chemical bonds to quantum theory) have all failed?

Why then do scientists go on attempting reductions? One reason, almost anecdotical, is because we can learn an immense amount even from unsuccessful or incomplete attempts at reduction. A more revealing reason is that scientists are strongly committed to extra-methodological beliefs which compel them, because of the architectonic and metaphysical presuppositions they embrace, to unify the existing complexity. This second reason is particularly applicable when the phenomena under study may be accounted for in radically different forms by competing theories since we know that the 'competition between paradigms is not the sort of battle that can be resolved by proofs' (Kuhn 1970).

The attempt to reduce complexity to elementary components as well as its foretold failure are but remnants of a conception of science that is being steadily challenged by new approaches and proposals highly incompatible with the positivist paradigm and its regulatory principles, such as reductionism.

Many of our competing points of view arise from the use of strongly 'polarized' concepts that make reference to a privileged direction in the flow of time. The concepts strike different emotional chords whether we are inclined to espouse explanatory models that are causal (events 'pushed' by the past and headed towards disorganization) or teleological (events 'pulled' by future explanations and adding order to an otherwise inert universe).

In order to escape this dychotomic way of reasoning which lies at the bottom of most of the contemporary disputes in the philosophy of science we must get rid of the lures of chronocentrism, that is, the deep association between chronology and causality and its associated European ethnocentric categories.

One interesting property of cybernetic feedback loops is that they point out the way finality or purpose appear in complex systems in an information/decision/action loop; information on the results of past actions is the basis for the decisions that will correct a present or future action. In as far as decisions are made to achieve an end, the consequent action is purposeful. A loop exhibiting such a trait reveals the occurrence of an intelligent act. In a feedback loop, cause neither precedes nor follows the effect. Causality and finality follow the entire circuit of the loop and neither can be distinguished from the other. With the discovery of cybernetic loops, circular causality was superimposed onto linear causality. In this kind of loop (not to be confused with a cycle) the arrow of time appears to close onto itself. The expression 'time passes' is, perhaps, erroneous. Time is balanced by something else.

Once the chronology of events is questioned, our logic falters since only chronology permits explanations by causes. We shall always face this kind of 'vicious circle' when looking for the origins of complex systems. It may well be that questions about origins can be answered only by means of arbitrary cuts which sever feedback loops in order to recover the familiar relationships of before-and-after, between cause and effect. This in fact is the core of the analytical approach (de Rosnay 1976). Not being able to consider all the inter-dependences of the functioning mechanisms in complex systems, we isolate those loops which we deem essential, opening them and patterning them according to the laws of cause and effect we find in the static and dynamical systems studied by physics (Varela 1975). This skewed reading of complex phenomena may well be a response to some kind of human habit stemming from the adaptive psychological meaning of 'after' and 'before'. Events seem logical insofar as there is a chronology: i.e. in so far as the arrow of time points toward increasing entropy. We have come to equate chronology with causality. That is why physics will accept only causal explanations, where improbability is given at the start, but rejected at the end.

The conventional direction attributed to the generalized vector of evolution leads to irreducible points of view which lie at the bottom of the most salient epistemological controversies in the recent history of science (materialism vs. spiritualism, Darwinism vs. Lamarckism, structuralism vs. phenomenology). One of the most remarkable by-products of the self-organizing revolution is the attempt to overcome this conceptual escalation by means of complementing opposing views. In a more productive vein the evolutionary standpoint imbedded in the self-organizing paradigm urges us to revise the normal epistemology endorsed by the positivists with its corresponding political and practical consequences.

Philosophers of physics such as Grunbaum and Reichenbach, have recently pointed out that both the principles of sufficient reason and the principle of causality originate from the adaptive direction of time. Phenomena are observable only when they occur in the direction toward which the life of those who observe them is also flowing. It is highly probable that we are capable of examining in detail only that which is in a state of transition towards decay and decomposition. If this is so, the fact that scientists generally seek, and find, certainty in the past would be much easier to understand.

The chronocentrism embedded in traditional thought is neither naive nor self-evident, it implicitly endorses metaphysical assumptions rarely acknowledged (Popper 1974). By pushing causal reasoning to its limit, we must rely on cosmological explanation such as that all negentropy is given from the beginning of time (as purported in the big bang theory).

I shall not summarize here the main points at which traditional positivist thought and the cybernetic paradigm clash (Skolimowski 1974). The cybernetics of self-organizing systems provide a theoretical clue to this new epistemological synthesis. To maintain the organization of a complex system one is forced to slow the increase of entropy into the system. The creation of information and organization results in holding time, preventing its loss. Conservation of time comes about through the maintenance of a balance between speed of organization and speed of disorganization in the world.

The increase in complexity is neither unavoidable, nor irreversible. All organization, no matter what its form, remains subject to degradation, to use. Human society is even prone to instant entropization through nuclear catastrophe. Even so, creative action compensates for the passage of time. Every original work is analogous to a reserve of time information.

In the complementarist view provided by the self-organizing paradigm, information and negentropy are no longer divided into two separate worlds; they are the hinges between the objective and the subjective. Superimposable and equivalent, information and negentropy possess opposite temporal poles.

Through observation we discover the world in a direction analogous to that of waves diverging from a source: the direction of conventional time. The universe is seen here exclusively according to its energetic, quantitative, material and objective aspect. Through creative action and in the richness of living experience we discover its other face in the direction of waves converging towards the centre.

What the self-organizing paradigm was able to show is that two basic entities lie at the end of such an epistemological journey, namely energy and mind — its intermediate forebears being matter and information. Nevertheless we insist in perceiving the world as if only two things existed: informed energy and materialized mind. If time can be conserved, freedom could be totally contained in the present. The universe would thus be a consciousness that creates itself as it becomes conscious of itself, that is, as it evolves.

One of the most important epistemological consequences of the findings of Von Foerster, Varela *et al.* is that the complexity of systems is much more a function of the observing systems than that of the observed systems. It is impossible to understand the social behaviour of man without taking into account the fact that knowledge and behaviour have both a biological and social basis and that all observation must be inscribed in a system of alternative perspectives. Given the relentless war about different paradigms that is currently raging in the social sciences, this approach is salutary since it enables us to realize that all the positivistic tenets which guided most of those unfortunate analyses are being currently debunked one after the other exactly there where they originated: in the hard sciences.

Work needs to be done to gauge the use of these new tenets in the factual research in the social sciences. In particular, a lot of new questions arise as to the relationship between the self-organizing paradigm, the future of rationality and the problems posed by political activism and the choice of a better world in which to live and let live. Much time and effort is being devoted to put together the bits and pieces of theory that are being produced in the most disparate realms of knowledge. As long as those contributions are particularly sensitive to their relative shortcomings and to the need of their mutual hybridation, they will all be equally welcome. This notwithstanding, the true test of these theories will not be in the realm of ideas — it will not consist in judgements made on the grounds of mere logical consistency. Their survival value will be intimately attached to the degree in which they help improve the living conditions of the submerged people in this world. For if these attempts to understand and transform the material conditions of the planet pass unnoticed, there shall be no more history — or anything deserving such a name; and then all theory will have been in vain!

Bibliography

Amsterdanski, S: *Between Experience and Metaphysics. Philosophical Problems of the Evolution of Science*, Reidei, Dordrecht, 1975.

Barnes, B. *Interests and the Growth of Knowledge*, Routledge, London, 1977.

Bloor, D. *Knowledge and Social Imagery*, Routledge, London, 1976.

Bourdieu, P. *Campo del poder y campo intelectual*, Bs As Folios, 1983.

Coward, R. and Ellis J. *Language and Materialism, Developments in Semiology and the Theory of the Subject*, Routledge, London, 1977.

De Rosnay, J. *Le mascroscope. Vers une vision globale*, Seuil, Paris, 1975.

Dupuy, J.P. 'Leessor de la premiere cybernetique 1943–1953' *Cahiers CREA* No. 7, Paris, 1986.

Ford, J. *Paradigms and Fairy Tales*, Routledge, London, 1976.

Foucault, N. *L'ordre du discours*, Gallimard, Paris, 1970.

Habermas, J. *Communication and the Evolution of Society*, Beacon, Boston, 1979.

Heims, S. *John von Neumann and Norbert Wiener, from Mathematics to the Technologies of Life and Death*, MIT, Mass., 1980.

Kuhn T.S. *The Structure of Scientific Revolutions*, Chicago, University of Chicago Press, 1970.

Lecuyer, B.P. 'Bilan et perspectives de la sociologie da la science', *Archives Europeans de Sociologie*, Tome XIX, 1978: 257–336.

Locke, J. *An Essay on Human Understanding* (1691), Verlag, Germany, 1963.

Merton, R. *The Sociology of Science*, University of Chicago Press, Chicago, 1973.

Popper, K. 'Scientific reduction and the essential incompleteness of all science', in F.J. Ayala and T. Dobzhanksy (eds.) *Studies in the Philosophy of Biology*, University of California Press, Berkeley, 1974.

Price, D.J.S. *Little Science, Big Science*, Columbia University Press, New York, 1973.

Rescher, N. *Scientific Progress: A Philosophical Essay on the Economics of Research in Natural Science*, University of Pittsburgh Press, Pittsburgh, 1978.

Skolimowski, H. *Problems of Rationality in Biology*, in F.J. Ayala and T. Dobzhanksy (eds.) *Studies in the Philosophy of Biology*, University of California Press, Berkeley, 1974.

Thompson, R. *Studies in the Theory of Ideology*. University of California Press, Berkeley, 1984.

Varela, F. 'Not one not two', *Coevolution Quarterly*, Fall 1975.

Varela, F. and Maturana, H. *Autopoiesis and Cognition: The Realization of the Living*, Reidel, Dordrecht, 1980.

Wilden, A. *System and Structure: Essays in Communication and Exchange*, Tavistock, London, 1980.

Part Two

Science and
Third World Domination

4

Science and Control
Natural Resources and their Exploitation

J. Bandyopadhyay and V. Shiva

How can modern science, which is defined as apolitical, neutral and universal be an ideological weapon in the control of resources? We shall argue that the characterization of modern science as apolitical and universal is itself an ideological act by which other systems of knowledge are rejected and subjugated without rational evaluation. Further, through this ideological act, control over resources is taken out of the hands of indigenous peoples and local communities and put into the hands of a minority. The 'experts' play a critical ideological role in this transfer by

1. creating a knowledge system which produces epistemological conditions for transfer of control;
2. creating technological systems which divert or destroy resources for commodity production;
3. creating a legitimization system when the transfer of control is challenged.

The challenge is emerging both from the knowledge traditions of the South as well as the alternative modernity in the North and South which puts sustainability of development above growth criteria and puts equity before profitability criteria. The conflicts between the dominant science system and an alternative science system thus express themselves as conflicts over natural resources. Science and technology are central components of contemporary political economy even while they are characterized as objective and interest-independent. This characterization is then used ideologically to legitimize a particular growth of political economy and control over natural resources by special interest groups.

Science and technology are evolving in directions that allow the control of natural resources to shift increasingly away from people who use them for sustenance and survival into the hands of groups who use them for profit. Precisely at a time when science is a primary source of economic, political and social accumulation and control, science and the social projects it generates and supports are presented as interest-free and as political instruments for human progress. And the sacredness that science enjoys in contemporary times makes it taboo to investigate its political and economic roots in social projects of accumulation of power and profits. People's movements to regain citizen rights to control over natural resources need to investigate these linkages and reveal the ideological role of science in control over natural resources both through the biased cognitive structures it creates and through the presentation of these structures as neutral and value-free.

Conflicts Over Natural Resources

The recent period in human history contrasts with all the earlier ones in its strikingly high rate of resource utilization. Ever expanding and intensifying industrial and agricultural production has generated increasing demands on the world's total stock and flow of resources. These demands are generated mostly from the industrially advanced countries in the North and the industrial enclaves in the underdeveloped countries in the South. Paradoxically, the increasing dependence of the industrialized societies on the resources of nature, through the quick spread of energy and resource-intensive production technologies, has been accompanied by the spread of the myth that increased dependence on modern technologies means a decreased dependence on nature and natural resources. This myth is supported by the introduction of long and indirect chains of resource utilization which leaves invisible the real material resource-demands of the industrial processes. Through this combination of resource-intensity at the material level and resource indifference at the conceptual and political levels, the conflicts over natural resources generated by the new pattern of resource exploitation are generally shrouded and ignored. The conflicts become visible when the resource and energy-intensive industrial technologies are challenged by the communities whose survival depends on the conservation of the resources threatened by destruction and over-exploitation. Or when the devastatingly destructive potential of some industrial technologies is demonstrated as in the Bhopal disaster.

Ecology movements emerging from conflicts over natural resources and the people's right to survival are spreading in regions like the Indian sub-continent where most natural resources are already being used to provide the basic needs for survival to a large population. The introduction of resource and energy-intensive production technologies under such conditions creates economic growth for a small minority while, at the same time, undermining the material basis for the survival of the large majority.

For centuries, vital natural resources like land, water and forests had been controlled and used collectively by village communities thus sustaining these renewable resources. The first radical change in resource control and introduction of major conflicts over natural resources induced by non-local factors was associated with colonial domination which systematically transformed the common vital resources into commodities for generating profits and growth of revenues. The first industrial revolution was to a large extent supported by this transformation of common resources into commodities which made South Asian resources available for the European industries.

With the collapse of the international colonial structure and the establishment of sovereign countries in the region, this international conflict over natural resources was expected to be reduced and replaced by policies guided by comprehensive national interests. However, policies have continued to be promulgated along the colonial pattern, and in the recent past a second drastic change in resource use has again been initiated by international requirements and the demands of elites in the Third World requiring natural resources. The most seriously threatened interest in this conflict appears to be that of the politically weak and socially disorganized group of poor people whose resource requirements are the lowest and whose lives are mainly supported directly by the products of nature outside the market system. Current changes in resource utilization have almost wholly by-passed the survival needs of these groups.

The use of natural resources for the production of paper conflicts with ecological demands on these resources for soil and water conservation as well as with the local people's requirements of forest products such as fodder, fuel, green manure, small timber, fruits, nuts, since the available land for biomass production is limited. The dwindling forest reserves have pushed paper industries into looking for newer bases of raw materials. As a result, in many parts of the country a rapid transfer of food-growing land to the production of commercial woods such as eucalyptus has taken place through what is commonly termed social forestry. This has decreased the potential for food production in two ways. Firstly, it has decreased the direct availability of existing cultivable land for much needed food production. Secondly, it has initiated a process of land degradation, particularly in the drylands, which reduces the long term potential for food production significantly. Similarly, the expansion of energy and resource intensive industries such as aluminium or steel puts further demands on land and water. For the mining activities that support these industries more land under forests or agriculture is being acquired and destroyed. The large hydro-electric projects that are set up to generate power, mainly for the growing industrial demands, in a similar manner destroy the potential for the production of food by submerging vast areas of fertile land in the river valleys. People's protests against such transfer of basic resources have so far been local and disorganized and as a result they have not seriously affected the formulation of national policies. With the advance of industrial growth in the years to come these protests are going to be more intense. These conflicts may not always be contained by ecology movements which hold out the possibility of resolving them in a just manner, but may get distorted into assimilating other social conflicts. Contemporary social complexity and instability create an urgency for understanding the difference between social responses that strengthen or damage the options for peace and survival.

Despite similarities, there is an important difference between the ways conflicts over natural resources were handled in the colonial period and the way they are handled now. Colonialism was characterized by transfer of and increased access to natural resources made possible by direct political and military interventions, whereas the post-colonial period is characterized by the subtle use of subsidies and an ideology of development that is shared by the industrially advanced countries in the North as well as the elites in the urban-industrial enclaves in the South. In the new political context of the post-colonial period the justifications of the resource utilization patterns are hardly different from those in colonial times, and the suppression of the conflicts generated by these transformations is rooted in the claim of the superior rationality of the sciences and the superior productivity of the technologies and economics on which the new modes of resource exploitation is based. Conflicts over natural resources must, therefore, be specially investigated in terms of the rationality and efficiency criteria employed to legitimize the present destructive patterns in the name of the development of the poor people in particular, or the nation in general.

The deteriorating condition of the natural resources has, in the recent past, created environmental responses at various levels; while the environmental consciousness of the elite in India can easily be traced back to the Stockholm Conference on Environment held in 1972. The ecological sensitivity of the rural people in India has always been a central element in their culture and consciousness; this has led to people's resistance to practices that are ecologically destructive.

About three centuries back the Vishnois of Kheri village in Rajasthan sacrificed more than 200 lives in a passive resistance to the felling of green trees by the royal forces from Jodhpur. Farmers and the forest dwelling communities in all parts of the country resisted the destruction of the forest resources by the British rulers through the forest *satyagrahas* in the 1930s. Among the other notable movements based on conflicts over natural resources in colonial India are the indigo movement of Bengal and Bihar, and the famous salt *satyagraha* initiated by Mahatma Gandhi at the end of the Dandi March that marked the beginning of the non-cooperation movement in the 1930s.

The intensity and range of ecology movements in free India has increased in reaction to the huge expansion of the industrial sector and the starting of development projects based on energy and resource intensive technologies which threaten the survival of the economically poor and politically powerless. The most important ecology movements in India in the last few decades have taken place around the conservation of the essential resources of soil, water and vegetation.

The Chipko movement, involving those living in and around the forests, is undoubtedly the most widespread and well known people's ecology movement in India. The economic activities connected with the livelihood of these people are based on the forest resources, which play a central role in the stabilization of the soil and water systems and thus maintain the agricultural productivity of land. The Chipko response was first sparked off in the Himalayan areas of the state of Uttar Pradesh and later spread to other mountain areas like the Western Ghats, the Aravallis and the Vindhyas. Movements to protect forests have also been characteristic of the tribal belts of the country, specially in Central India, notably in the Singbhum and the Hastar regions. These movements have been directed against the conversion of mixed natural forests to monoculture of commercial species like teak or tropical pine.

The destruction of forests and agricultural land has also been strongly resisted in those parts of the country where there are development projects like big dams or plans for urban settlements. Such movements are going on in diverse areas of the country like Tehri in the North, Sirsi in the South, Koel-Karo in the East and Inchampalli in Central India. The destruction of the life support system through the introduction of water-logging and salinity has been resisted by the farmers' movements in various parts of the country, particularly in Karnataka and Madhya Pradesh. Another significant movement based on conflicts over natural resources has evolved around the threat to the marine resources as well as to the survival of indigenous fishing communities posed by the reckless overfishing by mechanized boats and the ecological destruction caused by trawling.

Inspite of differing in their geographical location or the resource focus, all the people's ecology movements in India arise from and make visible conflicts over natural resources in a common way. These conflicts exist at three levels:

1. *Economic level.* The movements to resist ecological destruction and to ensure conservation of natural resources bring out the conflicts between two types of economic activity; one aimed at ensuring survival for the people in a sustainable manner through a genuinely collective mangement of the resources, the other aimed at maximizing economic growth for a few at the cost of the basis for the material survival of many.
2. *Technological level.* The movements based on the issue of survival do not merely question the control, management and distribution of resources. Their resistance against resource destruction also questions the manner in which resources get transformed, processed and utilized. These movements thus bring out the conflicts between the two groups of technologies, one aimed at ensuring survival by minimizing ecological costs and the other aimed at maximizing short term growth with heavy negative externalities in the form of resource destruction.
3. *Scientific level.* The people's movements emerging from the conflicts over natural resources are based on political analysis that is deeper than those

based on the politics of production or distribution. They raise issues in the politics of scientific knowledge and show that the dominant scientific expertise is geared to the objective of the maximization of growth. It is not capable of responding to the problems of survival by ensuring sustainable and ecological utilization of natural resources. The movements indicate that ecology provides the foundation for an alternative science which would recognize the diverse ecological processes in nature that relate the natural resources and determine their diverse material properties, thus creating a science for survival.

Ecology and the Politics of Knowledge

The dialectical contradiction between the role of natural resources in production processes to generate growth and profits and their role in natural processes to generate stability is made visible by movements based on the politics of ecology. These movements reveal that the perception, knowledge and value of natural resources differ for different interest groups in society. The politics of ecology is thus intimately tied up with the politics of knowledge. For subsistence farmers and forest dwellers a forest has the basic economic function of soil and water conservation, energy and food supplies etc. For the industries the same forest has the function of being a source of raw materials only. These conflicting uses of natural resources based on their diverse functions are dialectically related to conflicting perceptions and knowledge about natural resources. The knowledge of forestry developed by forest dwelling communities therefore evolves in response to the economic functions valued by them. In contrast, the knowledge of forest developed by forest bureaucracies which respond largely to the industrial requirements will be predominantly guided by the economic value of maximizing raw material production. The way nature is perceived is therefore related to the pattern of utilization of resources. Modern scientific disciplines which provide the currently dominant forms of perception of nature have generally been viewed as objective, neutral and universally valid. These disciplines are however particular responses to particular economic interests. This economic determination influences the content and structure of knowledge about natural resources, which, in turn, reinforces particular forms of resource use. The economic and political values of resource use are thus built into the structure of natural science knowledge.

When the dominant resource use is guided by vested interest or special interest objectives, it generates a partisan science which tends to be reductionist in character. Two central assumptions underlie this reductionist perception of nature: natural resources are isolated and non-interacting collections of individual resources; and natural resources acquire economic value only when commercially exploited. This approach to nature is reductionist on two counts. Firstly, it reduces nature to its constituent parts, and takes no account of the relationships between the parts, and the structure and functions of the whole system. Secondly, it reduces economic value to a man-made construction as something produced with technology and capital inputs for the market. Nature's work and the work of marginal communities which depend on nature's productivity are thus ignored and destroyed.

Partisan science tends to be epistemologically reductionist because maximization of special, vested interest objectives focuses on the exploitation of single resources. It must be narrowly conceived since it is inherent to its logic to concentrate on the special interest objective and to be blind to ecological and environmental costs.

Environment conflicts that emerge from the violation of the public interest through special interest groups must therefore not merely indicate the social and environmental consequences of narrow profiteering. A deep and sustained resolution of such conflicts in favour of the broader public interest must be based on the emergence of a different way of looking at nature. This would involve the creation of an ecologically-based public interest science which must be based on the recognition of relationships and interdependence among the various material components of nature. It must be able to see and assess nature's work and value it. And lastly, it must be able to locate how nature's processes support survival, not merely profitability.

Ecology provides an epistemological framework that shows that alternatives to reductionist science and technology are not merely possible, but preferable too, because reductionism fails to provide faithful accounts of nature. This cognitive failure of the reductionist sciences arises from the incapability of the reductionist to take into account properties that arise from relationships in nature. In this sense, the ecological foundations of an alternative science and technology differs from philosophies based on epistemological relativism. While epistemological relativism also allows for the possibility of alternatives, it denies the existence of materialistic criteria for the rational choice of alternatives. This is the limitation of the Kuhnian model, as well as the models arguing for plurality from a purely sociological or physiological perspective, not from materialist foundations. The ecological foundations for an alternative science and technology provide a materialist epistemology for evaluating the rationality of knowledge claims on the basis of their materialist adequacy in guiding action in the real and complex world. The rejection of the reductionist interpretation of materialism need not amount to an adoption of a materially vacuous philosophical position. The ecological perspective provides such a materialistic alternative to reductionism. The distinction between reductionist materialism and ecology is the difference between mechanical materialism and dialectical materialism repeatedly articulated by Marx. Engel's analysis of this distinction in his critique of Dühring reads exactly like a contemporary ecological critique of reductionist science:

The analysis of nature into its constituent parts were the fundamental conditions of the gigantic strides in our knowledge of nature which have been made during the last four hundred years. But this method of investigation has also left us a legacy of the habit of observing natural objects and natural processes in their isolation, detached from the vast interconnection of things, and therefore not in their motion but in their repose, not as essentially changing, but as fixed constants, not in their life but in their death, in contemplating their existence it forgets their coming into being and passing away, in looking at them at rest it leaves their motion out of account because it cannot see the wood for the trees. Dialectics grasps things and their concatenation, their motion, their coming into being and passing out of existence.[1]

In the world where relationships are an actuality, the denial of such relationships, and the multi-dimensional properties they give rise to, has created a reductionist world-view and knowledge system which is inadequate in functioning in the real world. The materialist criteria provided by ecology allow for the perception of such a failure of knowledge systems through the ecological instabilities induced by them. Reductionist knowledge is found to generate unreliable claims about the natural systems and processes on the basis of ecological criteria of materialist adequacy. The cognitive failure of reductionism arises because reductionist science has created ecological instabilities which in turn threaten survival.

On a materialist epistemology, systems of knowledge are simultaneously systems of action. Reductionist science leads to human transformation of nature which is successful in creating artefacts and generating exchange value, but which fails to maintain the essential life support systems on which human survival depends. Reductionsim is not an epistemological accident. It is a particular response to an economic need of a particular form of economic organization. The reductionist world view, the industrial revolution and capitalist organization are the philosophical, technological and economic components of the same process. Economic growth, the achievement of this economic organization, is materially based on externalizing the real costs of production, and on commercializing hitherto common resources to provide inputs to the production process. This entails large withdrawal of individual resources from the ecosystem in accordance with demands of the market, not in accordance with renewal capacity of resources or the needs of the people. Since it is the individual resource which generates exchange value through extraction, scientific knowledge of natural resources which gets created as a response to this economic system must necessarily be reductionist. Properties of resources which stabilize ecological processes but are commercially valueless because they cannot be exchanged in the market place are ignored, and in turn destroyed. Profits and commercial exploitation guide the creation of the context in which properties of the natural systems will be perceived and known. Scientific knowledge is not universal, objective and neutral as it is posited to be. It is always a particular response to a particular interest. When the interest is commercial utilization of resources for maximizing exchange value, the type of knowledge

system that is created is reductionist. Internalization of profits and external-
ization of costs is a normal consequence when nature is treated as if its individual components were isolated and unrelated, and the only components with
economic value are those that can be transformed into commodities. The basic
terms, concepts, and definitions have built into them the economic values of the
interest to which the knowledge is a response. In contrast, when the interest is
the sustainable livelihood of the people and the satisfaction of basic needs,
ecological knowledge is the response.

Ecology as a public interest science is central to a just resolution of environmental conflicts in the contemporary setting because it is science, not politics,
that is used as the explicit justification and legitimization of destruction, in the
name of progress. 'Science' is used as a final arbiter in all resource conflicts;
'scientific' is taken as synonymous with public interest. However, since dominant science is partisan, decisions based on it will serve the special interest
groups. Public interest science is a tool which makes explicit the political nature
of partisan science and makes it a factor located within environmental conflicts,
not a source of independent and neutral judgements *about* conflicts. Public
interest science, however, does not merely have a critical role in the politics of
knowledge and politics of the environment, it also has a constructive role in
generating new paradigms of science and development based on ecological
principles which ensure sustainability and justice. Probably the examplar of
public interest science is Rachel Carson's *Silent Spring* of 1963 which exposed
the destruction caused by the use of poisons in pest control and laid the foundation for alternative non-chemical means of control. In substance it was ecological, and in form it was different from the work of entomologists which, while
being critical of pesticides, had remained confined to debates among
entomologists and had failed to inform public debate and public policy. *Silent
Spring* as public interest science, as a technical critique of dominant partisan
science supporting pesticides, has helped the growth of the environment
movement.

Knowledge is power, and environment movements need the cognitive power
derived from public interest science. The Normada groups have displayed how
this power can be used to control powerful agencies like the World Bank.
Voluntary sector doctors working in Bhopal have displayed the power of public
interest science in the rehabilitation and relief to the victims of the gas tragedy.
To illustrate in detail how this politics of knowledge has been translated into
practice in ongoing environmental conflicts we shall discuss two situations in
which public interest science has been a source of effective environmental
action.

Public Interest Science in Environment Action in Doon Valley

It had been written of Doon Valley: 'The Doon is what is commonly called a
backward district, but so far as the comfort and well-being of all classes is

concerned, it is a matter for regret, rather than otherwise, that more districts are not in the same state of backwardness.'[2]

This prosperous 'backward' district moved towards underdevelopment with the beginning of limestone quarrying in its critical watershed areas. Limestone has been at the centre of resource conflicts in Doon Valley for nearly two decades. The beginning of the extraction of Dehra Dun limestone was for modest purposes like the preparation of lime for construction and whitewash. It later found use in the North Indian sugar and textiles industries. Finally, owing to its chemical purity (over 99 per cent) it is being used in the steel and chemical industries in far away places like Renukut and Jamshedpur.

The resistance to the extraction of limestone from this vulnerable ecosystem came in three phases. In the first phase, the local village organizations resisted the mining activities politically. This resistance was quickly interpreted as a block to national progress and the organization of villagers was subverted by converting them into co-operatives and providing them small leases. Without the support of science or the state, the villagers lost their campaign.

The second phase was characterized as a conflict between the state and the lessees. The UP government tried to stop a lease in 1977 on the grounds that it would affect the 'natural beauty and ecology' of the region. The court called on technical experts who were partisan scientists. They informed the court that quarrying in the lease areas 'does not necessarily affect the environmental and ecological balance in regard to water soil and other related factors'. Without counter arguments, even the state could not control mining in Doon Valley.

In the third phase, citizens groups in Dehra Dun and Mussoorie fought a similar case in the Supreme Court, this time informed by public interest science. The balance shifted, and the same expert who in 1977 had stated that quarrying was ecologically safe now stated of the same quarry that 'the lease area is situated right in the immediate catchment area of a *nullah* and is thus subjected to conspicuous denudation by flow of water. Rectification of the situation calls for a permanent closure of this mine'.

The emergence of public interest science in Doon Valley created a new countervailing power favouring the public interest. The ecological knowledge was generated with citizens' participation in an ecosystems study of Doon Valley undertaken by the authors for the Department of Environment. The study was completed in May 1983,[3] and in June 1983 it was used to file a public interest litigation against limestone quarrying through the Dehra Dun based Rural Litigation and Entitlement Kendra. The court acted as a public interest science laboratory where scientific ideas were tested, verified and developed into an opposing force challenging the power of partisan expertise. Public interest litigation backed by public interest science was successful in controlling the mining.

From an ecological point of view, the limestone in its fractured form provides the best and largest aquifer that can sustain the supply of water resources to the valley. The most efficient economic use of the mineral from this perspective which sees limestone in its relationship with other resources, is its conservation for the sustained supply of water on which all economic activities in the valley depend. 'Scientific' mining and 'scientific' geology in the reductionist framework is based on partial and incomplete knowledge of the diverse properties and functions of mineral resources. It is based only on the specific and particular properties which provide maximum exchange value of the mineral. But minerals have properties and functions beyond those that are commercially exploitable, some of which are only realizable *in situ*. Mineral extraction in the reductionist framework however is blind to the other functions, treats them as non-existent, and thus destroys them by maximizing benefits from the commercial exploitation of individual resources.

On 12 March 1985 a Supreme Court bench consisting of Justice P.M. Bhagwati, Justice A.N. Sen and Justice R. Misra, passed an order closing permanently or temporarily fifty-three of the sixty limestone quarries within the geographical limits of Doon Valley or the Dehra Dun *tehsil*. The bench introduced the order in the following words:

This is the first case of its kind in the country involving issues related to environment and ecological balance and the questions arising for consideration are of grave moment and of significance not only to the people residing in the Mussoorie hill range forming part of the Himalayas but also in the implications to the welfare of the generality of the people living in the country. It brings into sharp focus the conflict between development and conservation and serves to emphasize the need for reconciling the two in the larger interest of the country.

They justified the closure of mining operations on the grounds that 'it is a price that has to be paid for protecting and safeguarding the right of the people to live in a healthy environment with minimum disturbance of ecological balance and without avoidable hazards to them and to their cattle, homes and agricultural land and undue affection of air, water and environment'. With this order the Supreme Court of India has set a precedence in accepting a stable and healthy environment as a human right.

A second case is now in front of the Supreme Court against the Dehra Dun master plan and the location of hazardous and polluting industry in the valley. The earlier ecological study of 1983 was supplemented by a seminar on pollution in Doon Valley organized by the citizens' groups in June 1985 to generate the relevant public interest science for the case. The Rajpur community, which is the worst affected by pollution, has used the technical critique of the master plan and polluting industry to file a public interest litigation supported by a petition filed by the Rural Litigation and Entitlement Kendra. Like the case on mining, this case will open new avenues on citizens' control on plans for urban development and for location of polluting and hazardous industry.

Why is public interest science crucial to environmental action? Firstly, because as past experience has shown, in its absence partisan science prevails and swings public policy and court decisions in favour of vested interest groups. Secondly, it is only environment action backed by public interest science which has the potential of turning into an environment movement in which social and political events are replicated rapidly in a chain reaction. While the special features of an environmental context are unique, the scientific and philosophical principles defining environmental conflicts are not. Environmental action without public interest science is impotent. It cannot reproduce itself and has no potential or force to transcend its limits or grow beyond its immediate context. To turn into a political force, environmental action must be philosophically and scientifically enlightened.

Public Interest Science in Forest Struggles in India

Forest struggles in India can be divided into two phases — those that were a response to direct commercial exploitation, and those that were a response to commercial exploitation legitimized as scientific forestry.

Forest struggles have been a sustained response to commercial forestry. The earliest records of commercial exploitation are of a syndicate formed in 1796 by Mackonchie of the Medical Service for extraction of teak in Malabar for the demand of the government officials for shipbuilding and military purposes. In 1806, a police officer, Captain Weston, was appointed as the first conservator of forests in India in charge of Malabar and Travancore, to extract teak for the King's Navy indicating that policing, not science, was needed in the colonial forestry of that period. Indigenous trade was sealed and peasants were denied rights.

By 1823 the growing discontent of the forest proprietors and timber merchants chafing under the restrictions of the timber monopoly, and the outcry of the peasants, indignant at the fuel-cutting restrictions, came to a head. On the recommendation of the Governor of Madras, Sir Thomas Mundro, and with the consent of the Supreme Government, the conservatorship was abolished.

The Forest Act of 1927 generated a new response against the denial of traditional rights of local people. The year 1930–31 witnessed forest *satyagrahas* throughout India as a protest against the reservation of forests for exclusive exploitation by the British commercial interests. Villagers ceremonially removed forest produce from the reserve forests to assert their rights. The forest *satyagrahas* were especially successful in regions where survival of the local population was intimately linked with access to the forests as in the Himalayas, the Western Ghats and Central India. These non-violent protests were suppressed: in Central India, Gond tribals were gunned down; on 30 May 1930 dozens of unarmed villagers were killed and hundreds injured in Tilari village of Tehri Garhwal. After enormous loss of lives, the *satyagrahas* were finally successful in reviving some of the traditional rights of the village communities to various forest produce.

In post-independent India where commercial exploitation contrary to public interest could no longer be justified on grounds of revenue alone, 'scientific forestry' was increasingly used as a political weapon to justify over-exploitation. Isolated actions to stop the non-local exploitation were initiated. Chipko became an ecology movement in 1977 when its environmental action was strengthened by the public interest science captured in the slogan

What do forests bear?
Soil, water and pure air

created by the women of Henwal Valley in the Advani forest. They mocked the partisan forestry science in the slogan:

What do forests bear?
Resin, timber and profits.

The insight in these slogans represented a cognitive shift in the evolution of Chipko. The movement was transformed qualitatively from being based merely on conflicts over resources to conflicts over scientific perceptions and philosophical approaches to nature. This transformation also created that element of scientific knowledge which has allowed Chipko to reproduce itself in different ecological and cultural contexts. The slogan has become the scientific and philosophical message of the movement.

Forest movements like that at Chipko are simultaneously a critique of reductionist, 'scientific' forestry and an articulation of a framework for an alternative forestry science which is ecological, can safeguard the public interest and does not view forest resources in isolation. Nor is the economic value of a forest reduced to its commercial value. Thus, while for tribals and other forest communities a complex ecosystem is productive in terms of herbs, tubers and fibre, for the partisan forester these components of the forest ecosystem are useless, unproductive and dispensable. Two economic perspectives lead to two notions of 'productivity' and 'value'. As far as overall productivity goes, the natural tropical forest is a highly productive ecosystem. Examining the forests of the humid tropics from the ecological view, Golley has noted: 'A large biomass is generally characteristic of tropical forests. The quantities of wood especially are large in tropical forest and average about 300 tons per ha compared with about 150 tons per ha for temperate forests.' However, for the partisan foresters, the overall productivity is not important. They look only for the industrially useful species and measure productivity in terms of industrial biomass alone. Bethel, an international forestry consultant, refers to the large biomass typical of the forests of the humid tropics:

It must be said that from a standpoint of industrial material supply, this is relatively unimportant. The important question is how much of this biomass

represents trees and parts of trees of preferred species that can be manufactured into products that can be profitably marketed . . . By today's utilisa-
tion standards, most of the trees, in these humid tropical forests are, from an
industrial materials standpoint, clearly weeds.[4]

With these assumptions of partisan forestry science wedded to forest industry, large tracts of natural tropical forests are being destroyed across the Third
World. The justification is increased 'productivity' but the productivity
increase is only in one dimension. Overall there is a productivity decrease. The
replacement of natural forests in India by eucalyptus plantations has been
justified on the grounds of improving the productivity of the site; but while
natural forests and indigenous trees are more productive than eucalyptus in the
public interest paradigm, the reverse is true in the partisan paradigm of forestry. The scientific conflict is an economic conflict over which needs and whose
needs are more important. In such paradigmatic conflicts, dominant scientific
assumptions change not by consensus but by replacement. Which paradigm will
win and become dominant is determined by the political strength backing the
paradigms. The reinterpretation of basic terms in a shift from partisan to public
interest forestry implies major shifts in political economy. Partisan forestry
prescribed eucalyptus as a magical fast-growing species, for the quickest cure to
all forestry problems — those related to conservation, basic needs, and industrial requirements. Public interest forestry discovered that eucalyptus was fast
growing only in the context of industrial requirements. It had zero-growth for
basic needs of fodder and food, and actually had negative growth in the context
of soil and water conservation.[5]

Movements against indiscriminate planting of eucalyptus under 'social forestry' and 'wasteland development' schemes such as Mannu Rakshana Koota
or Save the Soil Campaign have grown with the support of public interest
forestry science.

In both the cases mentioned above, ecological perceptions of nature have
emerged from outside the reductionist partisan expertise. They have emerged
from the ecological perspective of the people whose survival depends on those
ecological functions of natural resources which reductionist and vested interest
has ignored. The evolution of ecological knowledge in general will depend on
the people's actions and movements because reductionist expertise is
epistemologically and politically constrained from evolving into a non-
reductionist framework. This dynamics of the evolution of knowledge from an
expert-dominated to a people-dominated process is according to Feyeraband
the only route to a free society:

> In a free society intellectuals are just one tradition. They have no special right
> and their views are of no special interest (except, of course, to themselves).
> Problems are solved not by specialists (though their advice will not be disre-
> garded) but by the people concerned, in accordance with the ideas *they* value

and by the procedures *they* regard as most appropriate . . . This is how the efforts of special groups combining flexibility and respect for all traditions will gradually erode the narrow and self-servicing rationalism of those who are now using tax money to destroy the traditions of the tax payers, to ruin their minds, rape their environment and quite generally turn living human beings into well trained slaves of their own barren vision of life.[6]

The evolution of public interest oriented ecological knowledge is however going to be resisted by the reductionist partisan expertise because this 'threatens their role in society just as the enlightenment once threatened the existence of priests and theologians.'[7]

The evolution of the ecological, sustainable and equitable utilization of natural resources in an alternative development strategy will also, quite naturally, be resisted by the vested interests who benefit from the existing reductionist, unsustainable and inequitable utilization pattern. This process has already been initiated in countries like India. At one level people's attempts at redefining development through sustainability and justice is resisted by the introduction of a false dichotomy between 'development' and 'ecology' which covers up the real dichotomy between ecological development and unsustainable economic growth. At another level the resistance is created by the rejection of people's perception of ecological destruction as 'unscientific', 'unproved' and 'unverified'. These attempts of experts and vested interests will work against human knowledge and public interest science, and in turn against the possibilities of human survival.

The growing conflict between the profitability imperative and the survival imperative will lead to the emergence of a politics of knowledge. It is in this sense that ecology as the foundation of an alternative public interest science and technology converges with ecology as a foundation for the politics of survival of the people.

The Real Experts

Development in a polarized society such as ours which depends mainly on a modernization strategy based on following the industrially advanced countries, slowly robs the majority of the people of their resource-base for survival in the name of an overall well-being of the people. It must necessarily exclude the people's opinion on development from the planning process which is left completely in the hands of experts and bureaucrats whose vision of prosperity cannot perceive how, as Susan George puts it, the other half dies. Nor do the experts and bureaucrats in one field recognize their links with other fields. This lack of co-ordination is a natural outcome of the misplaced belief in improving socio-economic conditions with the help of isolated technologies borrowed from a set up where they originated in a systematic and integrated fashion. The problem, in our view, is not merely that people have been left out of the process of planning for development. The problem really is that in a country with two societies, planning without people's participation, even of the most well-intentioned and competent sort, will necessarily develop one at the cost of the other in an irreversible manner. It is this irreversibility of development that permanently destroys resource bases that need thousands of years to grow either in terms of knowledge systems or natural resources.

The task of forming alternate development strategies thus becomes urgent, but we believe that these cannot even be imagined, let alone formulated in detail, without people's participation in the assessment of the present development plans and the generation of alternate ones.

Research, as understood conventionally, is a full-time activity of academic professionals, and their expertise in the respective field of specialization has guaranteed their monopoly on research as an activity. The restriction has so far not merely been on who does research, but also on what research gets done or recognized. Research possibilities get limited by the way in which the powerful groups of society can register their priorities on the research system. Because this mediation is very indirect and subtle in operating through the reward system, it becomes extremely easy to believe that research is an autonomous and socially neutral activity. Even if the impact of society and culture is recognized in social science research, it is rarely admitted as possible in research in natural science and technologies. The indirect mediation also implies that while the research system as a whole is guided by the priorities of the elite groups of society, individual researchers themselves are not committed to serving the interests of the elite. For them research is a freely and autonomously chosen activity which leads to autonomous results.

Added to the lack of awareness on the part of individual researchers of how their activity is unknowingly contributing to the growth of the knowledge system that serves the elite, is their lack of awareness of what research in which historical situation would work to the advantage of the weaker people. The myopia related to research possibilities is tied up very closely with extended training of the formal type and the requirements of specialization. A good illustration is the recent reaction of the specialists of the Punevik Supply Scheme to the call to farmers by Sharad Joshi. The call has not been taken seriously by the urban experts since, according to them, 'Milk is a perishable commodity and farmers have no means of preserving the huge quantities of milk. The only option open to them was to destroy it.' This negates the existence of all traditional milk product-preserving techniques.

Researchers thus end up reading only formal knowledge sources like publications and give up looking at society and nature. The research establishment, as it exists in India, has no room or mechanism for making good this gap from within. We have, after all, had no dearth of claims of all research being aimed at the needs of the people and removal of poverty, starting from economics and agriculture right up to sophisticated space technology.

The problem is only directly political in part. To a large extent it is also indirectly political through the built-in epistemological constraints on the modern research system. On the one hand it creates compartmentalized, unco-ordinated and fragmented expert knowledge and, on the other, it renders invisible the knowledge of the people involved in the real life activity at which research is aimed. However, there are two very good reasons for taking people's knowledge as an important element in research which tries to provide a more holistic understanding of the natural and social world. Firstly, assuming that the people are ignorant, it is they who know better than the experts, exactly where the shoe pinches. Secondly, people are really not as ignorant as the experts take them to be, at least not in matters related directly to their activities. Particularly for agrarian societies where the majority of the people are involved in primary production, their informal knowledge accumulated over centuries of practical experience has its own built-in reliability and viability. The whole life-style of the rural people in India is closely interlinked with the local eco-system and danger to it is obviously first sensed by these communities.

Most professional planners and bureaucrats are at best ignorant of the role of a stable ecology for the satisfaction of the needs of the rural population, and at worst they consciously contribute to the process of chanelling resources from the rural poor to the urban rich. In either case, their development plans, based on technological determinism, consciously support and encourage the development and use of technologies that tend to destroy the local ecology, and hence the sustenance of the material base for survival. Consequently, the traditional technologies on which the life-style of threatened communities is based, instead of being improved, get overtaken by ecologically unstable and socially irresponsible modern technologies. Lewis Mumford was probably addressing himself to these distinctive technologies when he wrote:

> From late neolithic times in the Near East, right down to our own day, two technologies have recurrently existed side by side: one authoritarian, the other democratic; the first system-centered, immensely powerful, but inherently unstable, the other man-centered, relatively weak, but resourceful and durable.

The weaker but ecologically stable technologies are, however, systematically threatened by the more powerful, ecologically reckless technologies which are projected as being more efficient and productive in some absolute sense. In the process, the traditional technologies are identified as unproductive and are marginalized in the development plans. Associated with this marginalization, the knowledge and skills of local communities are also rendered invisible. Professionals are the only ones viewed as having reliable knowledge. Their role in policy-making, therefore, gets more and more entrenched till an ecological crisis threatens the livelihood of vast rural populations which sets off organized opposition to development and technology policy. This opposition also takes a few steps in exposing the political base of technologies and the restricted nature of the knowledge of the experts who work on the development of these technologies.

The Right to Counter-expertise

Research carried out by de-professionalized intellectuals with the participation of the people can become a two-pronged tool for critical evaluation of science and development. On the one hand it strengthens the needs and wants of the common people by putting their feelings and views in a form which is easily understandable and hence respected by the experts and policy makers. On the other hand it exposes the restricted nature of expert knowledge and provides a platform for countering the political power at a level of expertise where no serious challenge to it has emerged in India so far.

The role of participatory research in science and technology assessment in supporting people's struggles cannot be underestimated. Firstly, it can help the people's movement grow at the down-to-earth level and establish more democratic decision-making in resource utilization around which most serious class conflicts are taking place in India today. Secondly, it can strengthen these struggles by taking their arguments to a level of theoretical sophistication that can demand serious attention and cannot be dubbed as political propaganda or anti-development moves. Critical evaluation of 'development' is not meant to be a block to progress. In fact it provides the only route to a meaningful progress for the people. After all, as Salomon has pointed out, people have a right to 'counter-expertise'. Participatory research is the vehicle to establish that right.

Notes

1. Frederich Engels, *Anti Dühring*, Foreign Language Publishing House, Moscow, 1947, p. 36.
2. H.G. Walton, Dehra Dun Gazetteer, Government Press, Allahabad, 1911.
3. J. Bandyopadhyay *et al.*, 'The Doon Valley Ecosystem. A Report on the Natural Resource Utilization in Doon Valley', prepared for the Department of the Environment, Government of India; 'Planning for Underdevelopment', *Economic and Political Weekly*, **19** (4) 1984.
4. J. Bandyopadhyay and V. Shiva, 'The Evolution, Structure and Impact of the Chipko Movement', in *Mountain Research and Development*, **6** (1) 1986.
5. V. Shiva, H.C. Sharatchandra and J. Bandyopadhyay. 'Social, Ecological and Economic Impact of Social Forestry in Kolar', IIM, Bangalore, 1984; 'Ecological Audit of Eucalyptus Cultivation', EBD, 1985; 'Eucalyptus in Rainfed Farm Forestry: Prescription of Desertification', *Economic and Political Weekly*, **20** (4) 1985.
6. Paul Feyerabend, *Science in a Free Society*, New Left Books, London, 1978, p. 10.
7. *Ibid.*, p. 79.

5

Science and Control

How Indian Atomic Energy Policy Thwarted Indigenous Scientific Development

Dhirendra Sharma

Even if we cannot be sure why the scientific revolution in our times appeared when it did and why in the western world, there is no reason to believe that some people are innately deficient in science and technology, or that any particular race or nation is peculiarly endowed with scientific temper. J.D. Bernal, the founder of interdisciplinary studies, now known as 'Science Policy', observed:

> All countries, even the poorest, are in fact far richer than they know, and the problem of raising their standard of living is a problem, essentially of learning to use the resources, both the natural resources of the territory and even more important, the human resources of its people. It is not a question so much of building as of releasing energies.

Indian science, in the pre-independence years, achieved great advances in fundamental sciences, without much support or encouragement from the government.

The gap between European and Indian science was not as wide as it is today; but then relations between science and technology were also not as close as today. The nations which were leading in technological fields, like the USA and Japan, were not necessarily considered leaders in science. Scientific activity depended mostly on the interest of individuals and only minor funding was needed to conduct research. In practice, scientific activities were open and universal, and publication of one's results was considered to be the most important function of a scientist. Indian scientists then could contribute directly to the advancement of science.

In the first half of this century Indians carried out pioneering scientific work with the emergence of researchers such as Nobel Laureate C.V. Raman, J.C.

Bose, K.S. Krishnan, S.N. Bose and Meghnad Saha. Their scientific excellence, however, was the direct outcome of personal dedication to the pursuit of scientific knowledge. Historically and collectively that period is epitomized by what is known as the Calcutta-Allahabad school of science.

At the end of the Second World War the relation between science and technology was fundamentally altered by the terrible experience of Hiroshima and Nagasaki. If science had lost its innocence in the destructive discovery of nuclear fission, the scientist had found a powerful means to explore to the farthest edge of reality in new technology. From now on science and technology became the Siamese twins which could not be separated. The alliance, which was primarily necessitated by urgency of the war, received unprecedented stimulus to open up scientific and technical organizations. In the fateful years of 1938–45, Fat Boy (the atom bomb) was produced. In the post-war years 'big science' involving massive techno-industrial and military establishments was born. Science was now intimately linked with technical advancement, their interdependence was total and they required a very high percentage of national resources. They received public prominence and national and political support; science and technology planning became synonymous with progress and development. Political leaders also saw popular advantage in outshining other nations through achievements in this field which also directly boosted military muscle. That was the period (1948–58) known to us as the cold war decade which stimulated war science and the military-industrial complex in the advanced countries. But India and other less developed countries were involved with problems of the restructuring of society after a politico-cultural domination of 200 years. Thus in this critical decade — these were the years of atomic tests, hydrogen explosions, aero-dynamics, electronics, and researches in frontier areas of chemistry and astro-physics — India (and China) were left behind.

But the realization of this limitation of the industrial techno-base was keenly felt by the Indian science community and a need arose to tame the atom for peaceful purposes. The first government of free India visualized a forward-looking atomic policy for the country. On 10 August 1948 the first Atomic Energy Commission (AEC) was established with a brief to take such steps as might be necessary from time to time 'to protect the interest of the country in connection with Atomic Energy', and they were told to promote research in their laboratories and to subsidize research in existing institutions and universities. The promotion of teaching and research facilities in nuclear physics in the Indian universities was encouraged.

In 1948, after the AEC had been set up and Bhabha had been appointed the first chairman, almost all science research activity was shifted from Calcutta to Bombay — the ancestral hometown of the Tata family to which Bhabha belonged. Even though efforts were already underway to advance research in new areas of science in existing universities and institutions, AEC was not inclined to accept the contribution of the universities. This led to serious differences of approach as to how fundamental research in India should be developed.

Meghnad Saha's school and almost all senior scientists in the country were keen to pursue science studies through open training and research in the universities. Meghnad Saha had felt that the new state installed in New Delhi after independence should not have monopolistic control over science in the country for he feared that the misuse of science might occur through governmental control, and that the free flow of knowledge would cease. This was the period when the science policy critics led by Albert Einstein, Bertrand Russell and J.D. Bernal had voiced concern about the use of science for evil purposes by nation states and their political leadership. Meghnad Saha therefore was opposed to the separation of fundamental research from the mainstream of science teaching in the universities. He opposed the creation of an independent atomic energy agency, and when the AEC was eventually established, he refused to be associated with the atomic establishment. Meghnad Saha observed:

> The whole difficulty here has been that the administrative policy with respect to the development of atomic energy has been extremely retrograde. From the very first there was a veil of secrecy about it. We were not allowed to talk about atomic energy. The Atomic Energy Commissioners never said what they were doing, what researches they were financing. Everything was under a veil of secrecy. This was extremely ridiculous, because other countries have imposed secrecy on atomic energy development simply because atomic energy was used to produce weapons of war. From the very first, we have said that we shall not use atomic energy for any aggressive purposes. Having said so, to have imposed secrecy on atomic energy work was not only the height of indiscretion, but the height of folly. Because if you analyse the work done in other countries, you find the atomic energy cannot be developed unless you enlist the services of thousands of scientists in your own country.

In 1958, Bhabha, in consultation with Jawaharlal Nehru, reconstituted the AEC in a manner that gave him a free hand in planning and executing his science policy. The government of India reorganized the AEC with a new resolution (no. 13/7/58-Adm. Bombay, 1 March 1958), this time, with 'full authority' to plan and implement the various measures on sound technical and economic principles and 'free from all non-essential restrictions or needlessly inelastic rules'. The Chairman of the AEC was granted 'full executive and financial powers' and he was also made the ex-officio secretary to the Department of Atomic Energy responsible *only* to the Prime Minister. He was empowered 'to overrule the other members of the Commission, except the Member for Finance and Administration' who only in a financial matter, could 'ask' to be referred to the Prime Minister. The AEC was further empowered 'to frame its own rules and procedure' and to meet 'at such times and places as may be fixed by the Chairman'.

Thus, under the reconstituted AEC, Bhabha secured personal autonomy within the formal constitutional framework of the country. In his person he combined the powers of both the Chairman of the AEC, and the ex-officio Secretary of Department of Atomic Energy; thus he represented the elected democratic government of India. Yet he was free from the rules and procedures of the government; he was free from all formal constraints as he was responsible only to the Prime Minister. He was Founder-Director of the Tata Institute of Fundamental Research, 99 per cent of whose funding came from the budget of atomic energy. He had thus secured a unique position in the country having direct access to Jawaharlal Nehru who by then had become totally dependent on Bhabha for scientific advice. In turn, Nehru was insulated from criticism by scientists working in universities and other institutions. Bhabha established close brotherly ties with the first Prime Minister, and, as reported by the official biographer of Nehru, Dr S. Gopal, he became only the second man in the country who could address Nehru as *bhai* or brother (the first one being Jayaprakash Narain).

With this new mandate to formulate his own rules and procedures, and with an open-ended budget, Bhabha adopted an aggressive policy of concentration of all big science research under the domain of the Department of Atomic Energy. Consequently, Indian universities were deprived of funds and denied their rightful role in the country's scientific advancement. Centres of higher learning could not purchase equipment or attract foreign-trained young scientists from abroad. Only the Department of Atomic Energy was authorized by an Act of Government to initiate, explore, plan and execute all nuclear studies and research; Indian universities were precluded from the emerging challenges of big science.

The Atomic Energy Commission responsible for formulation of national nuclear policy now has no independent scientists or economists or representatives of Indian universities as members. It is constituted only of those who are already engaged in the execution of and advancement of nuclear activities. Seven out of nine members belong to one ethnic and regional fraternity — South Indians — and six are South Indian Brahmins, and all of the seven have been students and/or associates of the Chairman of the Commission (Raja Ramanna). One member, Cabinet Secretary (Kaul) belongs to the Prime Minister's own fraternity: Kashmiri Brahmin. Since June 1986 he has been posted as India's ambassador to the United States. The ninth member (J.R.D. Tata) has been on the commission since 1962. Besides being the top industrialist of the country, he belongs to the ethnic fraternity of the founding-chairman of the AEC (Bhabha was nephew of Tata). And the Tata industries are directly involved in the construction works of the Department of Atomic Energy under the policy directives issued by the commission.

It was believed that by establishing an independent atomic energy organization the country could have been transformed from being industrially under-developed to developed within a short period. In total disregard of economic and social imperatives, the atomic energy programme was launched thus making India more dependent on external aid.

The reconstitution of the AEC Act in 1958 was indicative of a shift in the official perception and records suggest an increasing interest in nuclear science and technology for its non-civil application. Bhabha's role in this shift was decisive, and through reasons based on 'the diplomatic-strategic uses of nuclear energy', he swayed the thinking of Jawaharlal Nehru. Even though Nehru was publicly committed to peaceful uses of the atom, Bhabha did not evince enthusiasm for his concern for disarmament. In fact Bhabha refused to be associated with the Pugwash Movement, and no Indian scientist signed the famous Einstein–Russell anti-nuclear weapons declaration of 1957.

In 1958, Bhabha drafted a Scientific Policy Resolution (SPR) which was acclaimed by Parliament:

> The key to national prosperity, apart from the spirit of the people, lies in the modern age, in the effective combination of three factors — technology, raw materials and capital — of which the first is perhaps the most important, since the creation and adoption of new scientific techniques, can, in fact, make up for a deficiency in natural resources, and reduce the demands on capital. But technology can only grow out of the study of science and its application . . . It is an inherent obligation of a great country like India, with its great cultural heritage, to participate fully in the march of science, which is probably mankind's greatest enterprise today.

The 'march of science' required comparable advancement in social and economic spheres; it also demanded a national commitment to a just social order. But the official thrust focused on efforts to close the gap in the scientific fields rather than on serving the needs of the people. The period from 1960–70 saw the multifaceted growth of science and technology organizations, mostly on the pattern of the advanced countries. Even if we raised the populist stream of 'self reliance', we also begged, borrowed and bought technology from external sources.

In 1970 a conference was held to review the performance of the 1958 Scientific Policy Resolution by the (National) Committee of Science and Technology, under the Chairmanship of Dr B.D. Nag Chaudhuri. Participants included 130 scientists, technologists and educationists from different institutions and organizations in the country. The conference was of the view that the resolution was an 'admirable enunciation of the Government's faith in science and the role science must play in the transformation of our society'. But it also felt strongly that 'on several important counts, the implementation of the SPR had been highly ineffective. As a result many of the objectives of the SPR have remained largely on the paper'. The role of science in providing new conceptual frameworks and analytical tools for tackling social problems and in promoting a scientific temper in society was re-emphasized. The conference further observed. that there was 'widespread prevalence of feudal attitudes'; that education was 'confined to the periphery of our society'; and they noted the problems in 'the complex psychological and cultural implications of using English, a language alien to the vast masses of our people, as the vehicle for imparting higher education, performing R & D [research and development] and applying its results'.

The delegates emphasized the inadequate practical bias; the lack of an interdisciplinary approach; and the absence of a component dealing with the role of a scientist or a technologist as the agent of change in a developing society.

In 1978 (during the Janata government) a Review Committee on Post-Graduate Education and Research in Engineering and Technology was set up, under the chairmanship of Professor Y. Nayudamma. The committee found the state of education and research in the country highly unsatisfactory. It observed that little effort had been made to implement the recommendations made by the earlier review committees, and that the content of science and technology within Indian society and the quantum of research and development in scientific and technological activities was very low. They therefore recommended a considerably higher investment in science and technology education and research in order to meet the growing demands of the constantly changing social and economic institutions.

The committee consisted of representatives from public and private sectors, industry, research and development organizations, the University Grants Committee, the Federation of the Indian Chambers of Commerce and Industry, Institution of Engineers, government departments and educational organizations. They collected information from almost all post-graduate engineering and research institutions, and held hearings at Delhi, Bombay, Ahmedabad, Calcutta, Kanpur, Hyderabad, Coimbatore, Madras and Bangalore. They visited various institutions, and discussed the issues with teachers and students, and with representatives of industry and R & D organizations. Discussions with senior government officials directly responsible for the administration of R & D organizations and post-graduate institutions were held. The committee invited comments and criticism from the public through advertisements in national newspapers. And, in April 1980, before finalizing the report, the Chairman of the committee again held discussions with some senior policy-makers in the Union government, including members of the Planning Commission. The committee did not find the situation satisfactory, and it lamented the absence of reliable data on the patterns and trends in the utilization of post-graduate degree holders in engineering and technology.

Science and technology education facilities had expanded from six institutions in 1947 to seventy-four in 1980, covering post-graduate education and research in engineering and technology. About 350 doctorates, 2,700 masters in engineering (MEs), and masters of technology (M. Techa), and 16,500 graduate engineers were also produced. Nevertheless, the Nayudamma committee questioned official claims that India was among the first ten industrialized nations of the world with the third largest stock of science and technology manpower and called for vigorous measures for the creation of a sound scientific and technological base in the country by mobilizing adequate resources. Inadequate facilities, lack of recognition and incentives for younger scientists, incompetent administration and lack of employment opportunities were factors that mitigated against bright young people being attracted into higher education in engineering and technology.

Some of the radical steps recommended by the Nayudamma committee included the encouragement of mobility and exchange of faculty between academic institutions, R & D organizations and industrial establishments; and making industrial experience as well as a doctoral degree or equivalent qualification essential for teaching positions at post-graduate level in all national institutions of engineering and technology.

Working scientists have had little or no say in the decision-making and therefore have not been able to determine the direction of research or science policy. The present system in higher education and in most scientific establishments encourages nepotism. Admittedly, it is not fair to blame all failures on a few men whom history has elevated, but accountability must accompany those who enjoy and wield unrestrained power. The relevant questions we therefore must address are: why has the atomic energy investment failed to become the vector of social change? In the forty years of independence, claims to excellence in the midst of all-pervasive poverty sound hollow. We have witnessed the emergence of various types of science and technology organizations. But they have contributed little of excellence to science, nor have they helped eradicate ignorance and poverty. The country is still plagued with endemic diseases; our public medical facilities are perhaps the most neglected and ill-equipped and ill-managed in the world. Many diseases eradicated elsewhere are still prevalent in India, and knowledge of primary health, hygiene, and social medicine are absent.

Inspite of the Council of Scientific and Industrial Research which was to have made the country self-reliant, and offer import substitutes in areas of manufacturing and mining industries, imports in all major products in engineering have increased. The country has not produced its own small car model or any energy saving devices. No innovation or improvements have been made to the models and machines bought from foreign suppliers fifteen or twenty years ago — to the models of typewriters, bicycles or sewing machines which are bought by the million in India.

And, finally, the Atomic Energy Commission and the Department of Atomic Energy, which take almost 45 per cent of the science and technology budget, have not contributed 2 per cent power from the promised uranium to the national grid in twenty-five years.

The interrelationship of science and society is such that the problem of the acquisition of knowledge becomes inseparable from the question of social change. This protest and criticism of nuclear energy is, therefore directed not so much against nuclear sciences as against the power relationship associated with science and technology activities within and without nuclear establishments. The criticism is less against science and scientific research, but more against the manner in which science and public policy decisions are made in India and other Third World states. It is now imperative that a critical and open re-examination of past and present science and technology policies be made.

6

Science and Control

Sex, Race and the New Biology

Munawar Ahmad Anees

Nearly fifty years ago Virginia Woolf made a subtle observation about andro-centric science in these words: 'Science, it would seem, is not sexless. She is a man, a father, and infected, too.'[1] The present pathogenic state of western science and its gender-infestation has raised reductionism and determinism to the level of a scientific world-view. Both of these approaches are promoting many an absurdity in the name of objectivity and a value-neutral process of enquiry. This methodology, a logical inconsistency *per se*, is but an agenda for legitimization of banal political policies. Once again, gender/racial bias is being justified under the umbrella of scientific impartiality.

The reductionist argument rests on the assumption that the complex web of bio-physical and bio-social interactions, spanning molecules to morality, can be explained in terms of their constituent elements. That is to say, that molecular behaviour is reducible to atomic or sub-atomic level and that the social behaviour of organisms is amenable to such a reductionist treatment as well.

The high-priests of the pseudo-science of reductionism argue that the collective/social/holistic behaviour is the sum behaviour of the base units constituting the whole, implying that the unitary behaviour antedates the summative one. In other words, they deny the existence of any kind of inter-active behaviour at any level of organization and perceive unitary behaviour as additive and linear directed towards higher orders. Yet another implication of this argument is that holistic behaviour can be inferred and predicted from the unitary behaviour.

For a disciple of reductionism, bio-socially holistic and organismically integrated behaviour can be ignored and, therefore, collectivity can be unashamedly sacrificed at the altar of individualism. For him/her, individual-ism acts as the causative agent for any collective manifestations. In the same vein, altruism is viewed as the guidepost for it is the wonder touch of the selfish gene that perpetuates itself for maximization of the so-called inclusive fitness.

Reductionism is a one-way street, leading straight into the pit of determinism wherein biological determinism reigns supreme. The proponents of biologism consider all human characteristics, including social and behavioural traits, as absolute, discrete units that are evolutionarily and genetically determined. Biological determinism thus becomes biological inevitability — the innateness of what we are. Biologism views organisms as fixed, immutable entities. Their attributes are labelled as natural and as a natural extrapolation, summative bio-social behaviour is traced back to their genetic make-up.

Not all reductionism constitutes bad science. For instance, it is a useful exercise when applied to the deduction of one theory from another, as in the case of Galileo's theory from that of Newton. The deductive reduction remains valid insofar as the conceptual terminology for the two theories does not acquire new meanings as a result of reduction. Yet another logical step in reduction could be its limitation at operational level.

Contrary to some valid reductionist approaches, biological determinism altogether ignores the notion of levels of organization. These levels, whether sub-atomic, macro-molecular or supra-social, impart a certain structure as well as predictability to a given organization of matter. This structural/ organizational behaviour operates within an hierarchical system supported by various feedback mechanisms. If organic structures are reduced to an interplay of DNA and RNA, the notion of hierarchical pattern in the organic framework is lost with the attendant loss of power of predictability. Hence, biological reductionism fails to capture the 'elusive' essence of living organisms that does not lie in the molecular structures of protein units.

The classical logical empiricist analysis of determinism aside, it is the social and cultural context of science that gets cosmetic treatment of neutrality from reductionists. The evolution of scientific theory and that of societal order are inextricably linked together. To think of the two existing in separate, mutually exclusive niches is the sign not only of bad science but political naivete as well. In so doing, reductionists get engaged in the fragmentation of knowledge for they assert a separation between value and fact.

Biological determinism is not a novel idea. In fact, it has existed in the scientific establishment for quite some time but under different titles. At one time the 'science' of craniology was invoked to confirm the accepted racial bias of the times. Similarly, eugenics has been pursued in order to institute selective human breeding. What has brought this age-old absurdity to the forefront is the new synthesis of social sciences that purports to re-orient our entire method-ological approach to the study of human social behaviour. This fad is called sociobiology.

Sociobiology Revisited

The theoretical beginning of sociobiology can be traced to W.D. Hamilton's classic work on the genetic evolution of social behaviour in 1964. However, the term acquired an entirely new connotation with the publication of E.O. Wilson's *Sociobiology: A New Synthesis*.[2] An expert in insect behaviour, Wilson has attempted to establish sociobiology 'as the systematic study of the biological basis of all social behaviour'. The central theorem of sociobiology is that organismic behaviour is directed towards inclusive fitness which may be defined as their net genetic representation in the succeeding generations. In rather informal language, social behaviour is nothing other than the optimization by organisms of their selfish genes.

Following the dictates of Darwinian natural selection, sociobiologists argue that adaptive social behaviour evolves and becomes fixed in genetic apparati. Behavioural configurations which impart higher survival values for the individual are 'naturally' selected and this is how behavioural adaptations evolve over a period of time. The sociobiologistic catalogue of 'universal' human behaviour with adaptive values, therefore, includes racism, aggression, sexism, dominance, chauvinism, slavery and ethnocentrism.

It is obvious that sociobiologists are explicitly pushing the theory that human behaviour is genetically determined and that human socio-cultural evolution is a direct consequence of hereditary transmission of 'adaptive' human traits. The social phenomena are assigned natural/biological causes. This theory is in keeping with the zoocentric views about humankind, in spirit if not letter. It opposes the rejection of animality within human beings, if not at organic at least at the symbolic level.

In the context of zoocentric connection of sociobiology, John Wengle has pointed out that in this age of psychohistoric dislocation, sociobiology perhaps serves as a therapeutic philosophy by offering an anchor for the search for self-restoration.[3] But the same could be argued for Darwinism or even Social Darwinism. The psychohistoric dislocation to which Wengle refers is more characteristic of western culture than that of the more introspective eastern milieu. Nonetheless, he does recognize the integral need for a non-zoocentric view. He asks to what extent the sociobiologists are deceiving themselves in maintaining their own 'vital lie' that an intellectually and emotionally satisfying theory of human nature can be constructed by ignoring man's fundamental need to deny his animality.

Odd extrapolations from the Darwinian philosophical stock abound in the formulation of sociobiological theory. One example: the theory of natural selection states that genetic mutations are the substrata for the natural selection to act. That is to say that mutations that confer higher survival value upon living organisms are favoured by the forces of natural selection and thus evolutionarily adapted. The sociobiological postulates about social behaviour run into enormous difficulty when examined for their validity in the light of gene-regulatory behaviour at the molecular level. For instance, there is no evidence for any unitary genetic specificity for the selection of a particular behavioural trait in human beings. Molecular interactions of genes offer an infinite variety of genetic combinations that make predictions about behavioural inheritability an impossibility. The relationship between mutations and selection becomes all the more fuzzy if the recent argument for quantum leaps in organic evolution is introduced into this debate.

Behind these methodological inconsistencies, gross extrapolations of data on animal studies, fatal errors in logical typology, and outright fabrication of information (Arthur Jensen's racist research is a classic example) is the face of the reductionistic, deterministic pseudo-science of sociobiology that has provided fuel to social injustice, racial inequality, sexism and patriarchal dominance. By arguing for a biological ontology for gender, sexual and racial inequalities and attempting to maintain a status quo, sociobiology has become a political tool rather than a scientific methodology. Of the many socio-political ramifications of this pseudo-philosophy, what concerns us here is a brief examination of its strategy for perpetuating patriarchal dominance.

The reductionist idea of the gene as the primal agent for imparting developmental behavioural specificity has ordained that women can be immutably classified with certain traits that are innate to the feminine lot: passivity, masochism, dependency, and a nurturative tendency. Sociobiologists present the argument that these behavioural phenomena must have imparted adaptive success and owe their existence, therefore, to evolutionary selection.

In their sway of genetic determination of human behaviour, sociobiologists have reserved all sorts of stereotypes of the human female. Almost all of it is designed to maintain the male chauvinistic status quo. Thus writes Wilson: 'It pays males to be aggressive, hasty, fickle and undiscriminating. In theory, it is more profitable for females to be coy, to hold back until they can identify their males with the best genes . . . human beings obey this biological principle faithfully.'[4] Apart from endowing the human female with a remarkable ability to identify the best genes, Wilson has also decided that the so-called best genes are here to stay forever; they do not undergo mutation and are 'faithfully' selected by human beings!

The denigration of the human female extends to all conceivable areas of concern, from innateness of gender and sexual differences, to regulation of female sexuality, to job discrimination and outright sexual exploitation. Masculinity is celebrated and femininity is relegated: all in the name of 'objective' extrapolations from animal models where apes and birds are shown to indulge in prostitution, where masochism is encountered among insects and homosexuality is rampant among worms — all in Wilson's animal kingdom at Harvard University.

The classical 'double standard' of human sexuality comes fully alive under the patronage of sociobiologists. Polygamy, differential criteria for virginity, and marital infidelity are defended on the basis of crude 'scientific' observations.

For example, extrapolating from a numerical differential between female eggs and male sperm, a philosophy of reproductive strategy is developed where females 'hunt' for the quality and males for the quantity. It is common biological knowledge that a reproductively active human female produces nearly 400 ova in her lifetime as opposed to millions of sperms produced by males. This numerical edge of the male sperm is invoked by sociobiologists to justify male promiscuity, adultery and even violent expressions of their sexuality. The message is clearly stated by Barash:

> Genes that allow females to accept the sorts of mates who make lesser contributions to their reproductive success will leave fewer copies of themselves than will genes that influence the females to be more selective. . . For males, a very different strategy applies. The maximum advantage goes to individuals with fewer inhibitions. A genetically influenced tendency to 'play fast and loose,' 'love 'em and leave 'em' — may well reflect more biological reality than most of us care to admit.[5]

Barash wants all human males to confess that they are genetically endowed to rape the females of their species and there is nothing wrong with this innate biological tendency.

The genetic advantage of spermatic plenitude is presented as a survival tactic. Male promiscuity is thus biologically right for in nature an intense competition is going on to inseminate rather scarcely available female ova. Male reproductive instinct as well as sexual drive, under the intoxicating influence of the omnipotent selfish genes, therefore condone even a violent overpowering of the human female. In this context and only in this context, rape becomes one of the human male reproductive strategies. Animal data are once again used to demonstrate the biological and natural origins of this sexual act. Hence, the genetically blessed male aggression against females becomes the motive force behind patriarchal dominance.

Sociobiology and its cohabitants (reductionism and determinism) are active on many other fronts too. For instance, Steven Goldberg's glory in the androcentric hall of fame rests on expounding neuronal reductionism.[6] By creating a genetic-neuronal axis for the release and differential influence of sex hormones, Goldberg postulated the inevitability of the gender gap. In Goldbergian determinism, gender differences in power, wealth, status, dominance or subservience, sexual assertion or passivity are simply a matter of hormones.

Sexual Control and Gender Bias

Gender bias is a universal phenomenon, perpetrated over many centuries and by many cultures. For instance, during a recent study of the creed associated with sexual behaviour and human fertility among Muslims and Hindus in certain parts of rural Bangladesh, Profulla C. Sarkar of the University of Rajshahi observed the following:

> If a couple wants to have a male child, they have to maintain certain conditions: they must complete their dinner before ten o'clock at night and must sleep at least an hour before having intercourse, which must be complete before twelve o'clock. It should be noted that at the time of coitus the male partner should be alert about the maintenance of the flow of breath through the right side of his nose. If the flow through the right portion of his nose is stronger than through the left, then a male child may be conceived. On the other hand, if the couple wants to have a female child, they must maintain the opposite conditions. They must have coitus after twelve o'clock and the male partner must maintain a stronger flow of breath through the left portion of his nose.[7]

At face value, it does not appear that in this study the villagers made their wish known for begetting male children only. Nevertheless, the evidence in this folk tradition makes the male partner exclusively ascendant for his role in sex determination: it is the relative breathing pattern attained by him that ascertains the sex of the child. The villagers were, however, sceptical whether the male partner would be skilful enough to muster the strength in order to preserve the ordained breathing sequence!

The differential attitude toward producing children of one single sex only, in this case the *preferred* male sex, is neither an anthropological rarity nor a cultural trait distinctive of the rural population from Bangladesh. Nearly all human cultures, at one time or another, have demonstrated a preference for rearing male children only.

A folk (?) tale from the southern Slavic region gives an account of the creation of the female sex in these words:

> God absentmindedly laid aside Adam's rib when He was performing the operation recorded in the Bible. A dog came along, snatched up the rib and ran off with it. God chased the thief but only succeeded in snatching off its tail. The best that could be done was to make a woman out of it.[8]

The Hebrew vilification of the female sex is asserted through the daily prayers in this diction: 'I thank Thee, Lord, for not having created me a woman.' Saint Paul, the main architect of Christian theology, believed that female seductive power was so great that it caused even angels to sin. The ruthless practice of female infanticide in the pre-Islamic Arabic custom (*Jahiliyah*) had its roots in the symbolism of 'shame' personified by the very existence of a female child. The Papuans from the Torres Straits are also known to have practised female infanticide; but the Zulus are more merciful for while they slaughter an ox to celebrate the birth of a male child, baby girls are received with: 'Why should we kill an ox for a girl? She is merely a weed.'

The human female though now saved from infanticide, must encounter the inexorable realities of a male-dominated world in which she continues to suffer denial of equal rights and opportunities,[9] job discrimination and exploitation, rape, forced prostitution and genital mutilation.

There is at present an extremely tangled ethical and moral question facing human society caused by technological advancement in biology and manifest in the rise of genetic engineering. The discovery of a chemical substance Deoxyribo Nucleic Acid (DNA), essentially paved the way for today's phenomenal developments in biotechnology. Briefly stated, chromosomes are largely made up of the DNA substance; there is nuclear DNA as well that is known to interact with chromosomal DNA in many subtle ways; this helical macromolecule in its peculiar sequence of the four bases carries the instructions for genetic encoding of a complete individual. Thus, when looking at the molecular mechanisms of sex determination, and given the fact that the human female and male are equal partners in furnishing the genetic raw material for the foetus, any intervention or interruption in the structure or function of their respective DNAs would yield the 'desired' results — the introduction of a fundamental change in the genetic make-up of the human foetus. This seemingly simple statement harbours implications for the entire human race that are more nightmarish than the nuclear holocaust.

The technique of amniocentesis has been in medical currency for some time now. It was in 1963 that the first foetal blood transfusion was successfully accomplished through the use of this technique, considerably reducing infant mortality caused by the Rh factor. However, amniocentesis took a new turn when it was found that examination of the amniotic fluid can help in detecting the sex of the foetus. This opened a new 'sophisticated' route to female infanticide: examine the amniotic fluid through a series of cell culture techniques for clues to the foetal sex, and then abort if female! While the benefits of foetal monitoring in the detection and possible treatment of congenital disorders are abundantly documented, it is difficult to collate data on how much these methods have been employed for sexual selection. The drawback of amniocentesis is that it is insensitive to early foetal development, therefore sex monitoring is not possible till the early onset of the second trimester.

More recently, with the help of a DNA probe, it has become possible to detect foetal sex from a single drop of amniotic fluid. No doubt, the DNA probe could become a useful tool in the detection of a number of metabolic and sex-linked genetic disorders, but it carries also a great potential to hasten the selective foetal elimination. Similarly, another new technique, chorionic villi sampling (CVS), can be employed to draw a small sample of tissue within the first eight to ten weeks of pregnancy or even earlier to determine foetal sex. Thus, in comparison with amniocentesis, DNA probe and CVS techniques are proving to be more efficient in the detection of foetal sex in the early weeks of pregnancy with a greater promise for being used as the implements for abortions of females.

The production of a child without recourse to normal sexual intercourse must now be taken for real. No more limited to the flights of fancy or to the prerogative of certain animals to perform it through parthenogenesis, this mode of human reproduction appears to be gaining impetus. The method of artificial insemination was first experimented with around the time of American independence! It was not until 1942 that human sperm was put under deep freeze without any structural impairment. The first human child conceived through artificial insemination was born in 1954. Since that time two kinds of artificial insemination (AI) have gained currency: one that involves the sperm of husband (AIH), and the second where sperm is from an anonymous donor (AID). The legal, moral and ethical questions encompassing the birth of an AID-child continue to be debated.

In vitro fertilization is becoming an accepted medical practice. It was in 1978 that the first 'test tube baby' was born through the use of IVF technique. By the end of 1985, there were an estimated 1,000 IVF babies in the world. In the United States alone, more than 100 medical centres now offer IVF facility.

Embryo transfer (ET) goes a step beyond the now familiar IVF. In the traditional method, it involved fertilization of ovum by sperm in a petri dish and its subsequent implantation in the uterus. With embryo transfer, the fertilized egg can be retained outside the uterine environment for a longer period, stored under sub-zero temperature, then thawed for implantation. The effectiveness of new technology for obtaining ova from women's bodies is supplemented by ultrasonographic methods. Once outside the woman's body, these ova can either be subjected to AIH or AID, or a mixture of both. Usually, an embryo of up to sixteen-cell level has been used for inducing pregnancy. So far, embryos developing beyond this stage have not transferred. On the other hand, techniques for sustaining premature babies in incubators through a complex network of life-support systems may some day be utilized for maintaining these extra-uterine embryos. Thus, any combination of ova and sperm could be realized: it can take the shape of a normal AIH, or AID; or it may involve the donated egg combining with the husbands's sperm; or both ovum and sperm come through donations, are fertilized *in vitro*, stored, and subsequently implanted in a 'host' uterus. Yet another variation on the theme is the so-called surrogate embryo transfer (SET) or lavage: conduct an AID procedure on an ovulating woman and after a few days, extract the embryonic mass for implantation into a non-ovulating woman!

The embryo transfer came into the limelight with the story of an American millionaire couple who had had parenting experience from their previous marriages but were unable to conceive together. In 1981, three of Mrs Elsa Rios's ova were subjected to AID treatment by the Australian researchers. One embryo was implanted and the remaining two were frozen. After about ten days of implantation, there was a spontaneous abortion and Mrs Rios did not opt for a second embryo transfer. The couple took off for South America, adopted a child, but were killed in a crash of their private aircraft. This incident opened the question of the status of frozen embryos. Are these the legal 'children' of the

Rios family, with the right to inherit the family estate? What moral obligation does society has toward these entities? Do frozen human embryos have moral rights? Does the instant of conception bestow personhood upon these frozen masses? These queries can only exacerbate the already troubled world of gender question.

The issues of ethics and morality of AID, IVF and ET notwithstanding, all these procedures are an enormous physical, emotional and financial burden on the women whose bodies are being used as guinea pigs for these hi-tech ventures. In Australia, where most of these techniques were perfected, the success rate is reported to be a mere 13 per cent, i.e., the percentage of pregnancies carried to term, while the reported success range for Britain is 10–15 per cent. However, the proponents of IVF continue to make loud claims that IVF babies are more intelligent, healthier and socially better adjusted than the 'normal' children. An invitation to abandon conventional sexual intercourse in favour of IVF wizardry?

It was in 1982 that a single gene transplant in mammals was performed with success. In that experiment gene for growth hormone was isolated from rats. Next, ova removed from female mice were subjected to *in vitro* fertilization. During the IVF, the isolated rat growth hormone gene was injected into the ova and the engineered mouse ova were implanted back into the female mice. This resulted in the growth of mice to rat size — nearly double that of the average mouse.

In the case of human beings, the application of genetic engineering is usually considered under two categories: firstly, somatic cell gene therapy that is supposed to take care of a genetic disorder in any part of the body; and secondly, germ line gene therapy — that is the deletion, insertion or recombination of a genetic segment of the human reproductive tissue in order that genetic transmission to the offspring would be a completely tailored operation. It can be argued that somatic cell gene therapy comes as a great relief for the treatment of hitherto incurable genetic defects. However, at the present moment, genetic receptivity of human somatic cells remains relatively unexplored. The bone marrow is, perhaps, the only human tissue that welcomes foreign genes. On the other hand, it is not difficult to erect a scenario of how germ line gene therapy may be exploited in the near future when an assortment of biomedical and biotechnological tools are just waiting in wings. These formidable apparati include ultrasonographic imaging, artificial insemination, chorionic villi sampling, *in vitro* fertilization, embryo transfer, foetal surgery and other varieties on the rDNA theme. IVF has been successful in getting investment capital for its growth, without much forethought on preventive or adaptive measures for the physical and mental health of mother and child.

For the record: a human germinal depository, where the Nobel Prize winners 'donate' their sperms, has been in operation in Escondido, California, for quite some time. Their declared objective is to produce White only individuals — analogous to bionic woman or superman — through artificial insemination of carefully selected females. There have been reports that babies born out of these hi-tech extra-marital coituses are indicating, through their behaviour, that they have indeed inherited the Nobel genes! So far, the depository has had recourse to traditional AI methods. With the provision for rDNA technology for human germ line tinkering, the prospects for human engineering seem dreadful. After all, Nobel laureates too have their defects which would need to be eliminated from the stock of 'super' children.

In the midst of the unfolding drama of sex pre-determination, what is the Muslim response? One argument could be that much of the world-wide Muslim population does not have access to the kind of reproductive technologies present in the west. However, let us not forget that abortion policies in most Muslim countries are neither well formulated nor properly implemented. We would be amiss to assume that a covert female infanticide is not underway in these countries, particularly in view of a strong gender bias. On the other hand, there is a reassurance that the social status of the Muslim women is different from that of her western counterpart and many of the gender-specific problems that we have identified are not problematic.

The sexual paradigm in Islam is entirely different from that of Judaism or Christianity; it is sex-positive as opposed to sex-negative. This sexual self-assurance of a Muslim is not identical with carnality or lust; instead, it is governed by the macroparadigm of *tawhid* that serves as the guidance for gender-specific as well as marital behaviour. Next, the status of human beings as *khalifah* (vicegerent) of Allah on this earth becomes the progenitor of a world-view in which the biological significance of reproduction and life itself acquire an entirely different meaning. While Muslim thought does not deprive the sexual act of its essential dualistic nature — creation and recreation, it puts ethical and moral bars that prevent it from degenerating into an animalistic act. Moreover, the family structure as a manifestation of biological activity is so intricately built on the principles of love and mutual dependence that it ensures the social survival of a group. Above all, the *sunnah* provides a guiding light for the desired social coherence.

What has befallen western women in the wake of modern reproductive technologies should however be taken as the basis for a discourse on the future of Muslim women since high technology is infectious and, sooner or later, Muslim societies will face at least some of these instrument-related if not idea-related problems.

Technology or the Angel of Death?

Human reproductive technologies threaten to transform radically some of the fundamental attributes of human life. It is in the realm of embryonic human life that technological breakthroughs are raising new issues with serious implications for human rights. The question of technological encroachments into human embryological environment goes back to the 1973 US Supreme Court decision (Roe *v.* Wade) on the legalization of voluntary abortion. It was stated that: 'With respect to the State's important and legitimate interest in potential life, the "compelling" point is at viability. This is so because the foetus then presumably has the capability of meaningful life outside the mother's womb, albeit with artificial aid.' It was further argued that 'viability is usually placed at about seven months (twenty-eight weeks) but may occur earlier, even at twenty-four weeks'.

It is the key concept of viability of the human foetus that is prone to techno-logical intervention. In other words, since 1973, technology for neonatal care has made tremendous progress and neonatal medicine is now a part of high-technology medicine with its own expenditure in the USA exceeding $2 billion. Surgical intervention to correct foetal anomalies, care for premature birth and extra-uterine maintenance for genetic repair increase the viability limits of the foetus outside the mother's womb. Although the World Health Organization has established the limit of twenty-two weeks as a cut-off point between birth and abortion, nineteen to twenty week intervention for amniocentesis is not uncommon. That this limit will be crossed further as a result of technological gains is not a far-fetched idea.

A decade after its decision, the Supreme Court once again faced the question of legalization of abortion. Although the new challenge did not alter the older decision, the dissent of one of the judges, Sandra Day O'Connor, on the issue of foetal viability is quite illuminating:

The *Roe* framework, then, is clearly on a collision course with itself. As the medical risks of various abortion procedures decrease, the point at which the State may regulate for reasons of maternal health is moved forward to actual childbirth. As medical science becomes better able to provide for the sepa-rate existence of the foetus, the point of viability is moved further back toward conception.[10]

The focus of viability of the foetus, therefore, lies in the potential growth of technology that can sustain a flourishing foetus outside the mother's body.

The juxtaposition of neonatal medicine and the growing tide of anti-abortion terrorism (one needs to remember that in 1984 alone there were more than two dozen incidents of fire bombing and arson of abortion clinics all over America)[11] indeed creates an enigmatic situation. On the one hand, this high-technology is labelled as the deliverer of troubled foetuses, while on the other anti-abortionists — who like to call themselves pro-life as opposed to pro-choice — are increasingly exploiting the technology-based viability clause to raise such questions as: What is the definition of a human being? What is personhood? When does human life begin? Does the foetus have rights? Human rights?

A central question to be raised is how neonatal technology can act upon a twenty-two week old foetus under the pretext of salvaging the premature baby, while the same technology is an abortive agent for a similar foetus, who may be a victim of gender preference or moral quandary of the mother. Apparently, the question of viability leads to yet another question: Who makes the foetus viable, or who decides its status as such? That is to say, who determines the limits of viability? Is extra-uterine viability supported by sophisticated technology the only criterion that may be invoked for declaring abortion a crime no matter at what foetal stage it happens? Anti-abortionists will argue it from a human rights perspective — that foetus has rights too and neither the mother nor technology may violate foetal (human?) rights. Of necessity, this leads us into the elusive domain of defining personhood.

If, according to anti-abortionists, the conceptus is a human/person, and abortion is a violation of the 'right to life', then the pro-choice counter-argument is that viability or attainment of self-consciousness are not protectable rights. For them, personhood is an outcome of biological *and* social elements. It is not a static process, but a dynamic one where the biological and the social evolve interactively. In this definition of personhood, the conceptus develops *toward* a person and is *not* a person. Therefore, questions of viability and self-consciousness must be isolated from their biological perspective and integrated into the social milieu before even a consideration could be given as to whether an infraction of human rights is involved in the act of abortion. Furthermore, the constitutional freedom granted to women to abort legally is upheld by the pro-choice group on the strength of the fact that mother's rights take precedence over that biological entity that she carries in her womb.

Anti-abortionists claim that conceptus is a person and there is no time when it? he? she? is not a person. This is the developmental model of personhood that bestows the right to life upon the conceptus; to abort is to deny that right. Moreover, conceptus has the right to be sustained, nurtured and protected since fertilization has accorded the status of person to the conceptus. Thus, the Human Life Statute (US Senate S. 158, Section 1) reads:

> The Congress finds that present day scientific evidence indicates a significant likelihood that actual human life exists from conception. The Congress further finds that the 14th Amendment to the Constitution of the United States was intended to protect all human beings. Upon the basis of these findings, and in the exercise of the powers of the Congress, including its power under Section 5 of the Fourteenth Amendment to the Constitution of the United States, the Congress hereby declares that for the purposes of enforcing the obligation of the States under the Fourteenth Amendment not to deprive persons of life without due process of law, human life shall be deemed to exist from conception, without regard to race, sex, age, health, defect, or condition of dependency; and for this purpose 'person' shall include all human life as defined herein.

From the moral standpoint, abortion and the question of foetal rights may be entangled with a host of social, economic and political questions. But whatever may be the final outcome of this debate, one thing is clear — neither pro-choice persons nor anti-abortionists can hope to get a tax exemption solely on the basis of their definition of the beginning of life. For Internal Revenue Service, it begins at birth.

Once again, technology is beginning to open a window on the womb to pinpoint the molecular processes that ensue after the sperm and the ovum unite to form a zygote. Within three weeks of fertilization, the embryonic heart begins to beat and brain development starts. Within the next two weeks, the brain is divided into three parts and limbs make their appearance. By the sixth week, eyes begin to shape and the next week shows development of genital organs, musculature and teeth. Come eighth week and all organs are in place. The embryo is now known as a foetus. Ultrasonic trackings will show that at this stage the foetus starts moving on its own. A heartbeat, which may be regarded as the *sine qua non* of human life, first begins intermittently at three to four weeks.

The fact that the first appearance of detectable brain waves occurs at about twelve weeks after conception has given a new twist to the search for a definition of humanness. It is ironical that the combined force of western theology, philosophy and technology has stumbled in evolving a definition of life acceptable to divergent groups. On the other hand, the formulation of a definition of death had apparently very little by way of epistemological complexity.[12] The main criterion in the articulation of the *Uniform Declaration of Death Act* was the cessation of brain activity.

Even though the use of brain-death criterion presents its own biological, moral and ontological problems, its reverse configuration, brain-life criterion, is becoming increasingly a possible choice for critical redefinition. Consider, for instance, the arguments by John M. Goldenring: 'Whenever a functioning human brain is present, a human being is alive.'[13] He states further, 'Whether that being is in-utero or ex-utero, whether that person will die in the next minute, or at age nine weeks or ninety years is immaterial to this definition.'[14] He rejects the viability definition as artificial since it is dependent on a particular state of technology.

What are the implications of brain-life theory for either a pro-life or pro-choice stance on foetal rights? Goldenring argues that in Roe *v.* Wade the Supreme Court's stand was that the state interest in possessing co-equal rights for the foetus begins only if the foetus has a moderate chance of survival. Thus, it is only the use of the concept of viability and not a definition of humanness that is implied in the Court's verdict.

Assuming that his brain-life theory is free from attendant moral, philosophical, social and even biological drawbacks, he declares the eight-week gestation stage as the demarcation between humanness and otherwise. He states: 'One could allow abortion "on demand" up till the eighth week post-conception, but then require substantial reasons for later abortions which by the brain-life definition would cause a human being to die.'[15] Goldenring appears to be maintaining the position that abortions prior to eighth week gestation period are legitimate but those performed beyond this mark are not. He is actually leaning against the power of neonatal medicine to devise, in the near future, an abortifacient, probably a kind of progesterone blocker, that will arrest the hormonal function at cellular receptor sites. Once that drug becomes available, Goldenring opines, the brain-life theory will have little ethical problem with the question of abortion. Obviously, for him morality stops at the door of biochemical technology. His advice appears to be: wait till that happens and then shoot for a merry-go-round!

In contrast with Goldenring's covertly reductionist paradigm for a chemically-mediated morality, M.C. Shea offers a holistic perspective to the debate. She argues that '. . . a new human life begins when the newly built body organs and systems begin to function as a whole.'[16] According to Shea, mere cellular life of a human embryo does not impart to it the status of humanness until all body organs and systems start functioning as an integrated whole. A comparison with human death, where death characterizes the cessation of holistic functions of the human body, is relevant in this argument. Nevertheless, she uses the brain-life criterion in defending her stand on holistic definition of humanness. Thus, the beginning of a new human life resides in the development of a functioning brain which has the capability to co-ordinate bodily functions. This, in essence, is her criterion for a holistic approach to the definition of humanness. While Shea's holistic approach may have an intrinsic advantage over Goldenring's reductionism, she fails to differentiate between brain functioning of a born individual with that of an eight-week old conceptus. Perhaps the pro-choice argument that foetal viability without self-consciousness is untenable carries serious implications for brain-life theory.

Mary Anne Warren has identified the following traits as most central to the concept of personhood, or humanity, in the moral sense:

1. consciousness (of objects and events external and/or internal to the being), and in particular the capacity to feel pain;
2. reasoning (the *developed* capacity to solve new and relatively complex problems);
3. self-motivated activity (activity which is relatively independent of either genetic or direct external control);
4. the capacity to communicate, by whatever means, messages of an indefinite variety of types, that is, not just with an indefinite number of possible contents, but on indefinitely many possible topics;
5. the presence of self-concepts, and self-awareness, either individual or racial, or both.[17]

At last, we have someone addressing the issue from a human and moral standpoint. At the outset Warren refuses to subscribe to the idea that genetic humanity suffices for moral humanity. She argues that none of the above premises apply to the foetus. That is to say, these traits so incisively distinguish between the two contenders of humanhood (potential human, genetic human, ontological human, brain-life human vs. the actual human) that no further explication is required. Therefore, she proclaims:

> Neither a foetus's resemblance to a person, nor its potential for becoming a person provides any basis whatever for the claim that it has any significant right to life. Consequently, a woman's right to protect her health, happiness, freedom, and even her life, by terminating an unwanted pregnancy, will always override whatever right to life it may be appropriate to ascribe to a foetus, even a fully developed one. And thus, in the absence of any overwhelming social need for every possible child, the laws which restrict the right to obtain an abortion, or limit the period of pregnancy during which an abortion may be performed, are a wholly unjustified violation of a woman's most basic moral and constitutional rights.[18]

In introducing a 'moral' dimension into the debate on humanness, Warren has relied upon those criteria that in the long run, on a chronological spectrum, may be characterized as biological and social outcomes of humanness. Furthermore, we can argue that these traits do not become, *ipso facto*, moral traits. It would be a naive assumption to claim that consciousness, reasoning, self-motivation and communication skills fall within the purview of human morality.

Further critical examination of Warren's 'moral' criteria for personhood demonstrates that they are liable to be taken on par with the ontological criteria of anti-abortionists. This indeed is a glaring self-contradiction in Warren's argument. The pro-life stance is that the conceptus carries the potential for humanness and that very potential begs an anti-abortion position. Warren's criteria, therefore, face the challenge that the stated traits of personhood are inherently ontological in their biological as well as social contexts and cannot be accepted as a rationale for the denial of foetal rights.

Moreover, Warren's justification for upholding maternal rights over foetal rights raises the question of infanticide when she goes on to deny the right to life even to a *fully developed* foetus. Should this be understood as her lack of faith in the magical powers of neonatal medicine, or is it yet another manifestation of the selfish gene of radical feminism? It goes without saying that this attitude raises questions of human morality.

My contention is that high-technology is reshaping the philosophical, theological and social attitudes towards human life. Moreover, there is corroborative evidence that the massive technological invasions of both the process and the event of human reproduction have given rise to extremely vital questions bearing upon our fundamental concept of life, humanness and moral being. A redefinition of humanness in today's society is being sought under the long arm of high-technology. The forces of religious fanaticism, racial bigotry, hedonism, and scientism seem to be progressing towards an involuntary collusion that may one day suffocate the moral being at its foetal level of existence.

Notes

1. See, *Three Guineas*, Harcourt, Brace and World, New York, 1938, p. 139.
2. E.O. Wilson, *Sociobiology: A New Synthesis*, Harvard University Press, Cambridge, Mass. 1975.
3. John E. Wengle, 'Sociobiology: A Psychohistoric Note', *Journal of Psychohistory* 11 (3) 403–10, 1984.
4. E.O. Wilson, *On Human Nature*, Harvard University Press, Cambridge, Mass., 1978, p. 125.
5. D. Barash, *The Whispering Within*, Harper and Row, New York, 1979, p. 48.
6. Steven Goldberg, *The Inevitability of Patriarchy*, William Morrow, New York, 1974, p. 25.
7. See, Profulla C. Sarkar in *Eastern Anthropologist* 35 (2) 139, 1982.
8. Quoted in H.R. Hays, *The Dangerous Sex: The Myth of Feminine Evil*, Pocket Books, New York, 1966, p. 2.
9. See, for instance, some of the contemporary writings: Julia O'Faolain and Lauro Martines (eds.), *Not in God's Image: Women in History from the Greeks to the Victorians*, Harper and Row, New York, 1973; Andrea Dworkin, *Woman Hating*, E.P. Dutton, New York, 1974; and Marian Lowe and Ruth Hubbard, (eds.), *Woman's Nature: Rationalizations of Inequality*, Pergamon Press, Oxford, 1984.
10. See *Akron* v. *Akron*, 103 US Supreme Court 2481, 1983.
11. *Ms* 13 (9) 19, 1985.
12. See my review of *Redefining Death* by Karen Grandstrand Gervais, *Inquiry*, March 1987, p. 69.
13. John M. Goldenring, 'The Brain-Life Theory: Towards a Consistent Biological Definition of Humanness', *Journal of Medical Ethics*, 11, 198–204, 1985.
14. *Ibid.*, p. 200.
15. *Ibid.*, p. 204.
16. M.C. Shea, *Journal of Medical Ethics*, 11 (209) 1985.
17. Mary Anne Warren, 'On the Moral and Legal Status of Abortion', In: *Morality in Practice*, edited by James P. Sterba, Wadsworth Publishing, Belmont, California, 1984, p. 150.
18. *Ibid.*, p. 152.

7

Science and Efficiency
Exploding a Myth

Rakesh Kumar Sinha

Technology is sometimes described as the know-how of socio-economic trans-formation, with an ideological potential since it controls the production and distribution of all the goods in a society. Simply stated, the purpose of technol-ogy is to mobilize the available human and natural resources and with the help of the existing means of production utilize them to generate goods needed by society. The types of goods produced in a society — whether agricultural, industrial goods or services — are a function of the technology available in that society. One of the functions of technology is to improve the means of produc-tion available to it by using the knowledge, or science, already developed. The development of technology at the same time influences and accelerates its own further development. Science and technology thus develop in an interdependent manner and their development is an integral part of the development of the society.

Every society develops a technology consistent with its own conditions, resources and requirements. Traditional technologies are characterized by self-contained conservationism that sustains the society without jeopardizing its life support system. This is particularly true for technologies developed in Afro-Asian societies. The science from which this technology was derived could truly be called natural science since it developed from observation of nature. A major feature of traditional science and technology is, therefore, an effortless attach-ment to maintenance of a balanced ecological system. Natural resources are utilized at a rate not faster than one at which they are replenished.

So-called 'modern' technology has the dubious advantage derived from modern science of having access to results of controlled and directed observa-tions, especially in the laboratory.[1] There is in this new technology no attempt to maintain a balanced relation with natural processes, and it stands, more often than not, in contradistinction to natural processes. A society that develops on the strength of this type of technology does not limit itself to the mobilization of its own natural resources but attempts to conquer nature and in the process usurps the natural resources of other societies.

The history of the development of this post-industrial revolution technology is the history of imperialist development of European and North American countries at the expense of the colonial exploitation of Asia and Africa. Annihilation of the original civilizations of the Americas and Australia is an integral part of this history. And yet the glamour of modern science and technology has created an illusion of its desirability in the development of human societies. African and Asian countries, ever since independence, have been clamouring after a development process modelled on that of their ex-masters. In this way they are making themselves the agents of their own continuing exploitation.

The Influence of Modern Science

Modern science and technology has, no doubt, helped the countries of Europe and North America attain an unprecedented level of consumption. The standard of living is generally measured by the level of material consumption and on this scale, the industrially developed countries have scored high. Major new developments are taking place at regular intervals; new consumer goods are being developed; working hours are decreasing and conditions of work improving; leisure time and material comforts are increasing.

On the other hand the people of Asia and Africa are suffering ever-increasing poverty and destitution. They put in longer hours of harder work and in return get less than subsistence-level wages.

Providing food for its population is the major task in Afro-Asian countries. In less developed countries a major portion of the Gross Domestic Product (GDP) comes from agriculture, a sector that contributes only 3 to 5 per cent to the GDP in the advanced countries. Another important indicator of quality of life is that of services available in a society. In the USA, the services constitute the single largest sector (39 per cent of GDP), in Japan and Europe services equal the contribution of the industrial sector (30 per cent), whereas in India they constitute only 13 per cent of GDP.

This discrepancy is supposed to have happened because the Euro-American countries have achieved advanced industrialization while the Afro-Asian countries have lagged behind. The industrial revolution is based on the development of machines that increase the productivity of labour (output per worker). As a result of the increased output a surplus is created. Part of this surplus goes into improving the material conditions of the society and part of it is invested in further improving the means of production, the machines. It has been suggested by Marx and Ricardo that a machine replaces more labour than goes into its manufacture. The industrial development, once started, thus grows from strength to strength.

It is also suggested that the Afro-Asian countries can alleviate their misery only by following in the footsteps of Europe and America. For this the less developed countries will have to develop a scientific bias, change backward socio-cultural habits and create an industrial infrastructure. The contribution of agriculture to GDP should be reduced and that of industry increased. It is further suggested that to make up for lost time, Afro-Asian countries have to develop at a rate faster than Europe and America. For this they need the latest technology that gives the highest productivity of labour. To get this technology they need the help of multinational corporations.[2]

These theories have been picked up, and to some extent accepted, by the ruling classes of the Afro-Asian countries. In some countries they are trained by the ex-colonial master countries and are influenced by them. They are also in a hurry to obtain for themselves the luxuries available in the master countries. In the remaining countries, the ruling class is driven by visions of historical materialism that has become a most effective instrument of European industrialism. For one reason or another, the ruling classes in most of Asia and Africa are urging their respective countries towards the goal of modernization. The experiment is, however, leading to a disastrous conclusion.

Myths of Scientific Efficiency

India launched a programme of modern industrialization immediately on getting her independence from British rule. In the last four decades she has built up the third largest cadre of scientific and technical personnel in the world. India is exporting scientific and technical manpower to Europe and America in the form of doctors, engineers and management professionals. Within the country, a wide network of roads and railways has been built, large steel plants have been started, space and nuclear programmes have been established. Clearly in the field of science and technology India comes right after America, Europe and Japan. Yet India is one of the poorest countries in the world with a per capita GDP lower than that of most Afro-Asian countries. Even China, which has also embarked upon a programme of controlled modernization, has a lower level of GDP than some African countries.

In India industrial production has stagnated since the mid-sixties.[3] Ever since the new youthful government started a programme of launching the country into the twenty-first century with the help of multinationals, conditions have been deteriorating. More than half of the country is reeling under drought, goods are getting scarce, prices are spiralling and people in some areas have been forced to sell their children. This is an experience not only confined to India. All the Afro-Asian and South American countries are paying the price of modernization. The population of Sahel is facing total extinction as a result of letting multinationals control their farming.

What, then, is the reason for this contradictory picture of industrialization in Europe and America on the one hand, and Afro-Asia on the other? Evidently the conditions under which industrialization has flourished in Europe and America do not obtain in Asia and Africa. But the matter needs closer scrutiny. Let us consider the implications of GDP as a measure of efficiency of technology. Per capita GDP is, no doubt, an indicator of the material well-being in a society, it does not, however, say anything about the way in which this production is obtained. GDP is expressed in terms of its monetary value and, for the purpose of comparison, this is converted into one currency (say US dollars). This does not tell us anything about the material content (steel, cement, grains etc., that is, the type of goods and the amount of each). Countries produce these goods in different proportions and their prices are not fixed according to the efficiency of their production. In fact the same goods, particularly services, carry different values in different countries. This is sometimes explained in the patently racist terms of 'social cost of labour'. The pricing is, in fact, entirely political.

This is better understood when one considers that the GDP of the USA is considerably inflated by its armament industry. The arms and heavy machinery produced by industrialized countries are sold at enormous profit, resulting in a new flow of wealth from less developed to the more developed countries. Consider in addition, the trade that USA, UK and Germany carry on with South Africa which is clearly beneficial to them. Their trade with other Afro-Asian and South American countries is no less exploitative.

Economics of Industrialization

In the initial stages of the industrialization of Europe, goods produced by machines were of very poor quality and could not stand the competition from traditionally made goods. The machines were rendering workers jobless (inspite of the excess capital) and the living conditions of the workers and their families were so miserable that a number of attacks on factories were organized by the workers. They were brutally suppressed. But how was the problem solved?

The problem of widespread unemployment once again reached crisis proportions during the general depression of the 1930s. Einstein blamed the new technology for the world economic crisis. He believed that the problem was what he called 'false over-production'; that is, production higher than can be consumed by the given purchasing power of the society but less than that actually required by the society.[4] Einstein felt that the purchasing power of society was decreasing even as production was increasing because the new technology enabled all the goods required by that society to be produced by only a small number of workers. The remaining workers were unemployed and had no purchasing power. This according to him was the root cause of the economic crisis then facing the world.

It has been shown quite conclusively by economic studies carried out on industrialization during the nineteenth and the early twentieth centuries that the output/capital ratio has been decreasing with increasing industrialization i.e.

with increasing capital/labour ratio.[5] This implies that the output/labour ratio does not increase sufficiently to make up for the reduced number of workers. Or in other words, the total labour required to make the machine is not made up by the increased productivity of labour. If a machine, then, cannot generate any surplus, where does the excess capital available in the industrialized economy come from? It is obviously the result of colonial and neo-colonial exploitation of the Third World countries. What appears to be the increased productivity of labour in the advanced countries is nothing but the fruits of the labour of Afro-Asian workers transferred to these countries.

These are some arguments supporting updating of the technology on the logic that reduced employment will lead to reduced levels of consumption and hence higher savings. These can then be used to improve technology and achieve a higher rate of development. In fact the advanced technology further reduces the output/capital ratio and induces increased unemployment and destitution. The suggestion is patently flawed.

The economic parameters used thus create an illusion of efficient production associated with advanced science and technology when, in fact, only redistribution by means of expropriation and transfer of wealth from one region to the other is involved. A more neutral measure of technological efficiency is therefore needed. 'Energy efficiency' of technology is one way to measure. For this some technical clarification is needed.

Technology of Economics

That there is equivalence between matter and energy has been known to man for a long time. It is doubtful whether this knowledge can be used directly in measuring the efficiency of technology. However, a number of investigators have started probing the role of energy in the development of human societies.[6] The daily energy consumption of man has increased continuously with the development of civilization so that today a man uses sixty times the energy used by primitive man.

The use of energy efficiency as an indicator of the performance of technology is further justified when we consider its role in the process of production. Industrial processing is done to convert a given raw material into a given finished product. This processing may involve mechanical work, heating or chemical processing in any combination and each of these involves expenditure of energy. It must be recognized that work, heat and chemical energy are all forms of energy and may be converted from one form to the other. The value imparted to a given raw material as it is converted into the finished product is the work (or energy) usefully invested in it during processing. The energy actually spent is always more than the energy usefully invested in the process since part of it is always wasted. Energy efficiency of a process may be defined as the usefully invested energy as a fraction of the total energy spent.

A number of attempts have been made to estimate the energy efficiency of advanced technologies by considering the energy flow through the economy. Input of primary energy is in the form of coal, petroleum, liquid gas, natural gas, hydrothermal and nuclear energy. (Incidentally, even in an advanced country like the USA, the contribution of nuclear energy is less than one per cent of total energy consumed, all of it being used to generate electricity.) Twenty-five per cent of the energy is converted to electricity, more than half of which is lost in transmission. The remaining is put to end uses. The overall efficiency of energy utilization in advanced countries is between 35 to 40 per cent, the remaining 60–65 per cent being lost in different ways. This means that only 40 per cent of the energy consumed in the society is actually delivered for final commercial, industrial and household utilization.

If we include the efficiency of utilization of energy in the process itself, the overall energy efficiency is still lower (i.e. of the order of 25 per cent of the energy consumed). This efficiency depends on the nature of the process itself. If machines are used to carry out a process, a part of the energy spent will go in driving the machine itself and only a much smaller fraction will be usefully invested in the actual process. I shall clarify this with an example. If a plot of land is to be ploughed — let us say with a six hp tractor — then a part of the energy will go into the actual ploughing of the field but most of it will be spent in driving the heavy tractor. The energy efficiency of this process can be displayed as: energy spent in ploughing/energy spent in ploughing + energy used to drive the tractor + energy wasted. If a bigger tractor (say a twelve hp machine) is used for ploughing the same plot of land, the second term in the denominator will increase while the numerator remains unchanged. Hence the efficiency of a process decreases as the size of the machine used to perform the work increases. However, the same work is done in shorter time. The rate of doing work is power. Hence a machine does not save energy in performing a task, in fact it consumes more energy and delivers power. With the advancement of technology, the size and power of machines, the rate of doing work increases, but the energy efficiency decreases.

Since energy, usefully invested, is the real value of production, a scrutiny of the energy consumption and the patterns of its utilization will tell us something about the state of technologies of production currently in use. The total annual consumption of energy in the world today is about 9 billion tons coal equivalent (TCE). This works out to an average (per capita) consumption of 1.8 TCE/ year. More than one third of the total energy is consumed by Americans, who constitute only 6 per cent of the population, a per capita rate of 12 TCE/year. The advanced countries, constituting less than 30 per cent of the population, use up to 84 per cent of the total energy. The poorest 70 per cent of the population get only 15 per cent of world energy at an average of 0.4 TCE/year.

The advanced countries produce only 60 per cent of the total energy between them and the remaining 24 per cent is appropriated from the oil producing countries. If they do without this extra energy, the energy available to the less developed countries will be more than doubled. Third World countries should seriously review the policy of acquiring energy wasting modern technology and try instead to develop their own traditional, energy efficient technologies.

Efficiency of Traditional Technologies

The traditional technology of Afro-Asian countries was definitely more energy efficient than that obtained from advanced technology. I should like to consider two examples of this, one in agriculture, the other in the steel industry.

Providing adequate nutrition to its population is an important task for every society. In pre-industrial society it was the only major productive activity. Even today it is the largest single contributor to the economy of Third World countries like India.

In agrarian economies, vegetable production, manual work and careful husbandry of animal and human waste enables the input energy to be multiplied many times over. The input is mainly in the form of human and animal labour, seeds and manure. The output is the sum of vegetable and animal output. Modernization of agriculture reduces the output per unit of input energy as it replaces human and animal labour with machines of much lower efficiency. Further, it tries to extract a higher yield from the earth by supplementing its fertility with chemical fertilizers. Chemical fertilizers contain less energy than goes into their manufacture and deliver even lesser energy to the fertility of the land. Hence the yield of food-energy per acre of the land might increase but the increase in input energy is much higher in proportion. The ratio of output/ input energy contents, that is, the energy efficiency goes down. For example, in the USA, the output energy measured in GJ/ha-y is 9 and the input energy is 13 (with an input/output ratio of 0.7); in India the figures are 10 and 0.7 (thus making the ratio 15); and China has output energy of 281, input energy of 6.8 (making the ratio 41).

We can see from these figures that mechanization may ultimately lead to the case where even in agriculture the output energy is actually less than the input energy inspite of the solar subsidy. What is needed to increase food production in the Third World countries is not the so-called scientific, mechanized farming but more manual field work. The western approach to nutrition, emphasizing animal protein, is highly inefficient in terms of energy. Animals consume ten times the food energy they deliver. In addition, the food is intensively processed and then transported over long distances. Both these are energy intensive processes. The total per capita expenditure on providing food to the American citizens is 1.4 TCE per year. The energy content of this food, per capita, is 0.2 TCE/year. The total energy consumed in all industrial and other activities in the developing countries is only 0.4 TCE/year per capita.

Steel, the second example I wish to deal with, is the most important engineering material and an essential part of the industrial infrastructure. Steel production is a highly energy intensive process. The energy required (as delivered to the processing equipment) is 5.5 G Cal/Ton of steel. The energy losses prior to that are not counted, neither is the energy spent in constructing the plant and equipment. What is more, the energy required is very high grade: coking coal is needed for reduction of iron ore in the blast furnace; for converting the pig iron to steel, depending on the process used, high quality coke breeze, oil or gas may be used. For the oxygen convertors, high purity oxygen is required. For clean steel electric steel making is used.

Under the pressure of the 1973 energy crisis, the steel industry started looking for energy efficient, alternative steel-making processes and a number of traditional technologies were reviewed. It is a well known fact that the traditional steel-making practice in India was the most mature pre-industrial technology and used a direct reduction process. The practice could use low grade energy from charcoal and even wood and was highly flexible for it could be adapted to varying local conditions and ore composition. Steel was produced in batches of about 250 in a forty hour heat. The process was decentralized and a large number of furnaces were spread over the large steel belt. In the mid seventies, experts from all over the world searched the jungles of Madhya Pradesh for any artisans who could help them reconstruct the practice used. (Traditional steel making, incidently, had been banned by the government of India with the nationalization of the steel industry.) A number of scientists are now investigating the traditional Indian steel-making practice and trying to reproduce it in the laboratory.[7] It is difficult at this stage to prepare an energy balance for the process but there is little doubt about its efficiency and quality. Experts are still debating how the iron pillar of Delhi and the beams used in the Konark temple were forged and what the secret of their corrosion resistance is.

The Third World: A Case for Decentralization

Maximum production is achieved with a technology that has a capital intensity equal to the actual capital/labour ratio in a society. This may be defined as the condition of optimum decentralization. Here the industrial capital is nearly equally distributed amongst the workers. The instruments of production used must correspond, in their cost, to this situation. Any attempt to increase the capital investment per labour beyond this will go in the direction of centralization. This will decrease the total output of the society and at the same time, cause unemployment leading to all the accompanying problems.

The same analysis may be carried out in the framework of the technology of production. In this case, output is measured in terms of energy value added and input is also in terms of energy. The optimum conditions for maximum production are obtained when the total energy available for production is distributed equally among the workers. This is another way of describing decentralization. It should be recognized that aside from maximizing production, decentralization will also minimize trading between regions and reduce manipulation and exploitation. It also minimizes transportation that is the single largest wasteful activity in the advanced countries.

In conclusion it may be said that 'modern' science and technology operate on the logic of maximizing the rate of production of goods — that is, maximum goods produced in minimum time. This high power, centralized production is achieved at the cost of energy efficiency, with production consuming more energy value than it produces. Advanced countries make up this deficit by exploiting Third World countries. Third World countries pursuing the western model of industrialization are aiding their own exploitation. Since this untenable situation is destabilizing world order and creating international tensions, Third World countries would do well to review their human and material resources situation and develop a technology appropriate for their own conditions. This will be achieved by using a decentralized production process in which the industrial capital (or productive energy) is made available to all the workers in equitable proportions.

Notes

1. Sachchidanand Sinha, 'Science Technology and Appropriate Technology', Paper presented at the National Seminar on 'Science Technology and Transformation' organized by Socialist Discussion Forum, Jawaharlal Nehru University, New Delhi, 30–31, March 1986.
2. A. Emanuel, *Appropriate or Underdeveloped Technology*, John Wiley, Sussex.
3. I.J. Ahluwalia, *Industrial Growth in India: Stagnation since the Mid-Sixties*, Oxford University Press, Oxford, 1985.
4. Carl Seeling (ed.), *Ideas and Opinion by Albert Einstein*, Rupa Press, India, 1984.
5. Colin Clark, *The Conditions of Economic Progress*, London 1960; and H. Habakkuk, *American and British Technology in the Nineteenth Century. The Search for Labour Saving Inventions*, Cambridge 1962.
6. B. Elbek, 'World Energy Problems', in *Physics and Contemporary Needs*, Riazuddin (ed.), Plenum Press, 1979; D. Faude, 'Long Term Energy Systems and the Role of Nuclear and Solar Energy', in *Physics and Contemporary Needs, op cit.*; and M.W. Thring, *The Engineers Conscience*, Northgate Publishing, 1980.
7. B. Prakash and V. Tripathy, 'Iron Technology in Ancient India', *Metals and Materials*, September 1986.

8

Science and Health
Medicine and Metaphysics

Ziauddin Sardar

Some time ago, I visited my local chemist to make an on-the-spot survey of his stock. From what he had on the shelves, it seemed that people in Britain suffer a great deal from constipation, weight problems, indigestion and gas, headaches, allergies, colds, flu, tooth decay, ill-fitting false teeth, a mania for tanned skin, depression, premenstrual tension, insomnia and anxiety. I am quite sure that another chemist anywhere in Europe or North America would be catering for the same common illnesses.

Constipation is primarily caused by a lack of fibre in the diet. Weight problems are also largely food related, along with indigestion and gas, certain allergies and tooth decay. Constipation can be easily relieved by putting fibre back into one's diet, by eating whole, unrefined foods. But instead some two billion doses of laxatives are consumed every year in the United States alone. Depression and anxiety can be controlled by exercise, yoga, meditation and by slowing down the pace of life. But, to quote American statistics again, some four billion Valium pills, one and half billion Libriums, a billion Equanils and Miltowns and millions and millions of other mind controlling drugs are swallowed every year by Americans. These drugs may bring temporary relief, but they don't cure the disease; and the disease is life-style.

In *Diseases of Civilization* (Paladin, London, 1981), Brain Inglis lists heart diseases, cancer, mental illness, infectious diseases and iatrogenic disorders (illnesses induced by doctors and their treatments) as the main illnesses of western civilization. With the exception of iatrogenic disorders, all the illnesses are related to life-styles. For example, heart diseases are a consequence of affluence: they are the result of overeating, rich food, refined foods, stress, chemicals in the environment and lack of physical exercise.

However, life-styles do not only produce new illnesses. They can also radically transform old diseases. Diseases can be reactivated, or assume newer deadly forms. For example, in the early nineteenth century, polio existed in the USA as a mild childhood illness. It started to disappear in the 1920s as American cities began to clean and purify their water supplies. However, a few decades later it came back: this time, it could kill and cripple. It had now become a disease of affluence, the consequence of pure drinking water. Consider also herpes which has been with us in harmless forms for centuries as cold sores. But as genital herpes it assumes a newer more irritating shape: sexual behaviour has changed the epidemiology of the disease. Then, of course, there is AIDS (Acquired Immuno-Deficiency Syndrome).

AIDS has probably existed in Africa for centuries. There is evidence that it has been endemic to Africa for a long time; it has certainly been traced as far back as 1973. But the epidemiology of AIDS in Africa was, and to a large extent still is, quite different from elsewhere. While in countries like Zaire and Uganda up to 15 per cent of the population may have got AIDS as a mild childhood sickness, very few people actually died from the disease. A recent study in Uganda revealed that none of the children whose blood samples contained antibodies to the AIDS virus showed any signs of the disease itself. But the disease has, in the last two to three years, started to kill. Why?

In its newer, deadly form AIDS first appeared in the United States. And its fatal mutation is clearly connected to a particular life-style.

The AIDS virus is spread only through blood and semen. There are only two ways fluids from one body can enter another. The first involves the use of a syringe or a similar device. Thus haemophiliacs needing blood transfusions can become victims of AIDS if they are given contaminated blood. Or drug addicts using contaminated syringes can also acquire the dreaded illness. The second is by sexual intercourse: and this is the predominant way most victims of AIDS acquire the syndrome.

However, only certain sexual practices promote the infusion of contaminated bodily fluid from one person to another. The walls of the rectal lining, the mucosa, are only one cell thick. Thus during anal intercourse, homosexual or heterosexual, they are easily damaged and offer a ready entrance for the AIDS virus to enter the blood stream, carried by the sperm or by blood from an injury to the penis. However, even when no tissue damage is done, it is extremely easy for the infected sperm to enter the bloodstream during anal intercourse. The sole function of sperm is to penetrate the corona of the ovum and fertilize it; thus, sperms are designed for cell penetration. During ordinary intercourse, sperms cannot enter the bloodstream as the epithelial cells of the vagina are fortified against such an invasion. But the rectum has no such protection and sperm have no problem in overcoming the barrier and passing on the virus. Thus, despite propaganda to the contrary, AIDS is exclusively related to life-styles, a life-style practised by homosexuals; and is found in only those heterosexuals who widely practise anal intercourse. A recent Swedish study, which looked at families of fourteen haemophiliacs infected by AIDS virus from American clotting concentrate, revealed that only one of thirty-five individuals examined showed evidence of the virus, and she was the only female sexual partner who practised anal intercourse. Short of anal intercourse and intravenous drugs, AIDS is almost impossible to get.

So why has AIDS now started to kill its African hosts? The most likely explanation is that in its new form the disease was taken back to Africa by American homosexuals checking out hot spots in Uganda and Zaire. In that part of Africa use of needles for performing rituals is quite common. The new form of the disease spread quickly through contaminated needles and is now threatening entire villages. It is quite feasible for the life-style of one group of people to change the epidemiology of a disease and pass it back in the new form to people who were immune to the disease in its original form.

Life-styles are dictated by world-views. Certain world-views tend to encourage certain types of personal, social, and cultural behaviour, and discourage certain other types. For example, Islam takes a grim view of homosexuality and the Shariah prohibits casual sexual relations as well as anal intercourse. As such, chances of outbreaks of AIDS or genital herpes in a Muslim society which practises at least the outer manifestations of Islam are non-existent. But world-views do not only shape life-styles; they also shape the external environment within which these life-styles are pursued. And this external environment plays just as important a role in producing diseases as life-styles themselves. Many modern health problems can be traced to environmental problems. For example, the rise of infertility amongst men in the United States has been traced to toxins like PCBs which concentrate in men's reproductive organs: average sperm counts, which were 100 million in 1929 and 60 million in 1974, dropped to 20 million in 1978. The result: some 23 per cent of American men are now sterile; and the percentage is rising rapidly.

However, world-views are not only responsible for producing diseases and illnesses — both through promoting certain life-styles and producing an environment within which these life-styles can flourish; they also form the matrix within which attempts are made to cure these illnesses. Medicine is a sibling of world-view: modern medicine is a product of the world-view of the western civilization.

Modern medicine is completely true to the world-view of its origin. Reduction is its methodology; capitalism is its dominant mode of production; power and control is its prime goal. Thus, the human body is a machine made up of a number of different parts, the organs. Diseases are well defined entities which are responsible for structural changes in the cells of the body and tend to have singular causes. They are caused by germs, bacteria and viruses; and recently it has been accepted — only on the face of mounting evidence — that environment too is a causative agent. The body is attacked by these outside forces which cause breakdowns within the body. If these external factors are isolated and crushed, by chemical or surgical intervention, the body can be repaired and the patient cured.

But this reductive model, as Fritjof Capra has argued so forcefully in *The Turning Point* (Simon and Schuster, New York, 1982), has been successful in only a few special cases, such as acute infectious processes, and cannot explain the overwhelming majority of illnesses. The decline of the mortality rate over the past century owes almost nothing to modern medicine. The credit belongs, as recent research has shown, to pure or treated drinking water, pasteurized milk, indoor plumbing, closed sewers, improved nutrition, clean and safe work places and shorter working hours. In *The Role of Medicine: Dream, Mirage or Nemisis* (Nuffield Provincial Hospitals Trust, London, 1976), Thomas McKeown undertakes elaborate historical-epidemiological studies to show that medicine contributed little to the improvement of health in industrialized Europe in the late nineteenth century. After examining the possible causes for declining mortality, he finally settles for improved nutrition. A similar study in the USA, done by John and Sonja McKinlay ('The Questionable Contribution of Medical Measures to the Decline of Mortality in the US in the Twentieth Century', Milbank Memorial Fund Quarterly: Health and Society, Summer 1977), attributed the fall in mortality rates to the disappearance of eleven major diseases: influenza, whooping cough, polio, typhoid, smallpox, scarlet fever, measles, diphtheria, tuberculosis, pneumonia and the diseases of the digestive system. With the exception of the first three, all the other diseases disappeared almost entirely before medical intervention made an appearance.

But even if medical care had arrived before these killer diseases disappeared, it would have made little impact on the overall health of American society. According to Capra, 'the main error of the biomedical approach is the confusion between disease processes and disease origins. Instead of asking why an illness occurs, and trying to remove the condition that led to it, medical researchers try to understand the biological mechanisms through which the disease operates, so that they can interfere with them'. Thus, modern medicine does not ask the fundamental question of why diseases occur, but instead asks how they operate when they have occurred. Thus, on the question of the rise in infertility amongst American men, nothing has been done about the toxins in the environment which cause infertility. But enormous financial and intellectual resources have been put in finding ways and means of making infertile men fertile again; to the extent of the development of artificial insemination by donors.

The reductive methodology epistemologically removes society from medicine. If diseases and illnesses are external to the body, and sicknesses can be cured by isolating the diseases and exterminating them, then the role of society in both producing and treating sicknesses becomes irrelevant. In this way, western medicine simultaneously tries to identify and manage ill health, while concealing the origins of health and illness in social and economic relations. Thus, western medicine is ideologically oriented.

Nothing better illustrates the role of western medicine as capitalist ideology more than cancer. In western society, cancer is both an epidemic and big business. Since 1950, when it was declared public enemy number one, astronomical sums have been poured into research on cancer in Europe and North America. In 1985, the United States cancer budget was an estimated staggering $50 billion. The average treatment of a cancer patient cost $40,000. Yet, despite all the optimistic claims by cancer research organizations in the industrialized countries, virtually no progress has been made towards finding a cure. A few cancers, like childhood leukaemia, can be cured but these represent a very small percentage of cancer incidences.

There are three methods of treating cancer: surgery, which is used to remove cancerous tumours; radiation, which is used to kill cancer cells in the tumour; and chemotherapy, which is used to kill cancer cells in the tumour or throughout the body. All three methods have severe effects on the patient. However, after the treatment the chances of the patient's survival are still slim. Only one in three survives five years after treatment. Whether this constitutes cure is a matter of definition.

Cancer can be caused by a number of carcinogens which range from asbestos fibre, cigarettes, alcohol, various drugs including those used in the treatment of cancer itself, radiation, food additives and colouring, pesticides, nitrates drained into drinking water from fertilized fields, nuclear waste and a whole range of industrial pollutants. One can take precautions by avoiding certain foodstuffs and certain kinds of diet, but what is to be done about the environment itself which is overcrowded with carcinogens? Improving the quality of the environment and making it carcinogen free amounts to transforming the industrial culture itself. But even in avoidable areas capitalist society places pressures on individuals to consume cancer-giving commodities. Cigarettes and alcohol do not just produce vast profits for those who are involved in their production and distribution, they are also a major source of government tax. Thus, most governments turn a blind eye to massive advertisement campaigns designed to increase consumption of cigarettes and alcohol. Similarly, fast foods as well as food additives, which prolong the shelf life of processed or frozen food, are a cornerstone of consumer economies. And clearing the environment involves banning certain pesticides and industrial pollutants; it involves taking on the might of the petrochemical giants. An idea of how intrinsically cancer is involved with politics can be judged by the fact that the Board of Overseers of Memorial Sloan-Kettering Centre, one of the key cancer research centres in the United States, includes the chairman of the board of Exxon, the president of Exxon, a director of American Cyanamid, a director of Olin, a director of Consolidated Oil and Gas, the chairman of the board of General Motors, a director of Atlantic-Richfield, a director of Philip Morris and a director of Texaco. These people are not likely to permit research which will identify petrochemicals as the major cause of cancers.

However, capital does not only control the agents which cause cancer; it also controls the methods by which cancer is treated. The medical establishment does not permit the treatment of cancer except by its own methods. In the USA unorthodox methods of treatment, that is treatment not based on the three methods described above, is illegal. Doctors can be imprisoned for using alternative methods of treating cancer. In Europe, alternative therapies are mocked and ridiculed.

Just as violence is an intrinsic feature of the western world-view and the capitalist mode of thought, so is it also central to its medicine. Even the terminology is violent. Diseases are hunted down. War is declared on certain diseases. The cytotoxic chemotherapy that is used with cancer patients is literally referred to in medical slang as poison. Doctors treating cancer patients actually talk of poisoning. When patients are sedated or anaesthetics are administered to them, the process is referred to as slugging. One talks of killing pain; in the cytotoxic language there is the notion of tumour-kill, derived straight from the language of nuclear war.

No wonder then that a patient arriving in a hospital finds him/herself in the middle of a war zone. In this battlefield only the generals of the medical establishment are in control, patients are helpless victims who bring diseases for the doctors to fight and defeat. Thus an expecting mother becomes a helpless patient who is 'ill'. Pregnancy is not seen as a natural phenomenon but as a form of sickness that can only be cured in hospital. A world-view that places no premium on family life, indeed actively undermines family relations, is bound to see the home as a place unsafe for giving birth. In Britain, it is against the law to practise childbirth at home, unattended by qualified medical practitioners. And doctors who encourage natural child birth are sometimes disciplined; as the recent case of Wendy Savage demonstrated. Nature cannot be trusted to produce a normal birth; it has to be actively managed by technology. Once inside the hospital, the pregnant woman has no control over her body. She lies there helpless while obstetric technology takes over. Even though obstetric procedures often do more harm than good, it is not always obvious to the victim who is led to believe that home births are infinitely more dangerous. However, the most common danger to women in labour is haemorrhaging. The remedy requires plasma and sterile water but midwives are not allowed these supplies not because they cannot administer plasma drips or inject needles, but because handing even the limited amount of technology to the midwife means that the medical establishment undermines its own control and power.

Systems of medicine based on other world-views, of course, present serious threats to the power and domination of western medicine. On a very simple level, they present an economic threat: in the western world-view, both health care systems and diseases are commodities. Doctors, as Donald Gould writes in *Black and White Medicine Show* are not interested in health. 'By inclination and training they are devoted to the study of disease. It is sick, not healthy people, who crowd their surgeries and out-patient departments and fill their hospital beds, and it is the fact that the population can be relied upon to provide a steady flow of sufferers from faults of the mind and of the flesh that guarantees them a job and an income in harsh times as in fair.' Medicine is about income; and advances in modern medicines are not made with health but financial rewards, as well as prestige and fame, in mind. Heart transplants are a case in point.

Coronary heart diseases kill some 350,000 people a year in England and Wales. While heart transplants may make money, reputations and fame, they are bad medicine for they can never make more than an insignificant impact on the people who suffer from heart diseases. A transplanted heart has to cope with all the problems of organ disorder, particularly lungs, which lead to malfunctions in the replaced heart. Thus, even when it is accepted by the new host, the chances that it will acquire the problems of the old heart are high. Even when heart transplants are successful and the patients survive, which they seldom do beyond a year or two, the cost of the operations is huge. Gould estimates that if only half of the patients who lose their lives from the disease in England and Wales could be saved, the total cost would amount to the entire budget of the British National Health Service. Moreover, even if this money could be found, one would still require surgeons, nurses and the equipment necessary to perform some 500 heart transplant operations a day. Then, of course, there is the other side of the equation: where will the donor hearts come from? As Gould explains, 'the major contributors of healthy spare parts are young people badly injured on the roads who have been taken to hospital still alive but with irreparably damaged brains. Transplant surgeons and their supporters therefore have a vested interest in sustaining and even increasing the carnage wrought by motor vehicles. It is a matter of killing Peter in order to have the chance of a long-odds gamble on saving Paul.'

It is this kind of absurd logic and unlimited reliance on high technology which led Ivan Illich to claim that modern medicine is on a suicidal course. (*Limits to Medicine. Medical Nemesis: the Expropriation of Health*, Penguin, London, 1977). High technology treatments and profit and glory oriented medicine have combined to produce what Inglis calls the disease of western civilization: iatrogenic disorders or illnesses caused by medical treatment. Drugs that produce serious side-effects (chloromycetin, marketed as a broad antibiotic which induces aplastic anaemia — a fatal bone marrow disorder; thalidomide, marketed as a fertility drug which produced children with malformed limbs; and clioquinol, marketed as a treatment for upset stomachs but which produced 'subacute myelo-optic neuropathy' leading to paralysis and blindness, are but a few well known examples), surgery that is either unnecessary or leads to serious illness like the notorious operations to remove the 'foci of infection' from the gut or the jaw, or sort out 'slipped disc' problems; and treatments like radiation or chemotherapy which end up doing more damage than good — all these indicate that western medicine has itself become a disease.

The western medical establishment, however, is not too concerned about the disastrous impact of its methodology. It safeguards itself ruthlessly. The medical establishment guards its commodity, the way medicine is 'made' and the way it is 'sold', by having complete control and absolute power over its products. It has, throughout its history, ruthlessly subverted and systematically destroyed systems of medicine rooted in non-western world-views which may challenge its domination. How Islamic medicine was treated under colonization, and is still treated to some extent, illustrates this point well.

The main difference between western and Islamic medicine is metaphysical. Islamic medicine does not perceive itself as a commodity, but an obligation on society which must be fulfilled by religious dictates. This is why, even though western historians of medicine have tried to prove otherwise, classical Muslim doctors never used their medical practices as a source of personal revenue. Scholars like Ibn Sina and al-Razi did not make their living from their medicine, but from their scholarship. Most physicians were also philosophers — they combined their metaphysics with their medicine. While the western world-view epistemologically removes society from its medicine. Islamic medicine makes society its central focus. The primary methodology of western medicine is reduction. Islamic medicine, while acknowledging the importance of reductive reasoning and diseases — al-Razi's description and analysis of smallpox has not been matched for its sheer power of reasoning and reductive analysis — concentrates on synthesis and the whole person. Under colonization, the clash of western and Islamic medicine was thus a clash of world-views.

The encounter of the two systems of medicine in the Punjab provides a good example of this clash. Islamic medicine, locally known as *Yunan-i tibb* (Greek medicine) because the Muslim scholars traced their medical practices back to the Greek masters, was the dominant rational medical practice when the British arrived in the Punjab. From the early twelfth century, Islamic medicine had developed a strong indigenous base in the Punjab. It was promoted by the Muslim rulers of North India who supported medical libraries, medical schools, hospitals and prominent physicians. Although much of it was urban centred, the rural areas were not altogether neglected. For example, during the sixteenth century, Sher Shah, the Sultan of Delhi, made sure that a physician was stationed at all overnight stops on caravan routes under his control.

However, largely due to the turbulent century which preceded the arrival of the British in India and marked the decline of the Moguls as well as an overall stagnation in Muslim civilization, research in Islamic medicine had ceased completely at the time of the British arrival. The principal medical authority for the *hakims* was still Ibn Sina's *Cannons of Medicine*. On the whole, *hakims* tended to shun surgery leaving it to *jurrahs*, who were often barbers as well, and to *suthais*, who performed eye surgery. But the intellectual stagnation of Islamic medicine in the Punjab under the Raj did not mean that the system itself was not rational, or that it did not serve the health needs of the populace.

Indeed, as John Hume observes in his thesis 'Medicine in the Punjab: Ethnicity and Professionalisation of an Occupation' (Duke University, PhD., 1977), when the British first took control of the Punjab they were forced to admit, as reports of early observers indicate, that the cities possessed 'sophisticated' medical techniques and contained large numbers of men trained in medicine. The *hakims* were able to treat successfully most of the common maladies in the area and commanded high respect from the indigenous population. As the Punjab Civil Secretariat's Proceedings in the General Department recorded in October 1856, the *hakims* came from respectable families in the community, they were careful of their behaviour and 'above all their systems of practice are carefully adopted to the prejudices and practices of their patients'.

Even though British administrators saw Islamic medicine as a clear threat to their world-view and political control, initially they were forced into an alliance of convenience. Thus, in a number of emergencies during the middle of the nineteenth century, the *hakims* were used by the British administrators to assist in the treatment of disease. For example, in a programme proposed by the District Commissioner of Sialkot, Lieutenant Colonel T.W. Mercer, *hakims* were employed in a district-wide scheme to treat diseases, distribute sample medicines, primarily quinine, to act as sanitary inspectors, as registrars of vital statistics and to aid in the provincial vaccination programme. The programme was based on the assumption that *hakims* were socially acceptable to the local population (which had already rejected Indian doctors trained in western medicine and shown indifference to British doctors), they knew the minds of their patients and, in time, would come to appreciate the superiority of western, allopathic medicine. Mercer was not concerned with revitalizing Islamic medicine, he was seeking the most expedient solutions to the health problems in the area under his supervision. His declared intention was 'the gradual substitution of English medicine for useless native drugs, the attendance of the sick of all classes, to afford prompt medical relief, and ultimately, the subversion of the system of medicine as practised by the natives'.

The programme involved the selection of a *hakim* who was nominated by the local population, for a circle of thirty-seven villages. A central village was selected as his residence. The circles were organized in a *tehsil* which was administered by a graduate of the government at the allopathic medical school in Lahore. He was given the title of Hakim Ali to indicate that he was the chief *hakim* of the *tehsil*. The *hakims* proved more resilient and adaptable than Mercer could imagine. He was astounded by the success and popularity of his programme and believed it was due to the involvement of local people in all areas of the programme. This success led Mercer to start a training programme designed to train the sons of *hakims* in western medicine. Indeed, in 1870 Lahore Medical School started classes for the relatives of *hakims*. These classes emphasized anatomy and surgery — two areas where Islamic medicine lagged behind. Eventually, the College developed a programme for the training of *hakims* in the English system and awarded 'Titles of Oriental Medicine' to candidates proficient in western medicine in addition to *Yunan-i tibb*. No formal teaching was conducted in *Yunan-i tibb* but the candidates were required to produce a certificate of their competence from a recognized authority.

However, the oriental medical programme of Lahore Medical College was not allowed to mature. Right from the beginning the College faced opposition from western doctors on this programme. In 1882, when the Government of India proposed to set up a register of medical practitioners, this opposition reached its apex. The proposed bill was to confer certain rights on the medical practitioners: the right to sue for fees, the right to sign government certificates and the right to call oneself a registered practitioner. Under the system, both *hakims* and doctors trained in western medicine were to have equal status. But the medical establishment in the Punjab could not tolerate this.

The practitioners of western medicine opposed it for two reasons. On the one hand, they considered western medicine to be superior and more scientific than *tibb*. Promoting *tibb* meant promoting 'reactionary elements' and regressive tendencies in Indian society — all of which were seen to oppose British rule in India. On the other hand, they saw *tibb* as a professional and economic threat; *hakims* clearly commanded respect and trust from the local population and tended to undermine the sole authority and control of the medical establishment. Allopathic physicians objected to programmes such as Mercer's since they gave *tibb* some legitimacy and intellectual respectability. The use of *hakims* in rural areas as well as in training programmes which awarded titles in oriental medicine were seen by them as a government recognition of *Yunani-i* medicine. Not surprisingly, they wanted a complete ban on *hakims*.

However, allopathic practitioners were not content at simply raising their voices against *tibb*. They started an active campaign to discredit it, magnify its shortcomings and neutralize praise for the successes of *hakims*. The major weapon was the pages of the Punjab Dispensary Report, an annual document which was submitted via the Governor-General of India to the Secretary of State for India in London.

However, even a barrage of complaints against *hakims* and an active campaign to suppress *Yunan-i tibb* did not initially succeed. In 1869, the Punjab government replied to critics of *Yunun-i tibb* by pointing out:

> The most opposite of opinions have been expressed to its utility; by some, the benefits are said to be 'immense' while others declare it to be 'worse' than useless; the fact probably being . . . that its success or failure mainly depends on the amount of encouragement and co-operation it receives from the District Officer and Civil Surgeon . . . So far as the Lt. Governor can judge the establishment of *Hakim* arrangements has in many cases led to a large increase of subscriptions from the native community, and insofar as it has been voluntary, this forms an additional recommendation of the measure.

But this declaration of the Punjab government did not deter British doctors or Punjabis trained in western medicine from attacking *hakims*. In 1876, the government of Punjab made its last stand. It declared that while Islamic medicine was in many ways inferior to western medicine, medical science was nevertheless empirical and there was no doubt that *hakims* did some good as 'there is no reason to doubt that the native treatment of many diseases is to some extent effective and highly appreciated by patients'. But this was very much a last ditch stand. The following year all courses for the training of *hakims* were abandoned.

John Hume concludes that judging solely on the basis of scientific principles, the exclusion of *hakims* from the government medical services was a failure for scientific medicine. Many *hakims* were certified as proficient in both western and oriental medicine and these hybrid practitioners dominated, much to the dislike of western doctors, private practices in the cities and towns for many years to come. However, it was in the rural areas that the pro-allopathic decision had its most serious effect for allopathic practitioners were unwilling and unable to serve the rural populations and only *hakims* could cater for them. 'Thus by excluding the *hakim*, Punjab government decided, in effect, that if the choice was between poor quality medical relief or no medical relief at all in rural areas, then government must support no medical relief' ('Rival Traditions: Western Medicine and Yunani-i Tibb in the Punjab, 1849–1889' Bulletin of the History of Medicine (55) 214–31 (1977)).

While the British did at least recognize some qualities of Islamic medicine in the Punjab and tried to use it initially to their advantage, other colonial powers had little but contempt for Islamic medicine. For example, the French wasted no time in ruthlessly suppressing Islamic medicine in Tunisia as Nancy Gallagher describes in her study *Medicine and Power in Tunisia* (Cambridge University Press, 1984). At the beginning of the nineteenth century, a French doctor needed *ijaza* (permission) to practise medicine from the Muslim Chief of Physicians. However, by the end of the century, Hamada bin Kilnai, son of a former noted Chief of Physicians, was classed as *médecin tolere* and given a second class medical status by the French administrators. As Gallagher points out, this reversal was not based on any scientific consideration, or concern for the health of the local population, as both systems were unable to cope with the three epidemic diseases rampant in Tunisia at the time: cholera, typhus and the plague. However, both systems of medicine arrived at similiar solutions for fighting the plague. The French sought quarantine. The more established Muslim physicians advocated a similar course except they addressed the indigenous population in a language that it could understand. One group of Muslim physicians, led by Abd Allah Muhammad Bayram, a prominent Muslim scholar, saw prevention in 'abstaining from mixing'. He cited two well known hadith to support his position: 'No contagion, no evil omen' and 'flee the leper as you would flee the lion'. It was the political clout of one system of medicine which ensured that the other was relegated to a marginalized existence.

But the wrath of the colonial powers was not reserved for Islamic medicine only. Muslim countries, like Egypt, which tried to adopt western medicine and tailor it to their requirements were not tolerated either. Western medicine had to be adopted within its own frame of reference and world-view: it had to be accepted as a commodity, with class distinction at its root and capital at its base, and with power and control firmly in the hands of the western medical establishment.

In Egypt, Islamic medicine was replaced by western medicine by Muslim rulers themselves. During the first decades of the nineteenth century, Mohammad Ali Pasha, who was rather impressed by the achievements of the European powers, decided to establish the western system of medicine in the country. He went about the task in a systematic way founding, in 1827, the Qasr al-Aini School of Medicine at its original site at Abu Zabal. The idea was to produce a cadre of professional doctors to look after the health needs of the country. But Mohammad Ali wanted not just to produce Egyptian doctors but also a healthy, dynamic medical establishment which was self-sustaining, undertook medical research on the needs of the country and was health oriented. Even though he had to rely on European teachers, and despite the fact that most of the literature existed in European languages, he ensured that all teaching at the Qasr al-Aini School would be in Arabic.

To overcome the problems of Arabic teaching material, translations of medical texts into Arabic were extensively promoted; and as medical and scientific terms in Arabic had either become extinct or were not part of the language at all, a systematic attempt was made to develop Arabic scientific terminology by using classical Arabic medical texts or simply translating the particular medical term into Arabic. Mohammad Ali chose to administer the Medical School centrally both to ensure that its graduates specialized in the areas needed for the medical development of the country, and when qualified they worked in needed areas in accordance with government policies which placed great emphasis on provision of medical services for the rural areas. The School was open to all Egyptians with free education and training.

By the time Khediv Ismail came to power in 1863, the school had supervised the emergence of a highly qualified medical profession. Most of the teachers at the school were Egyptians who had also received some training in Europe. The school was undertaking considerable original research and published *Al-Yasub al-Tibbi*, which was the first scientific journal of its kind to appear in Egypt. A number of hospitals were opened and several mobile units were being used to fight epidemics. Teams of graduates of the schools, *hakims* and barbers were used to control smallpox and incidences of cholera were reduced substantially. Ismail also allowed doctors to work outside the government system and set up private practices.

However, the British occupation of Egypt in 1882 radically altered the development of the Egyptian medical establishment. As Amira el Azhary Sonbol writes in her dissertation, when the Qasr al-Aini School was placed in the hands of the British administrators in 1893, they began a systematic attempt to destroy Mohammad Ali's achievements ('The Creation of a Medical Profession in Egypt during the Nineteenth Century: A Study in Modernisation', Georgetown University, Washington D.C., Ph.D, 1981). Their first steps included cutting the number of students studying at the school, introducing fees thus limiting the profession to a certain privileged class which could afford to pay for medical education, abolishing Arabic as the language of instruction and making Secondary School Certificate a basic qualification for entry. Thus the medical profession was cut off from the vast majority of the Egyptian people.

But the British were not content with limiting medical education to a particular class of Europeanized elite. They were essentially against the whole notion of the Egyptain medical establishment being health orientated. Thus, they first changed the curriculum and shifted its emphasis from health to diseases and then prohibited Egyptians from specializing in any field. The number of years of medical training was reduced from six to four. Publication of the research journals was suspended. Moreover, they introduced a number of laws which made it easier for foreigners, particularly Europeans, to practise medicine in Egypt while at the same time making it impossible for Egyptians to do the same. Foreigners needed no qualifications to enter the Qasr al-Aini School. Egyptian doctors were not awarded a medical certificate which would permit them to go into private practice. Instead, the education and training at the school was recognized only in the government health service.

Mohammad Ali established the Qasr al-Aini School to serve the health needs of the Egyptian population. Indeed, before the arrival of the British administrators, health care was widely available in Egypt, both urban and rural areas being adequately catered for and patients treated free. However, under the British, health care in the rural areas was virtually eliminated. Moreover, even in urban areas western medicine was available only to those who could afford to pay. Under the British a number of hospitals were built in Egypt. But all of these were actually built by and for the exclusive service of various European national groups and Christian denominations — none of these was open for the general public. Thus the health care system in Egypt was destroyed and it still has not recovered.

As the years since independence have shown, the replacement of health orientated Islamic medicine in the Muslim world with profit-orientated, high-technology, western medicine has played havoc with health care systems in Muslim countries. Indeed, as recent experience in Pakistan reveals, only by upgrading the traditional medical systems is it possible to develop a health care delivery system for rural areas. Upgrading the Islamic system of medicine, and integrating it with the existing medical structure must be a major priority for Muslim countries. No other policy can lead to adequate health care for rural areas.

But in the long run, the re-emergence of Islamic medicine in contemporary times is the only viable solution for the health of the Muslim world. Medicine makes sense only within the world-view of the life-style it is trying to preserve. A civilization cannot hope to survive, physically, psychologically or intellectually, without a dynamic system of health and medicine based on its own world-view. Doing something positive about the wretched state of health in Muslim countries means giving Islamic medicine its true recognition and injecting funds, manpower and intellectual and physical resources it needs to acquire a contemporary shape. With appropriate resource and research base, Islamic medicine would not only be more than a match for western medicine, it may actually rescue mankind from a system of medicine and metaphysics determined to pursue a suicidal path.

9

Science and Health
The Redundancy of Drugs

Claude Alvares

Today's global, transnational-dominated drug industry seems to be facing a crisis of legitimacy and public confidence. In many ways, the situation is similar to fifty years ago when the present drug companies were just beginning their businesses, marketing patent medicines of mostly dubious value, and thereby raising the hackles of the medical profession — the slick advertising encouraged patients to bypass the doctor, and inaugurated a war.

It was Paul Ehrlich's discovery that some bacteria were selectively coloured by certain dyes, and therefore could also be selectively killed, that changed the basis of the drug industry's fortunes. It laid the foundation for the invention of modern medicine in a systematic manner, and enabled the companies to claim they were now securely wedded to a production system associated with modern science.

Over the past decades, however, the association of drug production with modern science has been calculatedly misused to generate a drug economy that has become not only exorbitantly costly and anti-human, but whose connection with human illness and social need has become increasingly tenuous. The depressing situation has called forth a major reaction from health professionals and consumer organizations all over the world. The pharmaceutical companies continue to receive polite, if not grudging, acknowledgement from the medical establishment, health activists and even the World Health Organization, for their contribution to public health, and much of this has recently been considerably devalued by the hard knocks delivered in the media and other forums: the drug industry *has* a bad name. A worldwide public is inclined to feel the industry is solely concerned with profits at the expense of health. The only consolation the companies have is that they have been joined in their corruption by most medical associations of doctors, and often, by consumers themselves. The 'pharmaceauticalization' of modern life is the result of a voluntary conspiracy between companies, doctors and consumers at large.

It was not too long ago, in fact, that John Wyeth was selling amidopyrine as a cure for tuberculosis. The names of individual drugs may have changed in our times (amidopyrine is now a banned drug in numerous countries) but the misfit between therapeutic chemical drugs and human illness has if anything progressively increased. In 1984, Social Audit and the Consumers Association published Charles Medawar's *The Wrong Kind of Medicine* which lists over 800 ineffective, inappropriately or extravagantly prescribed drugs in the UK. This year, the Voluntary Health Association of India (VHAI) with the assistance of the All India Drug Action Network (AIDAN) issued a list of nearly 8,000 banned and bannable drugs marketed in the country. August 1985 saw the publication of Andrew Chetley's *Cleared for Export*, an examination of the EEC's chemical and pharmaceutical trade. Earlier, in 1981, the International Organization of Consumer Unions (IOCU) published its own report, *Forty-Four Problem Drugs*. It seemed that all of a sudden the pharmaceutical companies' striking achievements were beginning to be questioned at source. People were no longer asking how much was being produced, but why so much of what was produced was dangerous, therapeutically ineffective, outrageously priced, or even downright worthless.

In the Third World there was an additional problem: not only were a large proportion of drugs produced of dubious value, or banned in the advanced countries, they were irrelevant to the health needs of the majority of the population: the pattern of drug production was at variance with the needs of the community. A major seminar of scientists and drug activists held in New Delhi in 1981 decided, for this reason, that multinational drug companies had really no positive role to play in the country's health policy if their present operations were subjected to criticism of relevance, a view directly contrary to the opinion of the London *Economist* (12 March 1983), that Third World countries 'need the drug multinationals more than the multinationals need them'.

India's is a classic instance of the paradox of drug over-production existing simultaneously with a shortage of essential medicines. The country's drug scenario illustrates how idealism can co-exist with an almost total lack of political will; how a powerful drug lobby can continue to perpetuate manifest injustice and 'drug colonialism', despite the presence of considerable scientific talent. It is a good illustration of how a campaign for essential drugs remains grounded due to lobby interaction between drug company representatives and government officials.

According to the Planning Commission, diseases like tuberculosis, gastro-enteritis infections, malaria, filaria, leprosy, rabies and hookworm account for 17.2 per cent of morbidity and 20.8 per cent of mortality in India. (It was estimated in 1980 that 18 million people suffer from filaria in India, 10 million from tuberculosis, and 7.8 million have gastro-enteritis.) The drugs required for the treatment of these diseases include chloroquine, streptomycin, diethyl carbamazyne citrate, metranidozol, chloramphenicol, dapsone, INH ethambutol and piperazine. The actual production of these essential drugs is now not only below consumer demand, but also much below installed capacity. Annual figures over the past five years indicate that the production of these essential drugs is rapidly declining, and that production targets have been scaled down. Pfizer, for instance, produced no INH or PAS in 1983.

In contrast to essential drugs, irrational drug combinations and trivial commodities like cough syrups constituted the bulk of the market. In 1980, for instance, essential or life-saving drugs comprised just Rs3.5 billion worth within an overall production total worth Rs12.6 billion. Such a distorted pattern of drug production directed and maintained by multinational dominance of the sector has had its own grievous consequences on the domestic industry, which has also indulged in the production of high-price drug combinations of little therapeutic value at the expense of cheaper, essential drugs needed by the majority of the population.

The consequences of this are obvious to any impartial observer: diseases like TB, filaria, and leprosy remain uncontrolled, and for the victims there is precious little relief. But pharmaceutical activity continues to increase sharply. Dr Mira Shiva writes:

Between 1952 and 1983 the number of production units grew by 300 per cent from 1,643 to 6,631; investment increased by 2,400 per cent from 240 million to 6,000 million rupees and bulk drug production by 1,800 per cent from 270 million to 3,250 million rupees. Inspite of this increase, drug production, distribution, etc., have not been keeping up with the health needs of our people. While on the one hand shortages of essential and life saving drugs have occurred, on the other, even rural markets have been flooded with costly brands of absolutely irrational drugs of doubtful therapeutic value and need.

The principal reason for this horrendous state of affairs is the conversion of health into a major new frontier for the expansion of capital. It really makes little sense for such a profit-oriented system to produce drugs essential to meet the health needs of the majority of the Third World, simply because the latter have little or no purchasing power and remain outside the effective market. The lesson therefore is quite clear and irrefutable: the present pattern of drug production dominated by giant multinational corporations is totally uninterested in meeting the health needs of the populations of the Third World.

The move to get governments to work towards a policy of ensuring essential drugs is basically an effort to reform such a wayward system. All medical experts recognize that only a few important drugs are required for ill-health. India however is reputed to have about 60,000 formulations in the market. Switzerland has 35,000. Yet both India and Switzerland restrict the number of drugs for which health insurance is available or reimbursements can be made. The UK's 6,500 preparations are seen as excessive by researchers like Charles Medawar.

However, the concept of 'essential drugs' is mistakenly believed to have originated with the WHO's report, *The Selection of Essential Drugs* published in 1977.

In 1973, for instance, the Chilean Medical Mission under Dr Salvador Allende decided that drugs should be limited to those with demonstrable therapeutic value. Allende was also keen to see that drugs not prescribed for Europeans or Americans were not available in Chile. The political repercussions of Allende's moves must have been serious, for within one week of the Pinochet regime taking over in September 1973, all those doctors who had participated in the new health programmes — which included emphasis on therapy based on community action rather than on swallowing pills — were assassinated.

Sri Lanka has worked with an essential drugs policy from 1959. The initiative came from another extraordinary person, Dr Senaka Bibile, who helped form the state pharmaceutical committee to launch a people-oriented drug policy. The number of drugs available was slashed from 2,100 to 590. These were purchased through international tenders and marketed under their generic names. It was the attempt to extend this system to the private sector in 1972 that led to active (and successful) efforts on the part of multinationals, their home governments and Sri Lanka's own medical establishment, to destabilize the policy. Likewise, in Pakistan, a proposal to restrict the production of drugs to essentials was browbeaten by the drug industry lobby in close association with the medical establishment.

Similarly, Brazil's state-run Central de Medicamentos used a restricted list of 347 drugs — but the target was not the general citizen, who remained captive to the private trade, but those receiving the official minimum wage or less. The regional co-operation efforts of the Andean countries also resulted in a limited list of 287 drugs. In India, in 1975, a committee appointed by the Government of India (later known as the Hathi Committee) concluded that a list of 116 drugs was sufficient for the country's basic health needs, and that these should be made available under their generic names.

It was the Mozambique government that took practical steps towards enforcing a policy concerning essential drugs. By 1977, the number of drugs in use in the country was reduced to 430, and private practice, one of the principal lobbies against judicious drug use, was abolished. In 1980, an essential drug list comprising 343 drugs was published and only such drugs could henceforth be prescribed.

The WHO initiative itself commenced with the World Health Assembly in 1975, which recommended the formulation of an Expert Committee to go into the question of a drug list. The Expert Committee eventually produced four technical reports on essential drugs (nos. 615, 641, 685, and 722). The definition of essential drugs was uncomplicated and unambiguous:

It is clear that for the optimal use of limited financial resources the available drugs must be restricted to those proven to be therapeutically effective, to have acceptable safety and to satisfy the health needs of the population. The selected drugs are here called 'essential' drugs, indicating that they are of the utmost importance, and are basic, indispensable and necessary for the health needs of the population . . . Drugs included in such a list would differ from country to country depending on many conditions, such as the pattern of prevalent diseases, the type of health personnel available, financial resources and genetic, demographic and environmental factors.

The burden of the technical reports was the drawing up of the *Essential Drug List* (EDL), comprising about 200 active substances which could be used to frame a standard formulary and adapted to the health needs of any given population. The technical report lists four principal advantages of such a list.

First, a reduction in the number of pharmaceutical products to be purchased, stored, analysed and distributed. Two, improvement in the quality of drug utilization, management, information and monitoring. In fact, a restricted EDL was a boon for many Third World countries handicapped by a rudimentary drug monitoring infrastructure, for it dramatically reduced the number of drugs that needed to be analysed and controlled. Third, stimulation of indigenous industry. Fourth, assistance to the poorest countries in urgent need of high priority drug programmes to solve their primary health programmes.

The Expert Committee also formulated four principles which could help as criteria to establish an EDL. First, the EDL should be a part of the country's health policy. Thus priority should be given to producing drugs of proven efficacy, safety, and covering the widest segment of the population, and necessary for the prevention and treatment of the most prevalent diseases. Second, only those drugs for which adequate scientific data are available from controlled studies should be selected. Third, each selected drug must meet standards of quality. Fourth, concise accurate unbiased sources of information should accompany each EDL.

In addition to the 200 drugs in the EDL, WHO also suggested a list of thirty complementary drugs for use in exceptional circumstances, or for rare disorders. Today there are available still briefer lists of essential drugs needed for primary or community health centres, and for emergency relief in disaster situations. All WHO contributions are itemized by their generic names.

We have already noted that a number of countries have been experimenting with limited drug lists, and in most of them, tremendous opposition was exercised to the idea from the company doctor cartel. The first country to legitimate its new drug policy on the basis of the EDL of the World Health Organization was also the world's poorest nation: Bangladesh. The new Bangladesh Drug Policy (BDP), promulgated in June 1982 took the world by complete surprise. Discussions on such a policy, however, had begun more than half a decade earlier. Inspiration had come not from the WHO report but from India's Hathi Committee. But the group of professionals led by Dr Nurul Islam and Dr Zafrullah Choudhury decided that for legitimation purposes the WHO report was a safer bet. The Hathi Committee Report was not as well known, and besides, none of its recommendations had been accepted by the Indian government itself.

The method employed by the Bangladeshis should prove an object lesson to other countries planning similar action plans. An Expert Committee was first appointed, and the deliberations of this Committee were held in secret: had any of the companies, or the medical association known what was afoot, rapid efforts would have commenced to neutralize any proposed course of action. The most sensational result of the Committees' deliberations was its recommendation that a total of 1,707 drugs be destroyed or removed from the market within a specified period of time (4,340 brands of drugs were on sale in the country at the time). The Committee also drew up a list of 150 essential drugs 'considered adequate for most therapeutic purposes' — with a complementary list of an additional 100-odd drugs for less general health needs.

The reaction to the BPD was instantaneous. While professional health and consumer organizations all over the globe, sick and tired of drug dumping, applauded and congratulated the President, General Ershad, the drug companies, led by the US Ambassador Mrs Jane Coon in person, mounted a massive counter campaign to have the policy changed.

The drug companies were clearly affronted with the suddenness of the decision, and the fact that they were not informed or consulted; they were also appalled that a poor country like Bangladesh, perennially seeking international aid, should now attempt to stand up to private enterprise. There was also fear that the more dramatic aspects of the BDP might lead to a chain reaction in other Third World nations.

What the pharmaceutical giants and their home country governments had to admit, however, was that this was one of those occasions when a group of smart health professionals from the Third World had presented the organized might of the western world with a *fait accompli*. The massive international support that poured into General Ershad's office, in the form of letters, telegrams, and media stories, strengthened the Bangladeshi bargaining position, and also, as General Ershad shrewdly realized, incidentally legitimated his regime in the eyes of the international public.

Evaluating the impact of the BDP four years later, Dexter Tiranti concluded that it was beginning to meet some of its avowed objectives: the indigenous manufacture of essential drugs has increased significantly, the country has saved scarce foreign exchange previously squandered by multinationals on trivial and therapeutically useless drugs, and the population has been safeguarded from the atrocity of known hazardous and dangerous drugs. Drugs have been selling cheaper. As the trends set in motion will consolidate, future scenarios can only produce more significant results.

What really upset the drug companies was not that the BDP decided in favour of essential drugs, but that it eliminated most of the inappropriate ones. Drug companies know better these days than to oppose an essential drug policy, they may even co-operate with a government that is concerned about such a list, in the interests of business. But what they cannot tolerate is 'unwarranted' restrictions that inflict damage to profitability. They would welcome an essential drug policy provided it permitted them to continue doing what they have in the past. The BDP cut at the root of that proposition. As a result of its implementation, the companies' super-profits were now replaced with moderate profits. Sold to other Third World countries, the formula could presage a vastly limited arena for market-oriented drug production.

The BDP, in addition, vindicated the theoretical correctness of the WHO essential drug list. At the same time, its results confirmed the opinions of concerned health professionals everywhere. It is not surprising that the Pharmaceutical Manufacturers' Association of the USA 'sponsored' an 'independent' study by a Sri Lankan lawyer, who not unnaturally found the BDP to be only a qualified success. The sponsored study in question, *The Public Health and Economic Dimensions of the New Drug Policy of Bangladesh*, written by D.C. Jayasuriya, appeared in September 1985, and sought to place the endemic problems faced by the ill-equipped drug monitoring and distribution set-up in Bangladesh at the door of the BDP! Its anxiety that other Third World countries would follow the example of the BDP is evident on every page; but the author lacked the courage to name the principal figure behind the BDP, Dr Zafrullah Choudhury, even while it kept alluding to his alleged self-interest in the developments involving the new policy (Dr Choudhury is a founder and trustee of the Gonoshasthya Pharmaceuticals, that produces essential drugs).

The Jayasuriya study was soon followed by another study, this time masterminded by the International Organization of Consumer Unions (IOCU), which commissioned Dexter Tiranti of the *New Internationalist* to make his own evaluation of the results of the BDP. *The Bangladesh Example: Four Years On* is a less hysterical, more evenly balanced account of what the BDP has achieved and what it has not. In fact, when Third World Network Coordinator, Malaysia's S.M. Idris, wrote a strongly worded letter to Dr H. Mahler, Director General of WHO, concerning the Jayasuriya study, the WHO officially responded that even the Tiranti study had not found the BDP entirely without weakness. WHO was playing games really. It is obvious that the Jayasuriya report was intended to destabilize the BDP, and delegitimate it in the eyes of drug administrators elsewhere, whereas Tiranti's study confirms the BDP's beneficial impact, and points out short comings merely to indicate that meeting them would strengthen the policy further and enhance its potential impact on Bangladesh consumers, and the rest of the world.

But what was the response of WHO itself all along? Their representative in Dhakka, Dr B. Sestak, remained totally non-commital and even the Director-General, Dr Mahler, on a visit to Dhaka in mid September, 1982, refused to comment on the new policy. WHO was perhaps taken aback by the hostile reaction of the Americans to BDP, and whether one likes it or not, the Americans still contribute 25 per cent of WHO's budget. When Dr Choudhury sought permission to reprint the EDL report, WHO remained unco-operative.

Today, things have changed. Mahler himself is more overt in his praise for the policy, and WHO is itself now irrevocably committed to the essential drugs concept. There is also the admission on the part of the multinationals that the BDP is here to stay. The WHO itself issues a regular *Essential Drugs Monitor*, which reports progress worldwide in the application of the EDL. Since the WHO launched its Action Programme around the concept in 1981, some eighty countries are now reported to be busy drawing up strategies for such necessary drugs. In May 1984, at the World Health Assembly (WHA) the Nordic countries proposed a resolution in favour of the EDL and the Action Programme: 116 countries voted in favour, only the USA against. The issue of essential drugs came up again at the WHA of 1985 and 1986 and continued to receive unanimous support. A WHO conference of experts on the use of Rational Use of Drugs held in Nairobi in November 1985, decided to intensify WHO's leadership role in the promotion of national drug policies based on the essential drugs concept. Also important for this meeting was a decision by Mahler to acknowledge officially the role of consumer organizations in making drug use more rational, a move that clearly upset the drug lobbies. At government level, both the Non-Aligned Movement and the Group of 77 have passed resolutions accepting the essential drug policy.

In Kenya, in a pilot project funded by SIDA of Canada, supplies of drug 'ration kits' containing thirty-nine drugs have increased the availability of essential medicines in fifteen rural districts. The Zimbabwe regime has also selected 376 essential drugs for use in the public health system and has decided not to make available foreign exchange for those who wish to import drugs outside the list. India is one major country whose drug situation remains well and truly hopeless. A 1982 request to the editor of the Indian *Mims* to delete drugs recommended for weeding out by Indian experts, and to indicate drugs included in the WHO list by using capitals or italics, evoked the editorial response that the relevance of the EDL was only for struggling poor countries, not India. Likewise, Argentinian companies attacked the concept of a limited list of essential drugs, saying that Argentina 'is not a country of the Third World'.

But such attitudes are changing. The rational use of drugs, or their restriction to those that are therapeutically effective, is no longer a Third World problem. Governments in advanced countries are increasingly anxious to cut the costs in health programmes, and drugs remain by far the biggest demand on the public health purse.

Both the UK and the USA, for example, have decided to cut costs by determining that some prescriptions begin using generic names. It is estimated that by 1987, 25 per cent of all prescriptions in the USA will be filled with generic drugs instead of 15 per cent as in 1983, leading to a saving of over one billion dollars. Norway, perhaps, has one of the best policies in this regard. It permits a list of 1,900 preparations, and keeps the list trim with a 'medical need' clause that applies to all prospective entrants: for a new drug to be registered, a company must demonstrate that it is more effective than existing ones in the market, and also safer and cheaper.

The drug companies have themselves begun to accept the WHO list as a fact of life; their only concern is that it remains applicable to 'poor countries'. Their paranoia is excited the moment government authorities seek to make the list exclusive, or, as in Bangladesh, reduce or restrict the number of drugs they wish the companies to produce. The companies continue to believe that their right to promote medicine is absolute. The WHO and health activists on the other hand, have a larger programme, not just concerned with essential drugs for the poor, but with the general 'pharmaceuticalization' of life itself, in poor *and* rich countries. This the drug companies refuse to see or accept.

They even attempt nowadays to collaborate with the WHO in the programme involving essential drug availability in Third World countries and cover drug colonialism with drug diplomacy. According to John Starrels in a study sponsored by the right-wing Heritage Foundation, a large number of pharmaceutical companies have agreed to make available 140 drugs and vaccines at 'concessionary prices' to the Third World. A US drug company group's agreement to supply the Gambians with the necessary essential drugs is cited. So is the Swiss drug company programme in Burundi. But the benefits of all that are clearly stated by Starrels: 'The West's readiness,' he writes, 'to engage international health projects designed to help poverty in the Third World (especially through the auspices of WHO) is but one aspect of what IFPMA's S.M. Peretz calls "enlightened self-interest". This includes, of course, the commercial benefit that the United States and other Western countries derive from sales to Third World countries with the wherewithal to pay for them.'

The very same Peretz will also claim elsewhere that 'the world needs a research-based pharmaceutical industry,' and that 'research can only be undertaken by firms whose drug prices cover research costs'. The world, adds Starrels, 'needs the contributions that only a market-orientated industry can provide'.

It is important to remember that the essential drugs campaign is basically a reformist campaign. It is a significant issue in a context in which drugs are meant to substitute for the absence of more elementary necessities like good drinking water and proper nutrition. It is also intimately connected, in the Third World, with poverty. In the absence of people-oriented development, an essential drugs list can only help mitigate the ill-effects of poverty, in place of actively helping to enhance the suffering of the victims, as drugs do nowadays. Further, an EDL is worthless unless governments actually do more to encourage the production and distribution of such drugs.

The de-pharmaceuticalization of life needs a campaign as consistent and forceful as the one that concerned breastfeeding. Most diseases are self-limiting, and often environmental conditions are the source of much ill-health. Improving such conditions is a much more effective programme of enchancing public health than using chemical agents. Also to be recalled for a future agenda must be the availability of more wholesome traditional nutritional recipes, and holistic traditions of interpreting disease. We must remember that a large number of the world's population still owe their allegiance to systems other than allopathy, and for them, this whole issue of the problematic nature of modern drugs is only minor.

10

Science and Hunger

A Historical Perspective on the Green Revolution

J.K. Bajaj

The history of modern agriculture in India begins in 1757. In that fateful year, the Indians lost the Battle of Plassey to the East India Company of the British soldier-traders. As a consequence of the defeat, the revenue rights of one district in Bengal — the twenty-four Pragannahs — were ceded to the Company. The foothold gained by the British in the civil and financial administration of India expanded rapidly. By 1765, large territories of India, particularly in the provinces of Bengal, Bihar and Orissa, had come under the control of the Company, and agriculture in India had become subject to British administration and its modernizing influences.

Before this conquest, agriculture in India was the focus of a traditional way of life. It was no mere economic activity, but the basic life-activity of the village people.[1] Its major function, if an integrated life-activity can be analysed in terms of functions, was to fulfil life-needs. The needs of the government, of the market, of industry were all secondary to the major function.

This independence of the traditional agriculture from external political or economic control was achieved through a social organization that left the village largely autonomous. The obligation of the village to the external political authority was limited to the payment of a small proportion of its produce as revenue. On the basis of the revenue records of those times Dharampal (quoted by Alvares 1979) estimates that the proportion of the produce payable to the external authority around 1750 was as small as 5 per cent. At the time of Jehangir this proportion was even smaller — about 4 per cent. Having met this obligation the village was free to organize its own political and economic affairs. In these affairs the village was so autonomous that who actually ruled at the centre was not of much concern to the villagers. Marx (1853a) quotes an official report of the House of Commons to the effect that:

The inhabitants [of the village] gave themselves no trouble about the breaking up and division of kingdoms; while the village remains entire, they care not to what power it is transferred, or to what sovereign it devolves; its internal economy remains unchanged . . .

Within the autonomous village, the cultivator was quite independent. According to the estimates quoted earlier, the cultivator paid perhaps another 25 per cent of his produce towards various heads of revenue. A large part of this 25 per cent however, went towards financing the religious, cultural, educational and economic activities of the village. This share of the produce was in fact often paid by the cultivator directly to the individuals or institutions responsible for the above-mentioned activities. The political aristocracy and the militia ended up receiving only 1.5 and 6.0 per cent respectively of the gross produce. Having received its share of the produce the aristocracy had no more rights on the land; in particular it had no right to separate the cultivator from his piece of land. Land was not the private property of the aristocracy — a concept yet to arrive in India!

The political-economic independence of the village and the cultivator was further secured through the independence of the village from external industrial and market influences. This independence does not mean that in the pre-British India there was no industry or no trade. In fact for the first hundred years of British contact with India the British traders dealt only in the manufactured goods of India. Up to 1757 they had to import silver and gold into India to be able to buy goods manufactured in India for Britain at that time had no manufactures to exchange (Alvares 1979). Even in 1840, Montgomery Martin, an early historian of the British Empire, could insist before a parliamentary inquiry that, 'I do not agree that India is an agricultural country; India is as much a manufacturing country as agricultural . . . her manufactures of various descriptions have existed for ages, and have never been able to be competed with any nation wherever fair play has been given to them . . .' (Dutt 1940: 129–30). Yet this vast manufacturing activity did not interfere with the autonomy of the village and the agriculturist, but rather was closely co-ordinated with agriculture. Textile manufacture, the most important industrial activity of pre-British India, was carried out almost entirely by the agriculturist in his free time. This close co-ordination between agriculture and manufacture, this 'domestic union of agricultural and manufacturing pursuits', was in fact, as Marx (1853a) noted, the pivot of the village system. It ensured the autonomy of the village by freeing agriculture from the demands of an external industry or market. That is why the spinning wheel (*charkha*) and the handloom — the basic tools of this union — became symbols of the independent cultivators and autonomous villages of traditional Indian civilization for both Gandhi and Marx.[2]

To the British, however, agriculture represented only a source of revenue that they set about collecting with great zeal. R.P. Dutt records that, 'In the last year of administration of the last Indian ruler of Bengal, in 1764–65, the land revenue realized was £817,000. In the first year of the Company's administration, in 1765–66, the land revenue realized in Bengal was £1,470,000. By 1771–2, it was £2,348,000, and by 1775–6, it was £2,818,000. When Lord Cornwallis fixed the permanent settlement in 1793, he fixed £3,400,000' (Dutt 1970: 114). With more and more money flowing into British coffers the village and the producer were left with precious little to feed themselves and maintain the various village institutions that catered to their needs. According to Dharampal's estimates, whereas around 1750, for every 1,000 units of produce the producer paid 300 as revenue with only 50 going to the central authority and the rest remaining within the village; by 1830, he had to give away 650 units as revenue, 590 of which went straight to the central authority. As a result of this level of revenue collection both cultivators and villagers were destroyed.[3]

How far agriculture lost its previous position of being the provider of the life-needs of the people, and became merely the source of British wealth is tellingly brought out by the communication sent by Warren Hastings to the Court of Directors of the Company, on 3 November 1772, a year after the great famine in Bengal that killed perhaps 10 million people. Warren Hasting reports:

Notwithstanding the loss of at least one-third of the inhabitants of the province, and the consequent decrease of the cultivation, the net collections of the year 1771 exceeded even those of 1768 . . . It was naturally to be expected that the diminution of revenue should have kept an equal pace with the other consequences of so great a calamity. That it did not was owing to its being violently kept up to its former standard. (Dutt 1940: 115)

The independent cultivator of yesteryear, who cultivated his land to fulfil his needs, had now become a tool to produce revenues that would fuel the industrial revolution of England.[4] This change in the function of agriculture from being a source of life in India to source of 'progress' in England brought in its wake untold misery. Irrigation works fell into dilapidation. Vast tracts of cultivable land decayed into a state of jungle. Industry got uprooted. Education was destroyed. Philosophical, scientific and literary activity came to a standstill. Culture stagnated. The story of that early plunder by the British and the consequent misery of India is well documented, though not so well known amongst educated Indians.[5] It should be remembered that the important point about that sad chapter of Indian history is not the immediate destruction and misery of that period. There had been plunderers before, and perhaps they had spread an equal amount of misery.[6] The British themselves soon realized that the type of destruction let loose by their early administrators in India was likely to be counter-productive and therefore, some semblance of order had to be restored. Cultivation was to reappear in the areas which had reverted to the jungle, some irrigation facilities were to be provided. Industrial activity, all of which had been moved from the villages to the cities of Lancashire and Manchester in the early phase, was to return in part to the Indian cities.

Education was to get reorganized, though only according to the patterns dictated by Macaulay — to produce lackeys, 'Indian in blood and colour, but Englishmen in tastes, in opinions, in morals and in intellect'. Philosophical, scientific and literary activities were to restart, though in the English mould. But, and this is the important point about that phase of Indian history, India was never to be the same again. The villagers and the agriculturists were never to become autonomous again. Their needs would always be subservient to the needs of the state, industry and the market — all of which were now severed from agriculture. Agriculture had become a mere economic activity; it had been finally 'modernized'.

The major instrument of this modernization, besides the use of naked force, was the system of landlordship introduced into India for the first time. Independent cultivators were not likely to put external economic needs before their own needs to eat and be clothed. A landlord, however, assured of his personal well-being, could be relied upon to produce and sell what the industry or the state needed. He could be relied upon to respond to the market, to divert land good for foodgrain to the cultivation of opium, indigo and so on, while famines stalked the country. Marx counted landlordism as one of the few regenerative forces introduced into India by the British. 'The Zamindari and ryotwari themselves,' declared Marx (1853b), 'abdominable as they are, involve two distinct forms of private property in land — the great desideratum of Asiatic Society.' Independent cultivators used to grow what they needed to live on rather than producing what was needed to 'progress'. They required the concept of private property in land personified in the landlord, to teach them that it is more important to progress, to industrialize, than to eat and be clothed.

How successful the British were in modernizing Indian agriculture can be gauged from the crop output data of the last fifty years of British rule. The central government had started publishing such data by the late nineteenth century. The period before the First World War was marked by a favourable world market in all export crops, and expansion in the domestic manufacturing capacity in textile and jute. Consequently we find Indian agriculture flourishing in this period with agricultural output rising at a rate faster than the growth of population. It is perhaps one of the best periods in British Indian agriculture, with per capita food availability hovering around 540g per day throughout this period inspite of substantial exports of rice and wheat. Then came World War I, followed by the Depression, and the World War II. Export markets contracted. Prices of agricultural products plummeted; Indian agriculture crashed.[7] While non-foodgrain production merely stagnated, foodgrain production started declining even when population was rising. Per capita food availability for the quinquennium ending 1946 was down to 417g per day inspite of imports. Interestingly the only crop that showed expansion in this period was sugar-cane which was granted protection through the imposition of new tariffs on import. The area under cultivation of this commercially favoured crop actually increased by about 40 per cent between 1930–31 and 1938–39.[8]

Thus, in this fifty-year period we see agriculture going up or down with worldwide economic forces. These forces, and not the needs of the people decided how much, and what, Indian agriculture would produce. Economics had won, life had failed.

Independence came to India at a time when agriculture was passing through a particularly bad phase: Bengal had had a major famine; per capita food availability was dangerously low (417g per day in 1946); rural indebtedness had been increasing alarmingly — according to the Central Banking Enquiry Committee, it had nearly doubled between 1929 and 1936. Cultivators were finding difficulties meeting their fixed liabilities such as rent or land revenue and many were turning into landless labourers. Partition of the country worsened the situation further, and the country faced an acute shortage of both commercial crops and food crops. Something needed to be done immediately to improve agriculture.

An obvious line of action was to concentrate on improving irrigation facilities which had been severely depleted with partition (only 19.7 per cent of the net sown area within the Indian Union was irrigated), and to take steps to put the cultivators back on the land and reduce rural indebtedness through land reforms. Some sort of land reforms had in fact become a political necessity given the aspirations that people associated with independence. Action on both these fronts was started immediately after independence. Between 1947–48 and 1949–50 the net irrigated area increased from 18.9 Mha to 20.2 Mha — most of the increase coming from the area irrigated by wells and other minor sources (NCAR, vol. 1: 221). This pace was maintained and an annual rate of increase of 0.67 Mha of gross irrigated area for the period 1950–51 to 1968–69 was achieved (NCAR, vol. 5: 43). Land reforms had been initiated by most states by 1950. These envisaged abolition of Zemindari, security of tenure for tenant cultivators and fixation of reasonable rents; later some ceilings on land holdings were also introduced. Though carried out in a half-hearted manner, these land reform measures continued to provide some relief to the cultivators right through the fifties and early sixties.

Agricultural production responded well to the restoration of some just order in land relations and to the slowly increasing irrigation facilities. Aggregate crop-output during the fifties increased at a rate faster than the population growth. Both the area under crops and the yield per hectare of almost every crop showed a rising trend.

However, Independent India also wanted to become 'modern' and 'industrialized'. It was important that agricultural production should respond to the needs of the market, and that food should come to the market for sale. Because, as the Report of the National Commission on Agriculture (NCAR, vol. 2: 14) noted, 'The entire industrial sector depends heavily on the supply of food from the agricultural sector. Since a sizeable part of the wages of the industrial worker is spent on food items, a sustained supply of food from agricultural sector is a necessary condition for stability in the industrial sector. . .' But the increased production was simply not reaching the markets or the industrial sector. The National Commission moaned that, 'The unique features of the food situation during the Second Plan period were the increasing demand for food grains and a steady decline in market arrivals despite higher production' (NCAR, vol. 1: 188). It may be that part of the reason for this phenomenon was 'speculative holding of stocks by the grain trade'. But that does not seem to be the only cause, since various experiments of introducing control in the food trade did not help matters and the urban industrialized sector had to be fed with increasing imports till the mid-sixties. It seems more probable that the general improvements in the land relations and irrigation that had led to increased production also improved the lot of the cultivators — and they simply ate more. This is what is likely to happen in a situation where the average per capita availability of food was low (around 460 g in 1960–61) and a large proportion of rural people (around 40 per cent in 1960–61 according to Dandekar and Rath, 1971) had insufficient purchasing power to buy the bare minimum of 2,250 calories of food per day. This tendency was further encouraged by the fact that agricultural production in the fifties and early sixties was by and large independent of inputs from outside the agricultural sector. As the National Commission noticed, production depended largely on the amount of labour a cultivator was able or prepared to put in. All inputs were farm produced (NCAR vol. 2: 9), so agriculture was becoming independent of the urban sector both on the input side and the output side. To the extent it was being freed from the yoke of landlordism it was once again showing the traditional characteristic of being self-sufficient.

Indian planners saw the solution as making those areas which already had a surplus in food produce more. This entailed concentrating resources in areas that were already well-endowed. In this scheme there was no danger of the producers consuming the increased produce as would happen if resources were allowed to flow to the deficit areas. This line of approach in fact was introduced into Indian agricultural planning rather early. Instead of spreading the efforts thin all over this country it was decided in 1950–51 to concentrate such efforts in compact areas called 'intensive cultivation areas which possessed assured water supply and fertile soils' (NCAR, vol. 1: 143). In 1959, the Agricultural Production Team of the Ford Foundation again recommended the intensive approach (*ibid*: 149). And with the visible failure of the Second Plan to get the food to the market inspite of increasing production, a new Intensive Agricultural District Programme (IADP) was launched in the closing years of the Second Plan. The programme was expanded in 1964 under the name of Intensive Agricultural Area Programme (IAAP) to cover more of the well-endowed areas.

The ostensible argument in favour of these intensive approaches was that resources spread thinly over a large area are lost leaving no appreciable effect on production; that only a package of practices involving concentrated doses of resources could be technologically effective; and that increased production achieved in these areas with improved practices would have a 'demonstration' effect in other areas. The latter argument carried no weight — there were just not sufficient resources to spread such intensive practices elsewhere — especially in areas which were to begin with not well endowed. As for the argument about the technological efficacy of an intensive package, the fact is that there were no agricultural technologies in use that could absorb and respond to intensive doses of resources.[9]

Traditional technologies, evolved in a more egalitarian context where the food needs of cultivators were more important than the needs of surpluses to support 'progress', were just not capable of absorbing more than their due share of resources. And within that context, there was little that the experts of the Ford Foundation could teach Indian farmers by way of possible improvements. Way back in 1889, Dr J.A. Voelcker, deputed by the Secretary of State for India to advise on the application of agricultural chemistry to Indian agriculture, had noted this perfection. He reported that:

> . . . it must be remembered that the natives of India were cultivators of wheat centuries before those in England were. It is not likely, therefore, that their practice should be capable of much improvement. What does, however, prevent them from growing larger crops is the limited facilities to which they have access, such as the supply of water and manure (quoted in Alvares 1979).

Therefore it is not surprising that the efforts of Indian planners to achieve increased production through 'improved' practices in areas which did have access to facilities like supply of water and manure, should prove abortive. In fact, the attempt was a complete failure. According to NCAR (vol. 1: 411) rice yields in the twelve rice districts and wheat yields in the four wheat districts under the IADP averaged only 13.3 quintals and 13.5 quintals per hectare compared with the pre-package average of 12.4 and 10.2 quintals. As against these marginal increases in yields, the added costs of the recommended packages were equivalent to 10 quintals of wheat on the average, and 10–12.4 quintals of paddy for most of the districts. The efficiency of the package for other crops was even worse.[10]

Thus the intensive package approach to agricultural development being tried out in India from the fifties had really nothing to do with technological efficacy. The policy in fact only expressed a political wish for a technology that would respond to these measures — a technology that would allow the concentration of resources and production in a few compact areas. The policy was asking for a technology that would achieve technologically what was achieved by the British politically through the landlords — namely, responsiveness of agriculture to the needs of industry and the market in preference to the life-needs of the cultivators. In other words, the developments sought for in the agricultural sector were not primarily to meet the needs of the rural population, but to provide the resources and capital needed for the industrialization taking place in the urban centres. What was needed was to break the independence of the rural sector and bring it into increasing dependence on the urban sector. There was a need for a certain technology to be introduced into the agricultural sector that would bring about such a transformation, but no such technology was available at the time the intensive approach policy was being formulated and implemented. By the mid-sixties, however, such a technology became available in the form of new 'miracle seeds' that had proved successful in Mexico. These seeds were genetically selected to absorb huge doses of chemical fertilizers, but since they had not evolved under natural conditions, they were susceptible to a number of pests and pathogens and they also required new sophisticated practices for irrigation, tillage etc. This was the ideal technology to make the policy of concentration of resources economically and technologically viable. At the same time it would make agriculture critically dependent on industrial inputs like chemical fertilizers and pesticides, and make the cultivator dependent upon the urban expert for knowledge of the correct agricultural practices, thus removing the 'dangerous tendency' of self-sufficiency in the agriculture sector. This technology was too expensive to be extended over the whole country, but all that was required was to make the surplus areas produce more. Acceptance of this technology would however involve imports of fertilizers and pesticides, and in the initial stages even seeds would have to be imported. Providentially, there was a widespread failure of the monsoon in 1965 and 1966 in India as well as in the rest of South Asia and Southeast Asia. The spectre of a major famine removed all hesitation about accepting the new seeds even if it involved massive imports. The ever-helpful attitudes of the Ford Foundation and the Rockefeller Foundation further encouraged the acceptance of the new technology and in 1966–67 the New Strategy of Agricultural Development was launched. Similar programmes were adopted in all of South and Southeast Asia at around the same time. The programme was declared an immediate success leading to what became known as the Green Revolution.

The Green Revolution

The new technology involving 'miracle seeds' and the associated practices was indeed successful in generating high yields, and in some areas the increase in yield could justifiably be characterized as revolutionary. This was amply borne out by a number of studies carried out to make a scientific evaluation of the response of different crops in different areas under the High Yielding Varieties Programme (HYVP).[11] However, my purpose here is not to provide an evaluation of the Green Revolution technology but to consider it as a breakthrough in the science of agriculture. I wish to evaluate the Green Revolution as an event in the growth of Indian agriculture for which it is not sufficient to assess the success of a few crops in certain localized areas but to look at the aggregate response of Indian agriculture to the event of the Green Revolution.

Aggregate Rates of Growth

I have taken the year 1967/68 as the dividing line for the Green Revolution, this separates my analysis into two phases: the pre-revolution phase runs from 1949/50 to 1964/65, the post-revolution phase runs from 1967/8 to 1977/78. The disastrous years 1965/66 and 1966/67 are excluded. There is some controversy about the correct dividing line for the two phases of post-independence agriculture, but most of the results under consideration here are quite independent of variations in the dividing year. An analysis with a different dividing line — 1960/61 — is provided by George Blyn (1979). Some of his data is used in this study. Keith Griffin (1979) focuses on 1965 as the dividing year.

While total agricultural production rose at a compound rate of 3.20 per cent per annum in the pre-Green Revolution period, the rate declined to 2.50 per cent per annum in the second period. The decline was visible in both the foodgrain output and non-foodgrain output; George Blyn, covering the slightly shorter period 1949/50 to 1973/74 (with the dividing line at 1960/61), finds an even sharper decline in the later period. Keith Griffin, analysing crop-output trends over all of the underdeveloped world, finds the trends declining after the Green Revolution in all the major regions except the Far East, where the growth rate is found to be practically the same in the pre-and post-Green Revolution periods. Thus it can be safely asserted that the compound rate of growth of aggregate agricultural production, as well as that of total foodgrains and total non-foodgrains, was lower in the post-Green Revolution phase.

The decline in the rate of growth of agricultural production is often explained as a consequence of the declining availability of additional land that could be brought under cultivation. In fact there is considerable statistical evidence to show that the decline in the growth rate of production must be to some extent attributed to the decline in the growth rate of area under crops. The total area under all crops grew at a rate of 1.60 per cent per annum during 1949–65 but the rate fell to 0.55 per cent during 1967–78. An interesting aspect of this trend is that throughout the period 1949–78, during which the food situation in the country remained precarious, the area under non-foodgrain crops rose at a rate much faster than the area under foodgrains. In the later period the trend rates of the area under foodgrains and non-foodgrains were 0.38 and 1.01 per cent respectively.

Though decline in the rate of growth of area does explain part of the decline in the rate of growth of production, it does not explain it all. In fact, the rate of growth of yield production per unit area itself declined. Thus while the aggregate yield rose at a rate of 1.60 per cent annually during 1949–65, the increase was only 1.40 per cent annually during 1967–78. Interestingly, on disaggregation into foodgrains and non-foodgrains, we find that while for foodgrains there is a slight decline in the rate of growth of yield, non-foodgrains show a slight improvement. Yet HYVP was supposed to have revolutionized foodgrains production! Further disaggregation of foodgrains into the major crops of rice, wheat and pulses shows more interesting features. We find the rate of growth of rice declining sharply from 2.09 to 1.46 per cent, and that of pulses which was already negative going further below, from – 0.24 to – 0.42 per cent. Only wheat shows an improvement in the trend rate (1.24 per cent in the earlier period, 2.53 per cent in the later one). Keith Griffin notices the same trend of increasing wheat production (except in Africa) and decreasing rice production all over the underdeveloped world.[12]

It is tempting to try to explain the decline in the growth rate of aggregate yields by referring to the law of declining marginal productivity. What it means in simple terms is that with the given technology and resources the productivity during the years before the Green Revolution had reached a saturation level, and without a technological change, maintaining the earlier rates of growth would have been impossible. If the new technology had not been introduced, the rates of growth of productivity, which admittedly declined a little after the Green Revolution, would have plummeted. Now to state that this law had started operating around 1964–65, one must show that by that period the possibilities of expanding irrigation and improving land relations, which were responsible for the increasing yields till then, had been exhausted in India. I shall later examine whether such a situation had actually arisen. For the present let us look at the statistical evidence, if any, in favour of the assumption that Indian agriculture in 1964–65 had reached the saturation level. If this had happened, we should be able to observe a declining trend in the rates of growth of productivity in the years preceding the Green Revolution.

During the Third Plan period (1961/62–1964/65), that is, during the period immediately preceding the years when the decision to implement the HYVP was made, productivity had reached an all-time high rate of growth. The rate of growth of productivity in this period was 2.7 per cent per annum, compared with the annual growth rate of 1.4 per cent and 1.8 per cent achieved during the First and the Second Plan periods. Thus the productivity graph, far from having reached a plateau, was actually moving upwards in the years before the Green Revolution. During the Fourth Plan (1969/70–1973/74), that is, during the five-year period immediately following the introduction of the Green Revolution technology, the rate of growth of productivity, however, touched an all-time low of 1 per cent. Thus it is obvious that the decline in the growth of productivity after the Green Revolution cannot be lightly explained by taking recourse to the law of declining marginal productivity.

There is no way to escape the fact that, notwithstanding highly visible increases in production and yields of a few crops in a few areas, both agricultural production and agricultural productivity in the aggregate showed a lower rate of growth after the Green Revolution technology was introduced. Even if one doubts the statistical significance of small changes in the trend rates, it is still impossible to maintain that there was any improvement in the growth rates of aggregate production and productivity. There definitely was no revolution in Indian agriculture with the introduction of the new technology.

It must be admitted that maintaining a growth rate of about 2.5 per cent per annum for aggregate production and above 1 per cent per annum for aggregate productivity over a period of about three decades is no mean achievement, even if the growth rates did decline a little in the later period as compared with the earlier period. My intention in pointing out this decline is only to establish that no revolution occurred in Indian agriculture with the onset of the so-called Green Revolution. What I want to criticize however are the special features associated with the attempt to achieve this growth through the new technology in the later period. It is to those features that I now turn.

Costs of Production

The rate of growth of production and productivity of Indian agriculture declined with the advent of the new technology. What is worse, however, is the fact that a high price had to be paid to achieve even this reduced rate of growth. The HYV technology is known to involve fairly high costs in terms of energy, in terms of depletion of soil-fertility and deterioration of the environment, and in terms of money.

A lot of data is available on the energy costs of the new technology of agriculture. And it clearly indicates that the HYV technology is energetically inefficient compared with traditional technologies. If all outputs from and inputs into agriculture are converted into equivalent energy units and output to input ratio is analysed, then the new technology invariably turns out to be inferior to traditional technologies. For traditional technologies the output/input ratio is often greater than one, indicating that these technologies are efficiently fixing freely available solar energy. For the new technologies this ratio is, however, always less than one. The difference in the energy efficiencies of the old and new technology can be as large as 50–250 times.[13].

The new agricultural practices are known to have a deleterious effect on the environment and soil-fertility. Chemical fertilizers change the flora and destroy the equilibrium of the soil. Consequently, more and more chemical inputs become necessary to get the same yield from a piece of land under this type of cultivation. This process of increasing chemical inputs year after year can even lead to permanent damage to the soil. Pesticides, an essential component of the new technology, form another component of the ecological costs. These pesticides have a way of being carried from food to man and other living beings, and form an almost permanent health hazard. All these ecological and energy costs of the new technology are important in any evaluation of the Green Revolution. However, I am now mainly concerned with the economic costs of this Green Revolution.

The new technology of agriculture is capital intensive. Since this technology depends critically upon industrial inputs like fertilizers and pesticides, it commits the nation to large investments in these sectors. Thus in nitrogenous fertilizer alone the indigenous capacity had to be increased from 0.37 mT of nutrients in 1967/68 to 2.23 mT in 1979/80.[14] Generation of 2.23 mT of nutrients capacity means (in 1980 prices) an investment of rupees 6,000 crores. Even such a heavy investment in fertilizers has not been sufficient to meet the fertilizer requirements of the Green Revolution, and in 1979/80, 1.3 mT of nitrogenous nutrients alone had to be imported. Besides, production capacity had to be generated for equipment like tractors and diesel-sets. In addition to this capital investment in the industrial sector, every farmer adopting the new technology had to invest capital in acquiring the necessary machines. This capital too often came through the public financing agencies. If the idea of introducing the revolutionary new technology was to provide new avenues of investments for the industrial sector, and not bother about the cost of food production, the Green Revolution technology has clearly done the job well.

Even more important than the capital costs are the actual unit costs of incremental production obtained through the HYV technology. It is difficult to put a uniform value on these costs since there is a lot of variation from place to place and year to year. Just to have an idea of the costs involved we can look at the evaluation studies of 1967/69 referred to earlier. From these studies we find that additional costs per quintal of additional wheat produced through HYV varied between Rs25 to Rs45. On the average the costs of fertilizer application per hectare alone were around Rs230, which at best would have produced an incremental response of 10 quintals. These costs look favourable given the 1968/69 wheat procurement price of Rs76 per quintal. However, it should be remembered that the price of wheat in 1968/69 had almost doubled from its pre-Green Revolution level, and that most of the inputs were heavily subsidized. These subsidies and price changes, in fact, make any evaluation of the economic feasibility of the new technology meaningless.[15] Once the decision to implement a technology is made, output prices and input subsidies can always be manipulated to make the new technology economically feasible. The high costs of production through the new technology can however be inferred from the rising prices and the fact that there is perhaps no country in the world where production through the new technology can be maintained without subsidies and price supports. In India it is perhaps an indicator of the high costs of HYV production that the procurement price of wheat, the major crop to come under HYV, has been rising at a much faster rate than that of paddy, which largely remained under traditional cultivation. And demand for higher paddy prices got some force only when the surplus Green Revolution farmers took to HYV cultivation of Rabi paddy.

It is commonly believed that the Green Revolution made India self-reliant in agricultural production. This belief is based on the impression that foodgrain imports after the Green Revolution declined substantially. In fact, however, the net amount of cereal imports in the decade before the Green Revolution (between 1956–65) at 43 mT were only slightly more than the net imports of 38 mT in the decade 1968–77 following the Green Revolution. It is true that the imports did not rise with the increasing population. But, as we have seen, the rate of growth of foodgrain production actually decreased after the Green Revolution, while the population growth did not show a corresponding decline. Under these circumstances, what could the declining cereal imports really mean? Imports of cereals in India have always been resorted to in order to feed the urban sector. Reducing the imports for this purpose became possible after the Green Revolution because more food started flowing into the government stocks, not because there was actually more food per capita to go around. The increased availability of food with the government was caused by a lopsided growth of agriculture on which we shall comment in the next subsection.

The important point to remember, however, is that decreased imports of cereals did not imply a decreased foreign dependence of agriculture. What was gained in terms of reduced cereal imports was lost in terms of increased imports of agricultural requisites, especially fertilizer. Before the Green Revolution, expenditure on imports of agricultural requisites used to be almost nil. In 1950/51, seven crore rupees were spent on this head, in 1960/61, the expenditure was thirteen crores. In 1970/71, this expenditure rose to 102 crores, and in 1973/74, it doubled to 201 crores. Then came the spurt in fertilizer prices, and in 1974/75, expenditure on fertilizer import alone stood at 532.5 crores. Thus the import dependence of Indian agriculture had in fact been rising quite fast. Let us look at this data in a different perspective. The price of nitrogenous fertilizer on a rough average remains around three times the price of wheat. In the decade 1967–76, on an average 0.72 mT of nitrogenous fertilizer was imported per annum.[16] This is equivalent to the import of 2 mT of wheat per annum, implying that the equivalent wheat imports in the post-Green Revolution decade had actually increased by 50 per cent. And we have not yet counted the imports required to build up an indigenous capacity in fertilizers and tractors and so on, which should also in fact be counted under this head.

Thus, after the Green Revolution, dependence of the agricultural sector on foreign inputs increased in diverse ways. While food alone had to be imported earlier, now a number of varied inputs had to be brought in. While the government had to depend on foreign countries for a large proportion of the new requisites of agriculture, the agriculturist had to depend even more on the government and the industrial sector. There was an increased external dependence all around.[17]

In addition to this dependence of the farmer on the government and of the government on foreign suppliers for tangibles like fertilizers, pesticides and seeds, an intangible, but not any less important, external dependence for knowledge of agricultural processes appeared. The farmer who till now was the expert on agricultural technology became ignorant in one sweep. He had to look to the university expert to acquire knowledge; and those experts themselves looked to the so-called international community of agricultural scientists to learn the latest on the new technology.

Disparities in Growth

Vast disparity in growth, from crop to crop and from area to area, was an inbuilt feature of the new technology. While a few crops in a few areas showed enormous increase in production and productivity, most of the crops and most of the cultivated areas in the country stagnated, and perhaps actually deteriorated.

Crop to crop disparity

We have already noticed that of the major foodgrain crops only wheat showed an increased rate of growth of production and productivity after 1967/68. To show this disparity of growth amongst various crops I have considered the absolute figures for the area, production and yield etc., of the three main foodgrain crops of India (rice, wheat and pulses) for every fifth year since 1950/51. In 1950/51, of the total foodgrain production of 52.58 mT, 21.81 mT was rice, 6.34 mT was wheat and 8.33 mT was pulses. In 1963/64, towards the end of the first phase of post-independence agriculture, foodgrain production had increased to 83.38 mT. Of this 36.17 mT was rice, 10.96 mT wheat and 11.34 mT pulses. The production of the three crops had thus increased at the same pace. In 1950/51, rice, wheat and pulses formed 41.5, 12.1 and 15.8 per cent respectively of total foodgrains production, in 1963/1964, their respective share was 43.4, 13.1 and 13.6 per cent. The relative importance of the three crops in the total foodgrain production of the country remained essentially unaltered, except for a small decline in the share of pulses. Interestingly, though the area under wheat increased at a faster rate than that under rice, the difference was made up by a higher growth of yield in the latter. In 1970/71, after the Green Revolution, however, we find wheat production jumping from 10.96 mT to 23.44 mT, while rice moved from 36.17 mT to only 41.91 mT, and pulses remained static. The share of wheat in the total foodgrain production rose from a mere 13 per cent in 1963/64 to 22 per cent at the cost of rice, pulses and other crops. While yield of rice and pulses remained almost unchanged, yield of wheat rose by 62 per cent. Of the 6.35 mha of additional area brought under irrigation 4.89 mha went under wheat. The same trend continued in 1975/76. Of 3.65 mha of additional irrigated area under foodgrains, wheat accounted for 2.84 mha; and of 6.41 mT of additional foodgrains, wheat accounted for 2.84 mha; and of 6.41 mT of additional foodgrains, wheat accounted for 3.88 mT. Output of pulses remained unchanged, while that of rice increased only slightly. All the benefits of growth thus went to the relatively prosperous wheat areas, while paddy growers, who formed the vast majority of the small cultivators were left to stagnate.

One reason for this imbalanced growth between rice and wheat is simply that western countries where the new technology evolved, are not rice-producers. Way back in 1820, Alexander Walker, while describing the failure of an experiment to introduce English agricultural technology in an Indian village, commented, '. . . It should also be well considered how far our agricultural process is suited to the cultivation of rice, the great crop of India, and of which we have no experience.' The problem is now solved simply by making the great crop of India the less important.

However, this is not the only explanation for the spurt in wheat production. It was also convenient to increase wheat production to meet the policy objectives which had in the first place led to the acceptance of this technology. There already were areas, almost surplus in wheat, and well-linked with the urban market economy.[18] By increasing wheat production therefore, it was easier to meet the policy objective of bringing more food to the urban market. Hence it seems no accident that out of the 10 mha of additional irrigation potential generated between 1963/64 and 1975/76, 7.73 mha has gone to wheat areas. What is more, the government has taken pains to provide a favourable market for the wheat growers. While wheat prices were maintained around the international market prices, the price of rice was kept substantially below the international price. For instance, as Keith Griffin notices, in early 1978 the ex-farm price of rice in India was $165 a ton, less than half the US price of $335 a ton, which also represented the international price, since USA is a major rice exporter.[19] On the other hand the ex-farm price of wheat at that time was $135 a ton, compared with the US price of $110 a ton.[20] The policy proved extremely successful. In January 1978, the country had 18 mT of surplus wheat, while about 300 million people in the country were below the poverty line, not having enough purchasing power to eat the food that was lying there.

The reasons for the decline in the growth of pulses are related to those that caused the spurt in wheat. Pulses, grown largely in rain-fed conditions, were not commercially viable. Some countries derived their protein requirements from meat procured through the expensive process of feeding good corn to cattle and pigs;[21] in largely vegetarian India, however, pulses formed the main source of proteins. Yet the availability of pulses per head per day continuously declined, from 64g in 1962, to 58g in 1964, 48 gms in 1971, 45 gms in 1976 and only 40 grams in 1979.[22] The solution was seen in trying to teach Indians to change their food habits and shift to commercially more profitable proteins. Let us give an example of the ridiculous extent to which the idea of changing the food-habits in a commercially favourable direction was carried. The Literacy House in India is a component of World Education Inc., a corporation that had, with the help of World Bank USAID and some other multinational agencies, taken up the task of preparing the rural masses in the Third World for the Green Revolution. This House in 1978 brought out an adult literacy primer, *Aao Charcha Karen*, wherein one finds the explicit message: 'Eating just rice has a bad effect on health; eat eggs to make up for protein deficiency.'[23] So, Indians were malnourished because they were vegetarians! If a commercializing society fails to produce vegetarian proteins they should learn to shift to other things.

Area to area disparity

Wheat and rice in India are traditionally grown in different areas. The fact that only wheat increased in production and productivity already gives an indication that the much vaster rice areas must have suffered stagnation after the Green Revolution. However, we can form a clearer idea of the type of disparities that arose in HYV and non-HYV areas.

Let us start with the assumption that all increases in yield in 1970/71 were due to the marginal productivity of HYV and of irrigation at the official yardstick of 0.5 t/ha, (that is, irrigating one hectare of land increases the output by 0.5t), which almost certainly is an underestimate. Now after subtracting the contribution of the marginal productivity of irrigation from the productivities, we find that from 1963/64–1970/71 productivity of rice (after subtracting the contribution of irrigation) rose from 827 kg/ha to 922 kg/ha and that of wheat from 634 kg/ha to 1,034 kg/ha. If we assign all this increase in marginal productivity of 1.12 t/ha for HYV wheat and 0.62 t/ha for HYV rice, then for wheat this implies that yields per hectare of unirrigated, irrigated and HYV-irrigated land were 634 kg/ha, 1134 kg/ha and 2254 kg/ha, respectively. The 1968/69 PEO studies based on field data from HYV areas gave an average yield of HYV wheat as 2,560 kg/ha (NCAR, vol. 1, Table 4.4). This means that our estimate of marginal productivity of HYV is a slight underestimate. It seems that the assumption that productivity in non-HYV areas remained unchanged is not entirely correct, it may have slightly declined.

If we consider figures for 1975/76 again and subtract the contribution of irrigation from the entire production we obtain the productivity of rice and wheat at 932 kg/ha and 1065 kg/ha, respectively. Interestingly they are not at all different from the corresponding figures of 922 kg/ha and 1,034 kg/ha in 1970/71. But the area under HYV rice had increased by 7.3 mha and that under HYV-wheat by 5.9 mha, between 1970/71 and 1975/76. Where did all the expected increase in production from these additional HYV areas go? Some increase in productivity over the 1970/71 figures can be observed if instead of comparing 1970/71 yields with 1975/76 yields, we make the comparison with 1976/77 yields to include an abnormally good year 1977/78 in the average. (Productivity, after subtracting contribution of irrigation of rice and wheat, then comes out to be 1,018 kg/ha and 1106 kg/ha respectively, with total HYV area under the two crops being 13.77 mha and 14.50 mha). Yet the marginal productivity of additional HYV rice and wheat does not approach anywhere near the earlier figures of 0.62 t/ha and 1.12 t/ha, which themselves seem underestimated. One way to explain this phenomenon is to say that as HYV areas were expanded, all the necessary resources could not be made available, and hence additional HYV areas did not show appreciable response to the new technology. Alternatively, one must assume that productivity in the non-HYV areas had declined to balance the increased productivity in HYV areas. In practice, both these processes are likely to have operated. Since bringing an area under HYV involves considerable expenditure it is not likely to be done unless there is some corresponding increase in productivity. What is more, the consumption of nitrogenous fertilizer almost doubled between 1970/71 (1.37 mT) and 1975/76 (2.4 mT) — and this increased use of fertilizer must have produced some response in the HYV areas. If the aggregate productivity still did not show any appreciable improvement, the only plausible explanation seems to be that as more and more resources got diverted to HYV areas, the productivity in non-HYV areas actually declined. Micro-level studies will be required to isolate the detailed causes of this phenomenon, but the aggregate trend of declining productivity in non-HYV areas seems unmistakable. And it is not very surprising. As prices rise all around and even ordinary inputs become expensive, those whose inputs are not protected by subsidies and those who do not gain by the increased prices of the outputs are likely to stagnate and deteriorate.

It is clear then that no revolutionary improvement in the production and productivity of Indian agriculture as a whole occurred with the so-called Green Revolution. If anything happened, it was that the rates of growth of Indian agriculture declined. What looked like a revolution was merely a spurt in the growth of a few commercially important foodgrains in a few areas which were already surplus. This growth too was achieved at a very high cost of resources, and at the cost of an enormously enhanced dependence of agriculture on external, often imported inputs. The increased costs pushed up prices all around and made the subsistence farmers — who were not protected by input subsidies and were not helped by higher output prices, since in any case, they had no surpluses to sell — even more impoverished. The yields in those subsistence farms consequently seem to have declined below the pre-Green Revolution levels. From the urban-industrial perspective, however, the change was truly revolutionary. With the growth concentrated in already surplus areas, more and more food flowed into the urban market, even though large numbers of people outside this sector still could not generate sufficient resources to get 2,400 calories of food. (That is the official poverty line for rural areas.) The improvement in the food availability in the urban-industrial sector was in fact so revolutionary that today leading economists can already advise resistance to the demands of surplus farmers for higher prices on the grounds that we do not need more food. It is now being declared that the country has already lost enormously by producing more food than is necessary, that the prices of foodgrains should now be kept low so that the surplus farmers are forced to grow more essential commercial crops (see the many recent editorials on this issue in the *Times of India*). And this at a time when 300 million people in the country are still hungry and living below the official poverty line! These are the achievements of this Green Revolution.[24]

Alternatives to the Green Revolution

Was there an alternative to the Green Revolution? The answer to that question depends upon what one expects a revolution in agriculture to achieve. If what is expected to be achieved is only a steady flow of food and resources to the urban market and the government stocks and industries, then the Green Revolution was perhaps the best way to achieve it. If, however, our expectation from a revolution in agriculture is that first of all it enables the millions of subsistence workers living below the poverty line to produce their essential requirements, then of course there would have been no question of even considering the Green Revolution technology. In some form, our subsistence farmers already had an 'alternative' to the Green Revolution technology. Even a cursory, but sympathetic, study of their agriculture (with the above objective in view) would have led to the conclusion that what was needed was not so much new technology but immediate action to remove the various resource constraints which were putting tremendous pressure on the agriculture — resources such as wood (fuel), manure, water, fodder, and of course land. However, any step in providing 'free' access to such locally available resources to the cultivators would have meant reversing the policy of achieving 'progress'. Let us consider, for instance, two of the major requirements for traditional agriculture: access to water and access of labour to land. As we shall see below, India had, and still has, a vast untapped potential of these resources.

Irrigation

Irrigation is the most important input for traditional agricultural technologies. It insures the farmer against the vagaries of the climate, opens up the possibility of multiple cropping, considerably enhances the employment potential of the land. (And it almost doubles the productivity of individual crops.)[25]

Costs of irrigation are difficult to work out because there is a large variation from area to area. For peninsular India, where irrigation costs are relatively higher, the Irrigation Commission in 1972 estimated the cost of irrigating one hectare to be roughly equivalent to the price of a quintal of foodgrains (NCAR, vol. 1: 437), which would give a return of five quintals of foodgrain. On the other hand, as discussed earlier, to bring one hectare of crop under HYV costs two to three quintals of foodgrains, and the return expected is about ten quintals. Thus in terms of economic efficiency, irrigation competes well with HYV cultivation. The possibility of developing this alternative also existed, for according to NCA estimates, our country has enough water resources to irrigate 110 mha of crops, whereas in 1965/66, the year before HYV crop was launched, the gross irrigated area was 32.2 mha.

Since the mid-sixties, some additional irrigation facilities have of course been generated. But this has been seen as only one of the inputs in the HYV technology and the stress has been on providing more irrigation in those areas which already had irrigation and had adopted new technology. This is obvious from the fact that of the additional 10 mha of foodgrain crops brought under irrigation between 1963/64–1975/76, 7.73 mha went to the major HYV crop, and only 1.44 mha to the vaster rice crops. Besides, the massive schemes of modern irrigation being launched in India have proved to be problematic in various respects.

However, one can conceive of alternative strategies for irrigation. In our country irrigation has traditionally been the responsibility of the community and the state. Traditionally, 'non-conventional' ways of generating irrigation using community labour and locally available materials have been used. Such a system of irrigation would be cheaper. More important, it could benefit small and marginal farmers to put their agriculture on a sound footing, instead of making it economically unviable as the Green Revolution technology has done. Such an irrigation would make all the difference between prosperity and hunger, between a living thriving culture and stagnation.

Access of Labour to Land

Besides irrigation, the other major prerequisite of traditional agriculture is labour. Productivity of this type of agriculture depends largely upon the amount of labour that the farmer is willing to or is capable of putting in. This fact is confirmed by the well-known observation that almost everywhere in the Third World, small farms, even those less than one hectare, on which labour is necessarily intense, are able to obtain much higher productivities then larger farms. The first series of farm management studies carried out in 1954–57 (NCAR, vol. 1: Appendix 4.1) brought out the fact that the difference between the gross output per hectare of the smallest and largest size groups was always more than 30 per cent except in UP and Maharastra (Akola and Amravati districts), where the districts studied were largely under cash crops, and in Orissa where the productivity was rather low irrespective of the size. In Tamil Nadu (Salem and Coimbatore) the difference was as large as 170 per cent. In Maharastra (Nasik) the figures are 109 per cent, in Andhra Pradesh (West Godavari District) and Punjab (Ferozepu and Amritsar) around 40 per cent. Similar data on other Third World countries (for example, Indonesia, Thailand, Taiwan) is available in Keith Griffin (1979). That the smaller holdings were able to utilize the available resources much better, is also clear from data found in the 1971 Agriculture Census quoted in NCAR, (vol. 1: Table 4.1 and 4.2). Out of 33.8 mha commanded by holdings of less than two hectares, 30 mha was sown, 7.7 mha of it more than once, giving a cropping intensity of 125. Holdings of size less than one hectare fared even better with cropping intensities of 134 and 123 respectively for unirrigated and irrigated land, while holdings of size greater than ten hectares sowed, achieved cropping intensity of only 109. Another study carried out at the ANS Institute for the Kosi Command Area in Bihar (Prasad 1972, quoted in NCAR, vol. 2: 37–38) shows the following. On the introduction of irrigation, whereas large farms (greater than eight hectares) irrigated during Rabi season, only 26.5 per cent of the area irrigated during Kharif, this ratio was 102.5% for small farms ranging from nought to eight hectares. Yet, according to the 1971 census, the most wasteful of farmers (of size greater than ten hectares) commanded 30 per cent of the total area, whereas small efficient farms (with an operational holding less than half the average size) commanded only 9 per cent of the total area.

What is therefore urgently needed is land-reform. Land to the tiller would not only result in an increase in agricultural production, but also the increase would benefit the small farmers, who need it most. The Green Revolution technology, however, is changing all this.[26] The small farms are being made commercially unviable, whereas the larger farms, with access to this technology, are producing more and earning profits. However, if the objectives are to improve the livelihood of our people, improving the access of labour to the land by redressing this skewed distribution through land reforms, and improving the availability of water, clearly offers a vast potential for a widespread and genuinely revolutionary improvement in agricultural production and productivity.[27]

This was an obvious alternative to the Green Revolution and was well known to anyone with any knowledge of agriculture. The National Commission on Agriculture (1974) itself recognized that 'small farms as a class are more efficient units of production compared to large farms when considered from the point of view of productivity and employment potential'. It also recognized that providing water to these small farms, 'would have by and large solved their problems'. If inspite of that a choice was made in favour of a technology that improved the fate of only already-surplus farmers and yet did not accelerate agricultural growth, then it can only be surmised that solving the problems of small farmers was not the most important policy-objective.

By bringing out the viability of the non-technological alternative, we do not wish to imply that in agriculture no technological change will ever be required. But it seems that technological changes which will emerge from well-fed farmers with a view to improving their own lot will have to be qualitatively different from the technological changes advocated by elite practitioners of the Baconian science of control with a view to commercial viability. The analysis above makes us agree with Lappe and Collins (1977) that, 'once it is manipulated by people, nature loses its neutrality. Elite research institutes will produce seeds that work perfectly well for a privileged class of commercial farmers. Genetic research that involves ordinary farmers will produce seeds that are useful to them'. And also, perhaps a genetic science that incorporates their view of nature. But then ordinary farmers in traditional cultures have been carrying out such research for centuries.

Notes

1. An early British observer of Indian agriculture, Colonel Alexander Walker, noted the following about agriculture in Malabar in 1820. 'In Malabar the knowledge of husbandry seems as ancient as their history. It is the favourite employment of the inhabitants. It is endeared to them by their mode of life,

and the property which they possess in the soil. It is a theme for their writers; it is subject on which they delight to converse and with which all ranks profess to be acquainted.' (Walker 1820)

2. For Gandhi, these were also the symbols of a resurgent India, of an India made free again through the independence of its agriculture and its villages.

3. Before moving onto the British phase of Indian agriculture, I wish to undo one prevalent misconception — that these decentralized village communities were technically inefficient. All available accounts of those times suggest that: the independent cultivators had achieved almost complete perfection in the art of agriculture producing 'the most abundant crops, the corn standing as thick on the ground as the land could well bear it' (Walker 1820); the decentralized manufacturers were able to produce the finest specimens of not only textiles, but also of steel; the village institutions had spread education so well that G.L. Prendergast, member, Governor's Council, Bombay, remarked in 1821 that '. . . there is hardly a village, great or small throughout our territories, in which there is not at least one school . . . there is hardly a cultivator or a petty dealer who is not competent to keep his own accounts with a degree of accuracy. . .'; this decentralized civilization was able to produce medical practitioners, astronomers, philosophers and artists of the highest order. For further details on these aspects of the Indian civilization, see Claude Alvares (1976), R.P. Dutt (1940) and R.C. Dutt (1970).

4. The revenues extracted from India after the Battle of Plassey have been recognized to be of critical importance in setting in motion the Industrial Revolution, by many observers. For details and references to some of the authors who have commented upon it, see R.P. Dutt (1940), pp. 116–19.

5. Both R.P. Dutt (1940) and R.C. Dutt (1970) give detailed accounts of this destruction. These books also contain detailed references to the historical accounts of this period.

6. There seems to be an important qualitative difference between the plunderers that visited India prior to the British. The earlier robbers, like the notorious Ghazani, looted the surplus accumulated in temples and with the aristocracy, leaving the life in the villages more or less unaffected. The legalized plunder by Hastings etc., and their hordes, on the other hand, ravaged every hut in every village.

7. The general wholesale price index for Calcutta (July 1914 = 100) which stood at 202 in 1920 declined to 173 in 1924 and 141 by 1929 and touched the rock bottom with 87 points in 1933. Indices of cereals, pulses and oilseeds in 1933 stood at 66, 84 and 74 respectively, (Vera Anstey 1949; quoted from NCAR, 1976; vol. 1, p. 128). Later with the outbreak of the Second World War, food prices increased reflecting general scarcity (NCAR stands for the Report of the National Commission on Agriculture, 1976).

8. For information regarding this period of Indian agriculture, see George Blyn (1966). Also see NCAR, 1976, vol. 1, ch. 3.

9. It should be noted that the thrust of the IADP and IAAP was not on introduction of new technologies, but on an intensive application of resources like irrigation, fertilizers, etc.

10. D.K. Desai (1969) and Dorris D. Brown (1971) have analysed the IADP programmes in detail.

11. This was how the programme to introduce new technology in certain well endowed areas was officially styled. The programme was monitored by the Programme Evaluation Organization of the Planning Commission during 1967–69. The relevant results on the yields of different crops in different areas under the HYVP have been gleaned from the various PEO evaluation studies and summarized in Appendix 4.2 of NCAR, vol. 1, by the National Commission on Agriculture. In Appendix 4.3 of NCAR, vol. 1, a summary of a study on the relative economic returns from HYV and local varieties carried out by the Agro-economic Research Centres at various locations in the country in 1968–69 and published by Ram Saran (1972) is also available. From these studies it can be said that HYV wheat fared rather well in almost all areas. The main kharif crop of rice, however, seems to have showed almost no response to HYV cultivation. This incidentally was the fate of the monsoon rice crop all over South and Southeast Asia. The studies also show a wide variation in the response to HYV from area to area.

12. The above statements of course refer to data gleaned from the statistics put out by the Government (NCAR 1976). There are somewhat different data available elsewhere in the literature (see for example Gail Omwedt (1981), Ranjit Sau (1981)), which employ either different sources or different base years etc. But from these data also, the same general trends are obvious: the decline in the rate of growth of aggregate agricultural production; no increase in the aggregate agricultural yield; marked decline in the aggregate yield of crops such as rice, pulses, etc.

13. A 1968 comparison of the energy efficiency of British agriculture as a whole with that of shifting rice cultivation carried out by Dayaks and Ibans in Borneo showed that while the efficiency of the former was only 0.20, that of the latter ranged between 14.2 and 18.2 (quoted in Caldwell (1979), p. 56). A more relevant comparison is perhaps the one carried out by Lockeretz et al (1977). They compared two sets of farms in the US corn belt that differed from each other only in the fact that one set used only organic manures and no inorganic fertilizer or pesticide while the other set used these inputs. They found that while the two sets of farms showed comparable economic efficiency the organic farms used 2.4 to 2.5 times less energy per dollar of output. Incidentally, the organic farms were also able to employ 12 per cent more labour, a commodity plentifully available. See also, for example, Reedy (1976).

14. These figures are taken from Economic Survey, GOI, 1980–81. Figures for 1979–1980 are provisional.

15. An idea of the level of subsidies can be obtained from the following:

Naphta, the major raw material for the production of nitrogenous fertilizer, is sold to the fertilizer industry at a controlled price of Rs900/ton while for other users the price is Rs2350/ton (1980 prices). The fertilizer produced is then further subsidized. While price support and subsidies are legitimate rights of the farmer if they must produce via the new technology, it should be borne in mind that these measures help only a miniscule proportion of Indian farmers, who use the new technology and produce for the market.

16. Data in this paragraph are taken from NCAR, vol.2, p.79, and Economic Survey GOI, 1980–81.

17. In conventional economics, this increased dependence will appear as development of new 'linkages' showing a positive effect on the overall economy. But objectively, what is really positive about loss of self-reliance of the agricultural sector?

18. Notice that in 1950–51 total production of wheat in India was only 6.34 mT. Incidentally, wheat is also the major grain traded in the international market. In 1974, under-developed market economy countries imported 31.2 mT of wheat and only 2.0 mT of rice. See Table 6.7 and 6.8 of Keith Griffin (1979). Also see his tables 6.1 and 6.2 to get a profile of the international wheat and rice trade.

19. Incidentally, before the Green Revolution, Asia was a net rice exporter. After the Green Revolution this region had become a net importer. In 1964, 181,000 tons of rice were exported from Asia; in 1970, 1,135,000 tons of rice were imported into Asia.

20. Part of the reason for the higher domestic price of wheat is perhaps to be found in the higher input costs of this crop because of the adoption of the new technology. (See the section on costs.)

21. The process is so expensive that non-vegetarian USSR imports about 50 mT of cereals every year to raise the domestic availability of foodgrains to about a ton per capita per year. Vegetarian India can feed itself with just about 1/5th of a ton per capita.

22. Triennial averages based on data in the Economic Survey, GOI, 1980–81.

23. Quoted from Ross Kidd and Krishna Kumar (1981). About the aims of the adult-literacy programme, launched by the World Bank, etc. in the late 1960s, in conjunction with the Green Revolution, the authors have the following to say: 'The purpose of the new programme was to cover all aspects of a peasant's life that would facilitate his initiation into a consumer society; aspects such as agriculture, health, sanitation, fertility and small-scale entrepreneurship. . .'.

24. Incidentally the salient features of the Green Revolution — decline in the aggregate growth, increased production in localized areas at the high cost of often imported resources, decline of production in less favoured areas and control of production by a small sector etc., are typical features of all modern technologies. The theory and practice of modern science and

modern technology was evolved in seventeenth and eighteenth century Europe. The driving concern of that evolution was clearly stated by Bacon, the prophet of the Scientific and Industrial Revolution, as simply power, power through control of nature, of production and necessarily of people. Resource efficiency, ecological efficiency, distributive justice, etc., were nowhere in the minds of the people who initiated this development. All ethical injunctions ensuring justice were in fact dismissed as obscurantist nonsense. The scientist and the technologist was to expend all his energies in increasing control — and hence profits. Justice and equality would, it was assumed, follow as a result of that singleminded search for power and control, through some inscrutable dialectical process. Resource efficiency, of course, was something about which the technologists of that era could not have cared much. All the resources of the colonies were there to be taken, almost free, till you could devise processes that would consume these resources efficiently or otherwise — within the mother country. It was under such conditions and such considerations that the science and technology that we call modern, emerged and it still carries its birthmarks with it. All the features of the Green Revolution that we have noticed are obvious manifestations of these birthmarks.

25. A comparison of data based on NSSO crop-cutting experiments for 1970–71 and 1971–72 shows that compared with unirrigated crops, yields of irrigated crops were higher by about 80–95 per cent for paddy and 105–115 per cent for wheat. According to a statistical analysis based on aggregate crop-production in the 1950s, quoted in NCAR, the differences in irrigated and unirrigated yields were 1.25 ton/ha and 0.46 ton/ha respectively, for wheat, and 1.45 ton/ha and 0.47 ton/ha for paddy. Official yardstick for the marginal productivity of irrigation is, however, 0.5 ton/ha. (NCAR, vol. 1: 437–8).

26. The Green Revolution, by making commercial cultivation with new technology economically more viable (at the cost of subsidies and price supports), seems to have partially neutralized the advantage of the small farms. Thus in Punjab (Ferozepur) the farms about 20.0 ha showed the lowest gross output per hectare of all sizes in 1954–57; in 1967–70 farms of 24.0 ha and above showed the highest output of all sizes. However, in most of the country the small farms still retain their advantage (NCAR, vol. 1: 431).

27. Such revolutionary change in agricultural productivity through improved access of labour to land and improved water control is not merely a pipedream, as was shown in Kampuchea during the few years of that ill-fated revolution. Using these two resources to the utmost, the Kampucheans were able to ensure 312kgs of rice per capita by 1977, in a situation where all experts had been predicting major famines. They had used only green and compost manures, vegetable insecticides and cattle power. Their belief in the workability of traditional agriculture was so strong that Khieu Samphan declared in 1977 that, 'The cattle and buffalo

are our closest comrades-in-arms in the national building campaign. If our cattle work hard we can build our country rapidly.' For an excellent review of the Kampuchean experiment, see Caldwell (1979) and reference cited therein, especially Hildebrand and Porter (1976).

Bibliography

Alvares, C. *Homo Faber: Technology and Culture in India, China and the West 1500–1972*, Allied, New Delhi, 1979.

Austery, V. *The Economic Development of India*, (3rd Edition), Longmans, London 1949.

Blyn, G. *Agricultural Trends in India, 1891–1947*, University of Pennsylvania Press, Philadelphia, 1966.

Blyn, G. *India's Crop Output Trends: Past and Present*, 1979.

Brown, D.D. *Agricultural Development in India's Districts*, Harvard, Cambridge, 1971.

Caldwell, M. *Kampuchea: Rationale for a Rural Policy*, Janata Pracharanulu, Hyderabad, 1979.

Dandekar, V.M. and Rath, N. 'Poverty in India', *Economic and Political Weekly*, 2 and 9 Jan. 1971.

Desai, D.K. 'Intensive Agriculture District Programme', *Economic and Political Weekly*, A83–90, 28 June 1969.

Dharampal. *Indian Science and Technology in the Eighteenth Century, Some Contemporary European Accounts*, Impex India, Delhi, 1971.

Dutt, R.C. *Economic History of India*, (1906) 2 Volumes, Publications Division, Delhi, Reprint 1970.

Dutt, R.P. *India Today*, Victor Gollancz, London, 1940.

Griffin, K. *The Political Economy of Agrarian Change*, (2nd Edition), Macmillan Press, London, 1979.

Hildebrand, G.C. and Porter, G. *Cambodia: Starvation and Revolution*, Monthly Review Press, New York, 1976.

Kidd, Ross and Kumar, Krishna. 'Co-Opting Freire: A Critical Analysis of Pseudo Freire in Adult Education', *IFDA Dossier* 24, July/Aug: Also Published in *Economic and Political Weekly*, 3–10 Jan. 1981.

Lappe, F. and Collins, J. *Food First*, Houghton Mifflin, Boston, 1977.

Lockeretz, W. *et al*. 'Economic and Energy Comparison of Crop Production on Organic and Conventional Corn-Belt Farms', in *Agriculture and Energy*, Academic Press, New York, 1977.

Marx, K. 'The British Rule in India', *New York Daily Tribune*, 25 June 1853(a); reprinted in *Marx and Engels* (1959).

Marx, K. The Future Results of British Rule in India, *New York Daily Tribune*, 8 August 1853(b); reprinted in *Marx and Engels* (1959).

Marx, K. and Engels, F. *On Colonialism*, Progress Publishers, Moscow, 1959.

NCAR. Report of the National Commission on Agriculture. 15 Volumes,

Chairman N.R. Mirdha, Ministry of Agriculture and Irrigation, GOI, New Delhi, 1976.

Omwedt G. 'Capitalist Agriculture and Rural Classes in India', *Economic and Political Weekly*, (26 Dec. 1981).

Reddy A.K.N. *Economic and Political Weekly*, Annual Number, 1976.

Saran, Ram. 'High Yielding Varieties Cultivation: Some Economic Aspects', *Agricultural Situation in India*, August 1972.

Sau Ranjit. India's Economic Development, 1981.

Shah C.H. *Agricultural Development in India: Policy and Problems*, Orient Longman, Bombay, 1979.

Walker A. *Indian Agriculture* (1820), reprinted in Dharampal (1971).

11

Science and Hunger

Plant Genetic Resources and the Impact of New Seed Technologies

Lawrence Surendra

The inroads into nature made by the demands of the industrial age have produced bizarre situations and problems with regard to the ecology and environment of both industrialized and developing nations.

With the growing pressure of population on land a steady deterioration of the ecology and environment is taking place in Asia, with very far-reaching consequences. According to the United Nations Food and Agriculture Organization (FAO), some 5 million hectares of forest are lost annually. Some 8 million hectares are burned and temporarily cultivated every year by approximately 200 million shifting cultivators, affecting about 300 million hectares.

Added to such pressure on the natural environment have been the distorted priorities for national development in many countries. For instance, the construction of massive hydro-electric dams in the 1950s and 1960s led not only to the inundation of vast tracts of cultivable land but to the accompanying destruction of centuries-old forests. This in turn led to massive soil erosion which has resulted today in many of the dams becoming choked with silt, rendering them useless.

One serious consequence of forest destruction has been the diminution of plant genetic resources, or PGRs. While wild genetic resources have been destroyed, the productivity of modern agriculture has come to depend on a very narrow genetic base. Thus, according to the International Union for Conservation of Nature and Natural Resources, 'every coffee tree in Brazil is descended from a single plant; the entire soyabean industry is derived from only six plants from one place in Asia; only four varieties of wheat produce 75 per cent of the crop in Canada, and one variety produces more than half; almost three-quarters of the United States potato crop is dependent on four varieties'. The Union goes on:

The desirable qualities such as high yields or resistance to disease that have been bred into these plant varieties are not permanent. The average lifetime of cereal varieties in Europe and North America is 5 to 15 years and new varieties must be constantly developed by cross-breeding, often with wild relatives of domestic plants. It is precisely these varieties which are threatened with extinction by the growing pressure on the earth's wild places.

However, strange as it may seem, the future of PGRs and ultimately the future of mankind's ability to grow grains, cereals and vegetables for its survival is not linked only to the ability of mankind to prevent the extinction of existing wild varieties but also to society's ability to control and supervise the intrusion of big corporate interests into the whole area of seeds and seed technology.

Today seed business is big business. In the rich industrialized countries of the North, the seed industry attracts an investment of more than US$50 billion and the commercial seed market of US$13 billion is mostly in private hands. The coming onto the scene of what is known as agribusiness coincided with the development of high-yielding varieties (HYVs) and the much acclaimed Green Revolution in developing countries.

The arrival of HYVs in developing countries induced in policy-makers and implementers what Prof. Gunnar Myrdal has called 'technocratic euphoria'. HYVs, for all their attraction to government planners and rich farmers in terms of bumper crops, more profits and the prospect of surplus available to provide export earnings, were not without some very serious drawbacks.

First, the improved HYVs came as part of a package which included agrochemicals such as pesticides and fertilizers. According to FAO estimates, Third World consumption of pesticides was set to rise from 160,000 tonnes in the early 1970s to more than 800,000 tonnes by the mid-1980s. Pesticides and herbicides become necessary to protect new varieties which tend to be vulnerable to diseases. It has been estimated that the increased use of pesticides causes 375,000 peasants in the Third World to become ill every year, of that a total 10,000 die.

The other important input is fertilizer, which is needed at the right time and in the right amounts. Fertilizer shortages mean crop losses. The shortage in 1974 is estimated to have cost the poor countries of the South 15 million tonnes of grain, enough to feed 90 million people.

All this means not only that many developing countries spend much of their financial resources paying for costly pesticide and fertilizer imports, but also that the spread of HVYs was a big boon for the multinationals, particularly those which are petrochemically based.

Some of the early criticism of the Green Revolution and the introduction of HYVs naturally was on the costs involved and the dangers of making national food production dependent on foreign multinationals and costly imports of agricultural inputs. Policy-makers weathered this criticism and went ahead enthusiastically with the active promotion of the Green Revolution strategy.

However, problems developed in other quarters. The acceleration of disparities in rural incomes gave rise to political and social tensions and greater hardships for tenants and the landless, as landlords now making greater profits resumed cultivating their land with machines.

Also, the self-sufficiency achieved by some of the grain-importing countries through the use of HYVs had adverse effects on vital foreign-exchange earnings in the traditional foodgrain-exporting countries such as Thailand. In large countries such as India, it also brought problems. In India, the creation of a new class of wealthy farmers in the wheat belt in Punjab and parts of southern and western India laid the ground for powerful political movements which continue to threaten the power centre. This has brought about a whole new regional dimension to Indian politics.

Side by side with such developments have been the devastating, negative consequences on the ecology and environment of the unscientific and unplanned spread of the new agricultural strategy and the hurry to reap the benefits of the new technology. Only recently has attention been focused by national and international agencies on these problems, but on the evidence of some of the partial, scattered studies available we may well be sitting on a volcano that could erupt with disastrous consequences.

All these problems may still be manageable, and indeed social and institutional changes, themselves triggered by the problems, may very well offer new solutions. But we now face a new and serious problem. Agribusiness, which has played a very powerful role in providing the two important inputs of the new technology, fertilizers and pesticides, has also moved quite logically into the whole area of seed technology. This has meant, in simple terms, not only stocking germplasm of new seed varieties, but looking for desired genes in order to make newer varieties.

But biotechnology as yet does not create new genes, it mutates existing ones. This means that seed germplasm has to be found wherever it is located. This of necessity has involved gene drain from the South to the North, and affects the world's pool of PGRs in two distinct ways. First, successful mutations of genes and the large-scale use of new varieties adversely affect the existing plant varieties in nature. Sometimes the effects are devastating in their reach and plant varieties can simply disappear.

For example, from 30,000 varieties of rice at the start of this century, India will be left with only fifteen by the year 2,000. New hybrid varieties of barley have annihilated 70 per cent of natural varieties in Saudi Arabia and Lebanon. Further, with every plant type lost, ten to thirty animal or insect species directly or indirectly dependent on it also disappears.

The second consequence of the South–North gene drain demands very serious attention and action. This is the heavy germplasm losses caused by commercial plant breeders and seed multinationals who plunder the germplasm of the South but do not use it at all or preserve it. Private firms exercise 'life and death' powers over germplasm under their collection and storage. For example, the

United Fruit Company may have control of about two-thirds of all the world's collected banana germplasm; it announced on 11 May 1983 that it may close down its conservation programme. As many as 700 rubber cultivars (cultivated varieties) collected from Southeast Asia, Brazil and Sri Lanka were held by Firestone Tire and Rubber Company. On 29 April 1983 the company announced simply that its germplasm research work had been 'suspended'.

Apart from the dangers of genetic erosion and the further narrowing of our base of PGRs, as we look to new plants to feed humanity in the future, control over major crop germplasm also could become a form of political control. About 55 per cent of collected germplasm is with the North. As in all environmental problems, where a long-term perspective is of the essence, neither governments nor industry seem terribly interested.

We face a situation where at one level earth's natural resources run the risk of both depletion and loss of people's control over them due to indiscriminate technological exploitation. At another level technology continuously legitimizes or at least seeks legitimization of its ruthless intervention in nature by claims such as rooting out hunger and starvation. In actual fact, however, and that is the sad irony of the situation, the spectre of hunger and famine looms larger precisely because of this kind of modern technological intervention. Therefore while hunger continues as a generalized phenomenon in the developing world, mass hunger caused by famines also seems to be set to become more often and more generalized. We are urgently in need of building a deepened community awareness of this web of captive relations that human beings, particularly in the developing world are being locked into by modern technology. People are in danger of losing control not only over their lives but over their very living environment and natural resources that sustain their life. It is in such a background that we look at the question of Plant Genetic Resources and the implications arising from the introduction of new seeds and seed varieties.

The PGR Debate — The Question of Science and Non-Science

Pat Mooney's 'The Law of the Seed' illustrates all too well the way an important topic like plant germplasm can become embroiled in international politics. While it is gratifying that germplasm work is receiving increased and deserved attention by the world community — and calls for increased support for germplasm work are frequent in Mr. Mooney's article — there are still some troubling aspects of the 'Law of the Seed'. Probably the most troubling is the fact that Mr. Mooney has written a polemic, and like all polemics, the discussion tends to be unbalanced and misleading. Two of the main targets of the article are commercial companies and Plant Breeders' Rights, both of which need not be defended here, since both are perfectly capable of doing so themselves.

The Law of the Seed is a bit like working your way through a detective story; it contains elements of intrigue and conspiracy, presents its 'good guys' and 'bad guys' and as a bonus includes some interesting background or historical information. The article even has its humorous moments. Mr. Mooney's characterisation of the IBPGR as a 'hybrid without parents' gives one reason for pause, especially in an article about seeds . . . Interesting and informative works on plant germplasm are always welcome. This one would have been more welcome had it been a little less polemical, a little sharper technically, and more balanced on its view of institutions working effectively for increased agricultural production in the Third World. (Donald L. Plucknett, Scientific Adviser to the CGIAR Secretariat, in *Development Dialogue*, 1985: 1)

Older Varieties often tend to be displaced by new strains and may become extinct, while wild rices may disappear when their habitat is disturbed. This phenomenon, which is called 'genetic erosion' is not due to the scientists' work of collection and conservation, but to the diversion of land for non-agricultural uses, destruction of national eco-systems and farmers' preferences at a point of time.

If officially approved and *standard scientific procedures* for seed exchange are described in terms of 'gene robbery', with the implication of political manipulation, it could be injurious to scientific work. It would be a great pity if such remarks discouraged rice research workers from exchanging and using genetic material from different environments. (Dr M.S. Swaminathan, Director, International Rice Research Institute, *The Illustrated Weekly of India* 29 June 1986)

I have taken these two quotes as a representative sample of rejoinders, replies or defence of existing institutions related to agricultural research in the face of critical articles by writers who have tried to bring the whole issue of Plant Genetic Resources to a wider, public debate. The first taken from Donald L. Plucknett's article, 'The Law of the Seed and the CGIAR — A Critique of Pat Mooney', is a defence of the Consultative Group on International Agricultural Research (CGIAR) to criticisms made by Pat Mooney in an earlier full issue he did on 'Law of the Seeds' for *Development Dialogue* (DD),[1] the Journal of the Dag Hammarskjöld Foundation. Pat Mooney in the issue of *Development Dialogue* in which Plucknett's article appears has also published a reply to Plucknett, entitled, 'The Law of the Lamb'.[2]

The second quote of Dr Swaminathan is taken from his rejoinder to an earlier article by Claude Alvares in· the same magazine called 'The Great Gene Robbery'.[3] Claude Alvares, in looking at the question of conservation of genetic resources in India particularly in relation to rice, had focused his attack on both Dr Swaminathan and the International Rice Research Institute (IRRI). Even if one does not totally agree with the tone of the article and its style, it nevertheless raises some very crucial points about genetic resources as the heritage of a people, about the 'modern science' of the Green Revolution and 'the schism that [developed] between indigenous science and international science!' Actually Alvares, talking of this schism makes a point in parenthesis that, 'in fact one could also argue that it was proof of the deterioration of science after it had given itself to modern agribusiness'. We shall come to this point later.

I have taken these quotes mainly to illustrate the image projected in these statements of scientists as people who, by the very nature of their activity in 'modern science', are aloof, objective balanced people and that there is a certain nobility about their work. Plucknett, in his criticism of Pat Mooney concludes by saying, Pat Mooney's article would have been 'more welcome had it been *a little less* polemical, *a little* sharper technically and *more balanced* in its view of institutions working effectively for increased agricultural production in the Third World'. The image conveyed is that the high priests of modern science are such perfectionists. We can then go on further to ask why cannot we let such perfectionists tinker with nature or whatever else science says is its proper concern? Dr Swaminathan talks of 'officially approved and standard scientific procedures', which he says if described as 'gene robbery' could be injurious to 'scientific work'. Gene robbery either officially sanctioned or, as Dr Swaminathan puts it, 'officially approved' (through CGIAR or the IBPGR, the International Bureau of Plant Genetic Resources) or the actual robbery from the wild reserves of the Third World, does go on,[4] but naming it so seems to violate the hallowed principles of modern science. The inference is that, if you dare to criticize these disastrous activities of modern science you can do so only within the parameters it has set. All else from the view of 'scientists' like Plucknett and Swaminathan, is 'non-science', 'polemical' and 'injurious to scientific work'.

Everyone realizes and recognizes that there is a debate and controversy over plant genetic resources and over how to deal with the unrelenting efforts of corporate bodies to privatize the genetic resources of the world. What is important to recognize first is that the controversy is not merely a debate between the Third World and First World, it is not merely a controversy between governments and corporations (where it is possible to make such a distinction) but it is crucially a debate, a fight over the terms of knowledge of science, of the fight against the one-dimensionality and reductionist notions of modern science, of what modern science considers 'science' and 'non-science'. It is a fight against 'corporatization' and for the principle of democratic access. These questions are closely and deeply interlinked. The fight to make the world genetic resources the common heritage of mankind for the benefit of all peoples, the fight against narrowing the genetic diversity of our plants and foodcrops and thus against mono-cropping, mono-cultures is thus a fight against larger processes of modernization. Processes that are relentlessly engaged in trying to wipe out diversity among cultures, knowledge systems and intellectual traditions. The fight is against modernization as a project, which as it is being thrust upon human civilization is nothing but a 'corporatization' of human societies, ecology, environment and nature.

Who Should Control the Seeds of the Future?

When the powerless say there is a problem and the powerful say there is not — as is the case in the genetic resources debate at FAO — the only thing absolutely certain is that there is a problem.

Pat Mooney[5]

It has long been realized that the breadbaskets of the North depend entirely on imported species of plants. By the beginning of the Second World War northern agronomists realized the 'breadbasket' of nations may be grain-rich but it is gene-poor and that western agriculture is wholly dependent upon the Third World. Over the decades Third World scientists and agricultural officials had been watching plant germplasm leave their countries. Recently, however, there has been the 'formation of an international programme intended to systematically remove this material as the threat of loss through erosion became better understood. This collection drive was led by IBPGR — an amorphous semi-UN, mostly autonomous organization that appeared both technocratic and aloof. Agricultural ministeries were becoming aware that the material they were donating had considerable value — and that it was disappearing.'[6] At the same time, as Pat Mooney notes, 'the nature of the seed industry seemed to be changing. Some of the old familiar names, Sutton's, Ohlsen's, Burpees, etc., had either vanished or been transformed by their absorption into the corporate world of transnational enterprises. The new seedsmen were tougher, more aggressive; they wanted more money for their seeds, and they were reluctant to share breeding information'. In this background, as Pat Mooney further points out, 'a sense of growing discomfort was ignited by the information that plant genetic resources were becoming a political weapon and that the outward flow of germplasm was almost entirely to the North, to the advantage of that region's high-tech genetics supply industry.'[7]

Describing these developments the *New Scientist* (6 June 1986) wrote that, 'The history of the world's plant species has leapt off the pages of botanical texts into the political spotlight' and went on to call it the germplasm war between North and South.

The struggle for the collection and conservation of plant genetic resources — and the political and commercial control of these resources is now taking place across the globe. The political debate is centred at FAO. Global responsibility for germplasm collection and conservation has been ceded to IBPGR. While there is much uncertainty about the work and effectiveness of both organizations, there is widespread agreement that time is running out.[8]

The countries of the South which have contributed most of the valuable food species, insist that plant germplasm is the world's common heritage and free to all. This communal gene bank in the view of the South would include the seed of hybrids and supercrops created in the laboratories of the North along with the developing world's own wealth of wild species. However, the industrialized countries which were so averse to the concept of the 'common heritage of mankind' in the context of the Law of the Sea work in the UN, have found 'the common heritage' concept in the context of the PGRs very acceptable. But the common heritage in the view of the North was to be applied only to the South's wild genetic resources, which the North's seed industry needs but not to the hybrids created out of these wild genetic resources. These could be got by the South if it were willing to pay the price. This willingness to pay of course ironically involves the pawning of the future of the South's food security to the North.

Both in order to protect its wild genetic resources as well as to preserve the diversity of earth's remaining genetic resources, the South has been fighting for an international gene bank and an international convention on genetic resources. The fight has also been over IBPGR, an organization over which member states have no control and which functions like a 'traffic cop directing South's genetic heritage to the North'.[9] It has hitherto remained a board 'without any membership or any legal identity, IBPGR has always donned the mantle of FAO in order to launch germplasm collection programmes in the Third World. Only because of its identification with FAO has IBPGR been permitted into many countries.'[10] The North, particularly countries with powerful profitable agribusinesses and seed enterprises, such as the US, most of Europe, Scandinavia and Australia, has all along resisted these attempts. Inspite of these attempts and the concerted efforts of private industry and agribusiness to push for plant breeder rights and the patenting of hybrids, the countries of the South have managed to make some modest gains on their FAO political battleground. In 1983 they wrested an International Undertaking on Plant Genetic Resources and the establishment of a high-level Commission on Plant Genetic Resources. According to Pat Mooney, who has been working on the issue of PGRs as well as monitoring the FAO debate quite closely, 'both the Commission and Undertaking are unassailable'. He sees the following proposals as those which Third World governments may wish to implement in the years to come. They are:

1. Expansion of the Commission and Undertaking to all forms of genetic material including animals and micro-organisms important to food and agriculture;
2. Strengthening FAO's financial support for genetic resources through the development of a World Gene Fund of US$100 million along the general lines raised by the Netherlands and Norway;
3. Renewed discussion on the development of an International Gene Bank and

a revised network under the auspices or jurisdiction of FAO and the commission;
4. A study of the opportunity for Genetic Co-operation among Developing Countries (GCDC) in the context of the expanding role of genetic raw materials in genetic engineering.[11]

Achieving these proposals is going to take some years and in relation to PGRs we are fast losing time. International seed company lobbies like the International Association of Plant Breeders for the Protection of Plant Varieties (ASSINSEL) are doing all they can to promote and protect plant breeder rights (PBRs), or basically the rights of private corporate institutions involved in seed hybridization and marketing. Their stake is increasing, considering that some of the world's leading chemical concerns are also becoming the world's seedsmen. This is because seeds, fertilizers, pesticides and pharmaceuticals have a common ground in intensive research related to genetics and chemicals. Add to this the smuggling of genetic resources from the South and the commerciogenic erosion continues. The point about commerciogenic erosion needs to be strongly underscored. As Pat Mooney says, 'This point needs strong emphasis: companies have no choice but to regard traditional cultivars as "competition" to their hybrid or otherwise proprietary (via PBR) varieties.'[12] The elimination of this competition by the elimination of traditional cultivars results in commerciogenic erosion.

Further, 'between 40 and 50 per cent of all types of living things — as many as five million species of plants, animals and insects — live in tropical rainforests, though they cover less than 2 per cent of the globe'.[13] At the same time, 'it has been estimated that 0.6 per cent of forest lands are being deforested every year: a phenomenon which is worrying in itself, since by the year 2000 a further 10 per cent of forest capital will have been eaten away'.[14] Modern science has so far managed to study only 1 per cent of the species of this fast eroding world's largest genetic reservoir.[15] We are equally aware that 'few scientists ever become as expert as the natives of the forest in distinguishing between the many hundreds of forest species'.[16] At the same time in terms of modern science and our dependency on it the situation particularly in the developing world is as follows.

According to the National Science Foundation, there are at most four thousand scientists in the whole world who are primarily concerned with tropical ecosystems. About half of these are taxonomists, whose concern is simply to name new species. One half of all tropical ecologists and taxonomists are North American and European (17 per cent in Europe, the rest in North America). Latin America has 22 per cent, as do Asia and Australia together; Africa has 6 per cent. The total number of scientific papers published each year on the environmental biology of the United States alone is greater than the total published on all aspects of tropical biology worldwide.[17]

It is in the light of such a situation, that the views of scientists like Plucknett and Swaminathan who would have us believe that all is safe in the hands of modern science and its handmaidens, the scientists, must be viewed. As Mooney comments in the context of the work of the IBPGR:

> Linnaeus would have been amazed. A science which can guesstimate the number of higher-order plant species to the nearest 50,000; a science that concedes that at least 65 per cent of the material in genebanks has not even basic passport data; a science that admits that between half and two-thirds of the genetic diversity placed in storage may have already been destroyed can still bravely tell the world that farmers and nature have produced (more or less) 110,000 wheat varieties (including landraces) and between 12,000 and 12,500 wild wheat types . . . and that 90 per cent of the cultivars and at least 75 per cent of the wild wheats are safely housed in genebanks. On the seventh day when God rested, IBPGR was obviously out counting![18]

In the face of this bewildering situation, it seems as if social scientists have also felt obliged to come to the rescue of modern science. In the June 1986 *New Scientist* article referred to earlier, Jack Kloppenburg, assistant professor of rural sociology at the University of Wisconsin, in putting forward the tired old argument that no region is genetically independent has said that no region can afford to isolate itself through 'a genetic OPEC'. He then goes on to argue that both sides in the battle over plant genetic resources are wrong. According to him, plant breeders never give away what cost them dear to perfect while at the same time wild species and those cultivated by the world's peasant farmers have intrinsic value, and like minerals or oil, should be paid for.

If one scans the media, particularly western media writing, on the PGR issue, one finds increasingly similar comparisons made between PGRs and oil. On one level, it is not quite accidental. Modern reductionist knowledge does not see an unextracted resource as a resource. There is no such thing as a resource that is a resource by not being extracted, used or made into economic value. Though such an approach has created problems in relation to resources such as oil. Oil must be explored, extracted to become a resource, but it is better to keep it below the earth or the seas for the more that is found the lower the price. Inspite of this contradictory fact even about oil as a resource, comparisons are made between oil and PGRs, though it is a well known fact that with the destruction of plant species, each disappearing plant kind can take with it anything from ten to thirty animal or insect species directly or indirectly dependent on it. PGRs are part of a complex ecosystem and are not just another resource like oil or coal. Yet, the kind of facile comparisons to oil and 'genetic OPEC' goes on, and involves a certain degree of mischief. I used the word 'mischief' deliberately, because it is mischief to suggest that all governments or scientific bodies in the Third World know the full extent of their PGRs and their economic value and are trying to form a cartel. Anyone knows this is not true and only creates many misleading conclusions. It is totally impossible to give any economic value that would have any meaning now or in the future to a very important and basic part of the ecosystem that we live in, like genetic resources.

In this complex background at a global level of the struggle to prevent the privatization and corporatization of PGRs and the controversy as to who should control the genetic richness and diversity in the Third World, we have to turn our attention to the issue of HYVs or hybrid seeds and their implications for growing more food, alleviation of rural poverty and related concerns in the Third World.

Introduction of New Seeds and Varieties

The controversy around the preservation of PGRs and who should have the right to exploit them is of course linked to the production of hybrid seeds. Not only is seed business becoming big money, but the growth in use of these hybrid seeds is also linked to the growth of the fertilizer and pesticides industry. As Lester Brown, once president of the Rockefeller Foundation has noted, 'Fertilisers is one item in the package of new inputs which farmers need in order to realize the full potential of the new seeds. Once it becomes profitable to use modern technology, the demand for all kinds of farm inputs increases rapidly. And only agribusiness firms can supply these new inputs efficiently.'[19]

Even though Third World Countries, particularly those in Asia, enthusiastically embraced the Green Revolution, especially to ease social tension caused by widespread food shortages, it was clear whom the Green Revolution was intended to benefit. Pat Mooney identifies them: 'The Green Revolution has been undeniably profitable for agribusiness. By the sixties, agricultural enterprises were in need of a new market to maintain their growth. Bilateral and multi-lateral aid program made expansion into the Third World financially possible. Twenty years later, major agrichemical firms have achieved a worldwide distribution system able to market successfully in Asia, Africa and Latin America. The Green Revolution was the vehicle that made all this possible.'[20] And the Green Revolution was launched on the promise of the miracle new seeds, the superior hybrids that modern science has created to replace the traditional variety of seeds.

The biggest implication in the use of the new seeds of HYVs has been the growth in fertilizer and pesticide production and consumption and the corresponding dependence of food production on them. Clarence Dias notes that:

By 1967 India was already paying out 20 per cent of its export earnings on fertilizers. When cheap energy vanished in the 1970s, the Third World found itself saddled with an energy-dependent agricultural system. Fertilizer shortages in 1974 resulted in a loss of 15 million tons of grain, enough to feed 90 million people. Ninety-seven per cent of the world's pesticides comes from the industrialized countries, but thanks to the Green Revolution, Third

World consumption is up 20 per cent and rising. Third World countries are increasingly finding that serious environmental problems are being imported along with their imports of fertilizers, herbicides, and pesticides and at an ever escalating cost in foreign exchange at that. The price of the seed itself has more than doubled since the early days of the Green Revolution.[21]

One could even go on to say that it was the Green Revolution and the HYVs that brought 'Bhopal' to India and very likely one cannot speak of 'No More Bhopals' without talking of putting a halt to this perverse revolution.

Apart from the fact that the Green Revolution emerged as a project of powerful corporate institutions as part of their planning for the future, what prompted the ruling elites of the Third World countries to adopt this technology? A major reason was that it initially came heavily subsidized by international, UN and other multilateral agencies.

The previously insurmountable obstacle of successfully selling a new variety to millions of smallholders — a hugely expensive marketing proposition among customers who could not afford to pay — now seemed solved as the World Bank, UNDP, FAO, and a host of bilateral aid programmes began to accord a high priority to the distribution of HYV seed. The companies could sell to their own governments or to a Third World government agency and let them bear the burden of distribution. Third World governments were prepared to heavily subsidize prices and also to force peasant farmers to buy new seed by attaching the use of 'improved' varieties to access to agricultural credit and other inputs including irrigation. From being a needful but uneconomic market, the Third World loomed as a vast and highly profitable one.[22]

All this notwithstanding, what were the internal pressures or rationale within a country for allowing these developments? The reasons were socio-structural and political. In the 1950s in almost all the newly independent countries of Asia, there was an urgent need to provide food, particularly cheap food in the cities; this was both out of a political consideration, *vis-à-vis* maintaining support for the government in the cities and an economic one of keeping wages low. The latter objective of low wages was to be achieved by ensuring cheap food prices, so that with low wages there could be higher margins on profitability to allow higher re-investments in industry. Of course this was part of the logic of industrialization and growth, a vicious cycle into which developing countries locked themselves. There exists much writing and discussion on it and this is not the place to go into it.

Efforts to achieve increased food production met with a serious structural bottle-neck. This was the highly unequal land-holding and social, hierarchical structures in the countryside which were an obstacle to more rationalized production and increase of yields both through productivity and increase in extent of land cultivated. A more basic fundamental approach would have involved a restructuring of land-holdings, a more equitable land-holding structure. However, restructuring of social class relations in the countryside has historically happened only during wars or through social movements. In newly independent countries trying to compress into three or four decades economic developments and levels of economic achievements that took almost a century in the industrialized countries, there was no political space for allowing social movements to bring more distributive justice and correct the lopsided nature of land-holdings and access to other natural resources in the rural areas. On the other hand, where such movements existed in the rural areas as in the case of the Huks in the Philippines or the Telengana movement in India there was no choice for the ruling elites but to crush them. This they did but the task of implementing land reforms remained. In India it was done partially, and it is in these areas that later Green Revolution technology was selectively targeted and introduced, and viewed from that limited perspective it was a success. Technology was seen as a way of circumventing the inherent problems and characteristics of the social structure and thus increasing food production. It was not meant to alleviate poverty and it is no accident that it accelerated and aggravated inequalities in the rural areas. This new miracle technology in agriculture seemed to hold all the answers, particularly since it came at a point in the 1960s when many countries in Asia were facing escalating social tensions, food riots and disillusionment among the populace with the general direction of economic development. The Green Revolution was embraced enthusiastically and promoted with great vigour, subsidies and all. Furthermore, the technology also provided space for top-down political 'mobilization' for its introduction and use, thus also having initially a depoliticization aspect to its introduction, in a situation (more often authoritarian and hierarchical) when all bottom-up, participatory mobilization was seen as suspect or not desirable. This was the socio-political background — at least in the Asian region — in which the adoption of the new seeds and HYV technology took place.

In terms of economics and increased food production, what is the record? Grain production did go up but at the same time required massive inputs such as fertilizers and pesticides that were getting more and more costly. So almost two decades later, independent rice scientists and economists who looked at the figures for the yields from the HYVs together with the inputs they needed, found them not as spectacular as they had been made out to be.

In relation to India for example, one analyst reports the following:

Starting from just five million hectares in 1970–71, over 18 million hectares or nearly half the area of rice has now been brought under the HYVs programme till 1982–83 . . . Therefore, this crop must have received a substantial share of the benefit of the overall increase in irrigation and the increase in the overall consumption of NPK fertilisers. However, compared to the increase in the area under HYVs and the increase in fertilisers and irrigation, the production of rice has increased to a lesser extent. During the period mentioned above [1970–71 to 1982–83], the production of rice has gone up from 42.23 million tonnes to 46.48 million tonnes. Assuming the production of non-HYVs did not experience any increase at all and all the difference in rice production was on HYVs land, we got an increase in production of about 4 million tonnes as a result of extension of HYVs programme to nearly 13 million hectares of land. In other words, an increase of 0.31 tonnes was achieved with HYV per hectare. This is a relatively small accomplishment which could have been easily achieved even without the expensive HYV programme and its infrastructure by making better use of village-based resources.[23]

Claude Alvares writes with regard to India again, that, 'A 33-number official working group headed by K.C.S. Acharya, Additional Secretary in the Ministry of Agriculture, has determined that the growth rate of rice production after the Green Revolution has been less when compared with the pre-Green Revolution period.' He goes on to add that, 'Millions of hectares of rice are now routinely devastated by BPH [brown plant hopper] and other pests — and no compensation is available to farmers who are induced to take to such "modernized" agriculture. Such pest infestations have been introduced into the Indian environment. The IRRI officials knew what they were doing, and they did it for the cheap objective of wanting to assert IRRI primacy.'[24]

In the case of the Philippines, home of the IRRI, the Rice Sufficiency Programme Campaign was launched by Marcos in 1967. Through various government information and dissemination campaigns, farmers were asked to adopt the 'miracle' rice. By 1970, 43.5 per cent of the total rice area was already being planted to HYVs and by 1968 rice imports were eliminated. But beginning already in 1971, rice production began to fall by an annual 5.3 per cent yearly resulting in rice importations again. Efforts were made again in 1974 through loans made available without collaterals to help in buying fertilizers, and in other schemes, to boost production. This was achieved till 1980, then rice production began declining again from 1981 to 1983 leading to the importation of 210,000 metric tonnes of rice in 1984 and another 100,000 just for the first quarter of 1985. Inspite of being the first recipient of advances in rice technology (or is it because it is the first recipient?) Philippines has one of the lowest per hectare yields in Asia.[25]

The picture is more or less the same for other developing countries in Asia and including other HYV crops such as wheat in countries like Pakistan. It is true that countries like Thailand, Indonesia and even the Philippines and India have or anticipate a rice surplus. But these surpluses have not entirely been due to dramatic increase in yields due to the adoption of HYV technology. The real picture in terms of production increases is similar to what we pointed out in the case of India and the Philippines. But we are not talking only about increased food production, we have to relate these figures for increased food production to other more important objectives like alleviation of hunger and poverty. The latter two objectives, especially the removal of hunger, is also the claim or stated *raison d'être* of the specialized agricultural and crop research institutions, like IRRI and others who are part of the CGIAR. When one questions their actual records or whether there are any positive implications resulting from the introduction of new seed varieties, the evidence in relation to the new technology or modernization of agriculture is really damning.

First of all, the production of food grains by this new technology which Filippino scientists refer to as the 'Cadillac' style of development (to emphasize how incongruous it is to the rural environment and to the socio-economic conditions and rice-farming practices of Filippinos), has led to decreased production of other crops which are sources of proteins, such as legumes. Clarence Dias estimates that 'half the Third World's protein comes from legumes. These legumes are being displaced both because of inability to companion-planting and also because of pressure on cultivable land. Thus, for example, per capita legume production in India dropped by 38 per cent between 1961 and 1972 because grain varieties were highly subsidized, making legumes less attractive and less profitable.'[26]

Apart from the actual physical displacement of crops as a result of HYVs, the decreased availability or even non-availability of crops other than food grains such as rice or wheat, the ability to buy even these food grains was further severely restricted due to increased pauperization caused by the new technology. Increasing poverty was simply the cause of hunger. This was true for almost all the countries of Asia. For example in the Philippines, malnutrition remains one of the top ten causes of child deaths and 'ironically enough the worst off children are those in regions which figure prominently as the country's food baskets: Southern Mindanao, where most banana and pineapple exports are grown; Central Luzon, the country's rice granary and Metro Manila the food manufacturing center'.[27] This was also the case in Pakistan, 'where agricultural growth was accompanied by increased poverty and unemployment, because of the nature of the agrarian situation into which the Green Revolution technology was introduced. It was a situation where there was a high degree of concentration of land ownership (30 per cent of farm area being owned by less than 0.5 per cent of landowners).'[28] A significant proportion of the poor peasantry suffered a decline in its level and quality of food consumption precisely during a period when overall food output was rising rapidly. One could cite one example after another in Asia.

In looking at the implications of the introduction of new kinds of seeds, plant varieties and chemical intensive agriculture, the list of negative, long-term implications seem endless. Particularly in looking at the impact in terms of costs and alleviation of hunger, we cannot escape the conclusion that in the final analysis it was a technical solution, a technological fix, to deal with problems that arose from other circumstances of social and political structures. The manner of introduction of this technological fix of course coincided with the vested interests of both internal/national actors and the external/transnational actors. In a way, this being the only consideration of those responsible for the introduction of the new HYVs and its technology, all other aspects became secondary or were not considered at all. We have already dealt with some of the consequences these HYVs have wrought upon the societies in which they were introduced. The devastating health, ecological and environmental damages caused by the chemical intensive agriculture that HYVs necessitate is yet another area. The studies and examples are so numerous and are becoming such commonplace knowledge that it is not necessary to catalogue them here. What needs to be further discussed is how to deal with the Green Revolution, how to block and reverse its effects. Doing so is the only way to do away with mass hunger and poverty in the developing world.

Where do we go from here?

In looking at the issue of plant genetic resources and its connection to the issue of seeds, the first link in the food chain and the interconnections, we have seen layers of negative implications arising out of the use of the science and technology associated with the Green Revolution. How does one cope with the situation? What kind of initiatives can be taken by social scientists, action-researchers, social action groups and by all those either directly involved with or concerned as a supportive organization for victims of 'modernization' projects such as the Green Revolution? I think the contributions in the recent ICLD publication on 'The International Context of Rural Poverty in the Third World — Issues for Research and Action by Grassroots Organizations and Legal Activists'[29] (hereafter referred to as the ICLD volume) addresses many of these issues, in particular, the editors' note on action implications of 'failed' rural development projects and the contribution by Clarence Dias and James Paul, 'Developing Legal Strategies to help Combat Rural Impoverishment: Using Human Rights and Legal Resources'. Dias and Paul have a concluding part on 'Developing Law and Legal Resources for Victims Groups', which has many very important recommendations which can be acted upon by organizations concerned or involved with issues around rural poverty. These recommendations, particularly with regard to the use of law and the building of legal resources, must be read in the context of some very valuable insights on law and the uses of law by social action groups made by Upendra Baxi in the same publication on 'Law, Struggle, and Change in India: an Agendum for Activists'.[30]

I would like to address some approaches to action on the issues of HYVs, plant varieties and chemical intensive agriculture, approaches or actions that could in broad terms be called alternatives. But first, I would like to preface my remarks with what I consider some very important perspectives provided by Reginald Green on alternatives in the ICLD volume. In particular, I share his view and would like to emphasize the dangers of looking in conspiratorial terms at all these developments in 'modern' science and technology as for example relating to the Green Revolution which we have examined.

Referring to some of the alternative approaches in relation to poverty that have emerged in the past and examining some of their limitations, Reginald Green identifies 'the tendency to slide into conspiracy theories and/or counter-simplification'. He writes, 'The justified moral outrage of some writers — for example Susan George in *How the Other Half Dies* — at the amorality leading to amoral results under the cover of technically determined, value-free public image models creates another kind of danger. The danger involves slipping into a perception of reality as a conspiracy rather than as a process of struggle. This is dangerous because it clouds analysis, especially of how holders of power (e.g. TNCs) might actually be constrained to act differently.' Contributions such as that of Susan George may have brought to the attention of a larger audience the issues involved in world hunger and poverty but they have hindered more than they have facilitated. There is a constant danger that in trying to popularize and reach to a popular audience, complex issues can be oversimplified. Reginald Green writes, 'Knowledge can be power and acquiring self-knowledge is a process which can benefit, especially in its early stages, from outside questions and comments. But this is true only so long as these questions and comments do not degenerate into overbearing rhetoric unrelated to the immediate context or become a procrustean agenda that reflects the priorities of the outsiders and not the people themselves.'[32]

Basically, I see three important elements in evolving alternative paradigms. One is a necessary, ongoing conceptual element. Two is the information/educational mobilizing element; and three is developing and sustaining the relationship to actual community initiatives *vis-à-vis* the issues thrown up. The three are interrelated and do not have any hierarchical (conceptually/theoretically) or chronological order in which they can be placed.

From a conceptual point of view, it is very necessary, particularly with regard to projects such as agricultural modernization projects, to show the totally irrelevent and artificial characterizations of knowledge made by modern science and the fake distinctions it creates between science and non-science, with regard to different traditions of knowledge. Such characterizations by modern science or knowledge systems should however be also situated within the context of the crisis that modern science or knowledge systems are currently facing. In the face of such a crisis one can then also see its inability, or rather the inability of the high-priests of modern science — the scientists — to come to grips with this crisis. We referred earlier to the response of scientists like Plucknett and Swaminathan to criticisms of the way in which modern science handles issues such as PGRs. These responses are also crisis responses and therefore our contribution must be to accentuate the internal crisis of modern science. Accentuating the crisis of modern science from its broad system level to the level of institutions and that of the individual scientist as an adherent of fragmented knowledge and ways of thinking, is also a way of strengthening different

streams and traditions of knowledge which are sometimes broadly characterized as traditional knowledge. Only then can we show the negative implications
of the assumptions such as the one that holds that modernization of agriculture
is premised on 'scientificity' and therefore constitutes a superior approach to
other accumulated knowledge and intellectual/knowledge traditions. This is of
course all part of a larger, more difficult conceptual struggle against the mono-
typing of knowledge and creating the climate and conditions for sustaining and
drawing upon alternative stocks of knowledge and achieving a legitimate status
for them.

The conceptual dimension or project is very important and also difficult
because the fight against modernization projects is done not in a sense
of a modern versus traditional framework which is a simple dichotomous
framework more akin to particular civilizational modes of thinking and to
which modern science is linked. It is necessary to view it as a 'corporatization'
project aimed at societies, ecology, environment and nature. Modernization
seen as corporatization is crucial in recognizing that corporatization is
the ultimate totalitarian project; and it is best exemplified by nuclear
power.

The informational/educational/mobilizing dimension is something that
both enriches and creates the possibility, as well as evolving out of the
conceptual dimension or project. Clarence Dias and James Paul make some
concrete suggestions for what they call strategic action campaigns:

> One step is to develop, share, and disseminate knowledge of the activity —
> the harms it has inflicted (or threatens) and the social impact and cost of
> these harms; wrongful practices associated with the activity which causes
> these harms or which, in other ways, violate the human rights of those
> threatened by the activity; law relevant to the governance and accountability
> of those who manage the activity, and legal remedies of people injured or
> threatened by it; and other measures which can be undertaken to prevent
> these outcomes (and strategies to secure them).
>
> Thus, strategic action campaigns have been directed at generating and
> sharing knowledge which different kinds of groups and concerned profes-
> sionals can use to induce more focused, reinforcing action — for example,
> the media; influential persons in differing professional circles; courts, legis-
> lative bodies, and other agencies which have powers to impose account-
> ability for wrongdoing; international organizations which can influence the
> development of law governing the activity.[33]

I now come to the third dimension, that of having a mutually creative and
interactive relationship to ongoing community initiatives. In Asia, there are a
number of initiatives at the level of rural communities, some of which are
particularly heartening since they have emerged not only in opposition to
modernization projects, but are also trying to develop real alternatives. There

have been initiatives to try and develop nationwide mobilization to promote 'sustainable agriculture'. In the case of the Philippines there have been attempts through a series of meetings of small farmers at the local, regional level culminating in a national meeting to identify root causes of the problems of rural poverty and find ways of promoting agriculture that are not overly dependent on HYVs and all the technological and other inputs that it entails.[34] In India there have been several initiatives at the local, state and national level. These initiatives counter what the editors of the Indian publication, *Lokayan*, describe as follows: 'A myth has been afloat that India has freed itself from the threat of famine through the Green Revolution. Yet the invisible, irreversible decay of our living earth through ill-conceived, inappropriate technologies is symptomized in the numbers of people leaving the land to scrape out a living in the cities.'[35] The same issue of *Lokayan* reports on three conventions, one on organic farming, one on rice and indigenous rice varieties and one on indigenous breeds of cattle, all very important and significant efforts to, as Lokayan puts it, 'reviving sustainable agriculture'. There are similar efforts beginning in Malaysia, in Thailand and other countries.

In many of these countries there are also emerging groups that are concerned about the issue of seeds and the whole question of genetic resources. It is true that some of these activities tend to be narrowly focused on certain issues and have a kind of repetitive campaigning that tends to lull rather than activate. But these fragmented types of activities are also taking place side by side with, and are being influenced by, the activities of ecological and environmental groups who are increasingly trying to develop holistic approaches.

We stand at the crossroads of some very exciting possibilities in working against the evils of modernization as corporatization. Activists, action researchers and concerned scholars from law, people's science, ecology, developmental activism, social research have the possibility of moving away from fragmented issue-oriented activism to not only developing holistic approaches to the problems of the poor and marginalized but also to generating new knowledge, strengthening useful aspects of traditional knowledge systems and thus contribute to what I would like to call alternative stocks of knowledge — alternative to the mono-typing of knowledge by reductionist modern science and technology and thus are an important aspect in the preservation of diversity. This must always be at the heart of any efforts taken to deal with modernization processes that seek to wipe out genetic diversity, for as Pat Mooney says, 'The diversity of agriculture and human culture are bound together.'

Notes

(Much of the information with reference to the background section of the paper is drawn from the Open Letter on Plant Genetic Resources circulated by Clarence Dias and Upendra Baxi and from a popular article that the author of this paper published on PGRs in the *Far Eastern Economic Review*.)

1. Pat Roy Mooney, 'Law of the Seeds — Another Development and Plant Genetic Resources,' *Development Dialogue*, **12**, 1983, Dag Hammarskjöld Foundation, Uppsala, Sweden.
2. Pat Roy Mooney, 'The Law of the Lamb', *Development Dialogue*, **1**, 1985.
3. Claude Alvares, 'The Great Gene Robbery', *Illustrated Weekly of India*, 23 March 1986, Bombay.
4. See Catherine Caufield, *In The Rain Forest*, Picador, London, 1986; and also *New Scientist*, 6 June 1986.
5. Mooney, 'The Law of the Lamb'.
6. Mooney, 'Law of the Seeds'.
7. *Ibid*.
8. *Ibid*.
9. *Ibid*.
10. *Ibid*.
11. *Ibid*.
12. *Ibid*.
13. Caufield, *op. cit*.
14. *The Vanishing Forest — The Human Consequences of Deforestation, A Report for the Independent Commission on International Humanitarian Issues*, Zed Books, London, 1986.
15. *Ibid*.
16. Caufield, *op. cit*.
17. *Ibid*.
18. Pat Roy Mooney, ' "The Law of the Seed" Revisited: Seed Wars at Circo Massimo,' *Development Dialogue*, **1**, 1985, Uppsala.
19. Lester Brown, *Seeds of Change*, Pall Mall Press, London, 1970 (quoted in Ibon Facts and Figures, no. 163, Manila, 31 May 1988).
20. Mooney, 'The Law of the Seed'.
21. Clarence Dias, *Reaping the Whirlwind: Some Third World Perspectives on the Green Revolution and the 'Seed Revolution'*, The International Context of Rural Poverty in the Third World, International Center for Law in Development, New York, 1986.
22. Mooney, 'The Law of the Seed'.
23. Bharat Dogra, quoted in Claude Alvares, *op. cit*.
24. Alvares, *op. cit*.
25. Ibon Data Bank, Facts and Figures no. 163, 31 May 1985, Manila. Also study on HYVs and Green Revolution in the Philippines (forthcoming), ACES Foundation, Manila, Philippines.
26. Dias, *op. cit*.
27. Ibon Data Bank, Facts and Figures, no. 195, 30 Sept. 1985, Manila.
28. Akmal Hussein, 'Economic Growth, Poverty and the Child', paper presented to the Harvard Law School Conference on 'Who Speaks for the Child', 11–12 April 1986.
29. International Center for Law in Development (ICLD), 'The International

Context of Rural Poverty in the Third', *World Issues for Research and Action* by Grass Roots Organization and Legal Activists, New York, 1986.

30. Upendra Baxi, in publication cited above.
31. Reginald Green, ICLD Publication, *op. cit.*
32. *Ibid.*
33. Clarence Dias and James Paul, ICLD Publication, *op. cit.*
34. Aces Foundation, Bigas Conference Report, UP at Los Banos, Manila, August 1985.
35. *Lokayan, Special Double Issue: On Survival*, New Delhi, October 1985.

12

Science and Development

Trends and Outcomes of the Transfer of Technology in the 1980s

David Burch

As with most issues in the social sciences the question of aid and its meaning to the donor and the recipient is characterized by sharply diverging views. At one extreme, there are those observers who argue that aid, whether bilateral or multilateral, involves a transfer of real resources from rich countries to poor, and represents a foregone opportunity for consumption or investment in the developed countries in order to generate development in the Third World. At the other extreme there are those observers who argue that aid programmes largely serve to advance the interests of particular groups within the donor country, and only secondarily selected groups within the recipient country. Seen from this perspective, aid may only serve to further exploit Third World countries by creating indebtedness and a technological dependence, in return for new resources which confer benefits on the few and costs on the many.

While some validity may attach to each of these extreme positions, I suggest that it is the latter position which is the most persuasive when it comes to explaining patterns of allocation of aid and the distribution of benefits which flow from such allocations. In order to substantiate this argument, I propose to document the major determinants of British aid policy over time (and that of other countries where relevant), and to undertake a general analysis of the distributional implications that flowed from such policies. In particular, the study will focus upon the impact that recent British aid policy has had on the choice and transfer of technology, and the acquisition (or loss) of technological capabilities by the Third World.

A full understanding of contemporary issues cannot be achieved without the historical perspective which, by exposing and explaining continuities and/or discontinuities in policy, enables the researcher to establish generalizations about the nature of the social world. Accordingly, the following section considers the historical background to Britain's aid programme.

Historical Development of British Aid

In considering the historical development of British aid, it is possible to distinguish four distinct phases of policy, each of which held different implications for the choice and transfer of technology (Burch 1987). The first phase commenced in the early 1920s and lasted up until the beginning of World War II, during which time aid programmes were initiated as a response to the mass unemployment and depressed economic conditions experienced during the 1920s and 1930s. Under the terms of legislation passed by Britain at this time — the Trade Facilities Act of 1921, the East African Guaranteed Loan Act of 1926 and the Colonial Development Act of 1929 — finance was made available for the purpose of funding 'development' projects in the colonies which would result in the stimulation of economic activity in key sectors of the UK economy.

As a consequence, aid in this period was dominated by a major programme of colonial railway building, mainly in East and Central Africa, which was designed to stimulate the employment of labour and capital in Britain's steel and heavy engineering sector, as well as to bring colonial producers of low-cost raw materials in closer contact with their UK markets. The first of these programmes, financed to the tune of £6.6 million under the East African Guaranteed Loan, was openly justified by the fact that,

> . . . approximately half the capital sum would be spent in Great Britain, on rails, bridging materials, rolling stock, etc, which at this time would provide work for the engineering industries of Great Britain and so lessen unemployment charges (Comd. 2387, 1925: 182).

The significance attached to colonial railway building as a means of stimulating economic activity in Britain can be gauged from the fact that some 60 per cent of the total of £17.9 million made available under the East African Guaranteed Loan of 1926 and the Colonial Development Act of 1929, was allocated to railway construction. But of course, it was by no means inevitable that aid policies designed to favour British domestic interests would also have beneficial effects in the colonial territories where projects were established. In fact, policy at this time imposed significant costs on the colonial territories, adding a financial burden at a time of serious economic hardship. Railways seldom paid their way and usually involved the colonial territories in substantial interest repayments and operating losses, thereby drawing off resources from other important projects and programmes.[1] Moreover, the commitment to capital-intensive technology implied by the choice of railways rather than roads (Kenya spent only £60,000 to £80,000 per annum on roads during the 1930s) meant that development could be concentrated only in a limited number of regions. The areas around the railway could be intensively developed, while those further away were starved of resources because of the drain imposed by the choice of railways over other alternatives. Needless to say, railways as a form of transport were — and still are — exclusive, in the literal sense of that term. Most roads, built to serve the interests of one particular group, may also be used to advantage by other groups; railways, on the other hand, put a price on access that would exclude the poor in many instances. Finally, there was the creation of a technological dependency in the expansion of the colonial railway system. The technological capability required to manufacture and maintain railways was — and still is — absent from most of the peripheral countries, a fact which did much to assure future markets for metropolitan producers.

All of these consequences — the technological dependence, the concentration of resources in areas close to the railway and the relative neglect of those further away, the patterns of regional imbalance that flowed from this and which are today reflected in social inequalities along a number of measures — could have been avoided. But, as Brett (1973: 297) has argued:

> . . . capital was only forthcoming on terms determined by metropolitan interests, who benefited from the use of capital-intensive technology which provided employment of British workers, rather than the labour-intensive technology which might have been used to provide a territory-wide programme of rural animation based on local road-building, agricultural extension and research.

In other words, the choice of technology at this time, dominated by the need to stimulate the employment of men and machines in Britain, led to a capital-intensive strategy of colonial 'development' and precluded the adoption of an alternative, labour-intensive strategy based on 'development from within', which might have stimulated a greater number of new and different economic activities leading to an increased emphasis on local technological capabilities.

The second phase of Britain's aid policy lasted from the end of World War II up until the mid-50s, and although British interests were again uppermost in the aid policies formulated over this period, such policies were dominated by quite different considerations from those that had prevailed in the 1920s and 1930s. In the immediate aftermath of the 1939–45 conflict, post-war reconstruction of a shattered and deeply-indebted economy was the top priority for Britain and indeed, all of Europe. Consequently, aid policy at this time was determined by the need to assist in this reconstruction, by encouraging increased colonial production of low-cost (and underpriced) consumption goods and raw materials, which could either be used directly by Britain or which could be sold on the world market in order to earn the scarce US dollars that Britain needed to re-equip its domestic industry and repay its war loans.

The problem for Britain was that increased colonial production implied increased capital investment, and since capital was the factor of production in short supply in the post-war years, Britain was reluctant to make available to the colonies the capital resources required for domestic regeneration. Under these circumstances then, colonial 'development' could only take place on the basis of labour-intensive techniques, and in the second phase of British aid there was a tendency to reverse the priorities which had been established in the first phase. For example, under the Colonial Development and Welfare (CD & W) Acts which gave effect to post-war aid programmes, the largest single item of expenditure in the ten years 1946–56 was road construction, which was given clear precedence over capital-intensive railway development in the provision of communications and infrastructure facilities. In fact, only 0.1 per cent of CD & W funding totalling £1,200 million was spent on railways over the period, compared with 16 per cent allocated to roads.[2]

However, this shift in emphasis was not symptomatic of a general move towards a strategy of colonial development based on labour-intensive technologies, but was a temporary expedient adopted by Britain as a way of recruiting the colonial territories into the process of British post-war reconstruction, without the necessity of making available large amounts of scarce capital. That this was the case is clearly demonstrated by the policy changes which occurred from the late 1950s, as the problems of post-war reconstruction were overcome and new issues emerged, which were to give rise to the third phase of British aid policy.

The problems faced by Britain from the mid-1950s were many and varied; there were the difficulties of adjusting to the resurgence of Japan and West Germany as industrial competitors, especially in areas such as shipbuilding and engineering which had been traditional British strengths; there was the dismantling of the imperial economic system which, as a consequence of long-standing US pressure, resulted in the full convertibility of sterling against the dollar, and allowed countries and territories locked into the restrictive sterling area to establish new trading relationships; there was a growing movement for political independence among the colonial territories, and the chance once again to establish new trading relationships which did not focus on the old imperial relationship; there was the decline in commodity prices over the 1950s, which reduced the purchasing power of the colonial territories and led to fewer imports of manufactured goods from the UK; and there was the continued use of sterling as a major trading currency, which rendered the British financial system (upon which industrial strength depended) prone to sharp fluctuations and uncertainties as confidence in sterling weakened.

The result was a growing lack of competitiveness and a substantial reduction in Britain's share of world and sterling area markets. This resulted in increasing levels of surplus capacity and growing unemployment in industrial sectors, especially in the depressed regions in the north of England, Scotland and Northern Ireland. For although the loss of Britain's market share was reflected in most industrial sectors, a number of industries, often associated with the depressed regions, fared especially badly, i.e. shipbuilding, aircraft and railway locomotives and rolling stock. The railway industry was the most seriously affected, losing nearly half of its total share of world markets between 1954 and 1962 (Beckerman *et al.* 1965).

Once again these issues began to be addressed directly in Britain's aid programme, and increasingly policy was oriented towards the use of aid to absorb unemployed resources of labour and capital. Numerous policy documents and statements alluded to the belief that this was a justifiable use of Britain's aid resources, and the content of British aid began to reflect these concerns. Priority areas for inclusion in the aid programme were shipbuilding (accounting for some 10 per cent of known British capital aid expenditure over the period 1966–72), railway locomotives and vehicles (7.7 per cent of known British capital aid expenditure 1966–72), and textile and leather machinery (5.0 per cent of known British capital aid expenditure). Other sectors, although drawing on a range of 'newer' industries, were also given priority when, at particular points in time, they faced difficult trading conditions and were threatened with declining markets and significant job losses, e.g. steel plant and equipment, electrical power machinery and switchgear, fertilizers, tractors (Burch 1987).

Once again then, the composition of British aid came to accommodate the needs of British domestic interests, and in the process resulted in priority being given to the transfer of capital-intensive technologies. The example of locomotives and railway construction is particularly interesting since it provides strong evidence of the continuities in British aid and the recurrent values and interests which, even under conditions of change in the composition of aid, shaped policy over time.

It should not be assumed that the British experience was in any way unique; it is clear that Britain stands as an exemplar of the patterns and programmes adopted by other industrialized countries. Thus the Depression years of the 1920s and 1930s not only laid the basis for Britain's initial aid programme, but also provided the impetus for the creation in 1934 of the US Export-Import Bank, an institution designed to fund US exports and to stimulate activity in depressed sectors of the domestic economy (Feinberg 1982). Furthermore, in the post-war period, all of the European powers with colonial territories established 'development' programmes aimed at post-war reconstruction in the developed centres. Similarly, Britain was not alone in increasing its aid and redirecting its programmes in the late-1950s as a means of maintaining old markets and penetrating new areas in order to sustain and expand domestic economic activity. With similar goals in mind, Japan and West Germany both introduced new aid and export credit programmes at this time (White 1965; Little and Clifford 1965), while in 1959 the US government introduced stringent new measures to tie its aid to the purchase of US-produced goods. The result of this particular move was quite dramatic; the percentage of US aid spent within the US increased from 40–50 per cent in the period before 1959, to some 90 per cent in 1966. Moreover, in the period 1959–65, the US increased its exports to fifty-one countries by $2,160 million; of this amount, $1,260 million (60 per cent) was attributable to US aid, while by 1965, nearly 5 per cent of all US exports were funded by the Agency for International Development (AID). In the case of particular commodities, this percentage was even higher; in 1965, some 30 per cent of fertilizer exports, 24 per cent of exports of iron and steel products, and 30 per cent of exports of railway transport equipment were financed by AID funds (Hyson and Strout 1968). It is no coincidence that these sectors were also priority areas for UK aid.

In summary, it seems clear that to a large extent the composition of bilateral aid programmes was determined by the need to solve particular problems in the donor country, rather than to generate 'development' in the Third World. In this process, particular preferences came to dominate the choices of technology that were made available at any particular point in time. In most instances, such choices would have coincided with the interests of one or the other of the dominant social groups within a Third World country, which is why these issues have apparently generated little conflict between donor and recipient. But the consequences for the poor of the Third World were often very different, and were frequently devastating. Nowhere perhaps, is this more dramatically revealed than in the case of the Green Revolution, and the growing commitment to agricultural mechanization that accompanied this programme. The widespread introduction of tractors, stimulated by the demands of large farmers in the periphery and the availability of tractors under aid, created widespread unemployment amongst agricultural labours in the Third World, and rendered many small farmers and tenants marginal to the new system of capital-intensive agriculture (Burch 1987).

It should be clear by now that this historical review of British and other aid is of more than academic interest, but is of critical importance for an understanding of the contemporary developments which culminated in the major global recession now being experienced. This recession has initiated the fourth and latest phase in British aid policy, resulting in a significant shift in the composition of aid as Britain's programme was redefined in order to protect domestic interests in a period of stagnation, uncertainty and weakness.

Overseas Aid and Global Economic Recession

Since the early 1970s, the world economy has become increasingly unstable and unpredictable. The collapse of the Bretton Woods Agreement in the early 1970s; the greatly increased cost of oil from 1974 and 1978; the growing condition of depression and the adoption of protectionist measures by both developed and less-developed countries; increasing long-term unemployment arising out of depressed economic conditions and the introduction of labour-saving production technologies; the growing indebtedness of many countries of the periphery and the likelihood of major defaults in repayments to private lenders in the developed centres as oil and other commodity prices collapsed and external markets for manufactured goods contracted; continuing high rates of inflation in certain countries and a persistent weakness and instability in key currencies — were all factors that confirmed that the long period of post-war growth and relative stability had come to an end, and that the world economy had entered a critical phase.

Under these circumstances it was inevitable that there would occur significant changes in the nature of the relationship between aid donors and aid recipients, although the thesis that official resource transfers from the developed to the underdeveloped countries have largely been replaced by official flows has, according to some sources, been greatly exaggerated (DAC, 1983). But of course, there have been other significant changes in the source of aid flows and the terms upon which aid has been made available, although it is debatable whether, in overall terms, such changes have worked to the benefit of the recipient countries. It is arguable, for example, that changes in policy initiated by the donor countries, e.g. the growth of export credits, have been designed more to assist the developed centres weather the worst effects of the recession, rather than help the worst affected underdeveloped countries in a period of economic stagnation and decline.

This is naturally not the view that is adopted by the developed capitalist countries which provide the bulk of official development assistance to the Third World. Instead, in a re-affirmation of the old assumption that the relationships between the centre and the periphery are not in conflict or antagonistic, western capitalist nations have argued that the current crisis could be resolved only by a recognition of a mutuality of interest and interdependence between the North and the South.

This position was the standard western response to demands from the peripheral areas for a New International Economic Order, and was a central theme in the report of the Independent Commission on International Development Issues (1980: 66–8, 617), also known as the Brandt Commission. This report argued that growth was necessary in the industrialized North in order to expand the markets for southern goods and create a suitable climate whereby the North could assist the South. At the same time the North needed the South, and especially the markets for the manufactured goods of the North, upon which so many jobs depended. Of course, the key problem was finance, and the Brandt Commission argued for the provision of a large volume of financial support, both private and public. The recycling of oil revenues in the post-1974 period had served this function and had done much to mitigate some of the worst effects of the depression. However, there was growing evidence that the recycling process of the 1970s would not be repeated, and that governments would need to take positive measures to ensure that the flow of capital to the peripheral areas was maintained.

There were several interrelated and contradictory issues associated with this course of action. For example, a large number of the peripheral countries had become deeply indebted to multilateral institutions such as the World Bank, and to public and private lenders in the metropolitan centres over the late 1970s, and had, by most criteria, become bad risks when it came to further injections of funds. But clearly, without such transfers the peripheral countries would not be able to purchase the manufactured goods which they required, and which the centre needed to sell. In addition, a policy for the massive transfer of funds to the periphery in order to stimulate economic activity in the centre ran counter to the policies individually embraced by many western countries as a means of reducing the impact of growing inflationary pressures throughout the 1970s, e.g. reduced public expenditure. However, the Brandt Commission argued that the problem of inflation generated by the transfer of resources could, under certain circumstances, be contained. Again, this implied a conscious attempt to link aid to the underdeveloped countries (UDCs) to the absorption of surplus capacity in the industries of the developed centres:

> . . . export orders from developing countries would not be as inflationary for the North as demand generated domestically by public expenditure, if these orders went to sectors of industry which have excess capacity (Brandt Commission 1980: 67–8).

And as part of its recommendations for an immediate programme of action the report suggested that:

> . . . the present predicament of the world economy can be resolved only with a major international effort for the linking of resources to developmental needs on the one hand and the full utilization of under-utilized capacities on the other (Brandt Commission 1980: 254).

This of course was what Britain and, increasingly, other donors had long practised in their aid programmes, but in the context of the 1980s, this policy took on a new urgency and significance. As a consequence, Britain and other developed countries undertook a number of new initiatives in aid policy in order to counter the possible loss of Third World markets and a worsening of the economic depression. Taken together, these initiatives facilitated an increase in real aid resources whilst avoiding inflationary pressures, and offset the cost of concessions by ensuring that the commercial returns to western companies were enhanced even further.

The first of these initiatives involved a measure of debt-forgiveness on past loans under aid which implied, amongst other things, a modest contribution to the recycling problem. Britain initiated this policy when in 1978 the government cancelled around £1,000 million of outstanding debt, involving repayments of some £60 million per annum, owed by seventeen of the poorest nations of the periphery. However, the cost of cancellation was to be met out of the unallocated sums expected to arise out of the future growth of the aid programme, in effect converting loans into grants. Although it was expected that this would marginally reduce the commercial benefits to the UK in the short-term, the financial institutions in the City of London welcomed this action, since by shifting to the UK exchequer the cost of servicing the official debt of the poor countries, it helped to ensure that the private debts incurred by these countries could be repaid. A number of other countries followed Britain's lead, including West Germany (£1,100 million debts written off), Japan (£240 million) and Switzerland (£53 million) (*The Times*, London, 13 July 1978, 1 August 1978, 4 August 1978 and 26 September 1978).

The other major initiative revolved around the rapid growth of export credits, and the practice of combining low-interest export credits with loans or grants from a donor country's aid programme, i.e. the offering of 'mixed credits'. Western countries have, for many years, made available low-interest credits as a means of enabling Third World customers to purchase capital and other goods over an extended period. In general terms, developed-country governments made available subsidies which constituted the difference between prevailing market rates and a fixed-rate of interest provided at less than the market rate to certain overseas customers. In order to avoid cut-throat competition between the industrialized countries as they sought much-needed overseas orders in a period of world economic recession, the member countries of the Organization for Economic Co-operation and Development (OECD) introduced in July 1976, a 'consensus' arrangement under which governments established maximum lengths of credit and pre-delivery payments, and a fixed common minimum rate of interest. However, partly as a means of circumventing these agreements and stealing a march on competitors by further subsidizing the purchase of capital and other goods by the Third World, the practice of combining subsidized export credits with overseas aid emerged. The concept of mixed credits was pioneered by France in the mid-70s, and such facilities are now widely used by most aid donors. The available data up to 1983 revealed that project financing packages involving the use of mixed credits amounted to an average of US$3.5 billion per annum between 1981 and 1983. France accounted for 45 per cent of this total, followed by Britain (22 per cent), and Italy and Japan (some 9 per cent each). Eleven other countries offered mixed credit deals in the period 1981–83, and although these involved relatively small amounts, it is anticipated that the practice will grow as more and more countries turn to such programmes as a defence mechanism. As the world recession led to a reduction in lucrative contracts from Third World countries, the competition amongst western producers intensified, with the result that the aid component in mixed credits rose from 27 per cent in 1982 to 37 per cent in 1983 (*South*, No. 52, February 1985).

The use of mixed credits has been criticized on a number of grounds. For example, from the perspective of the industrialized countries the policy may be self-defeating in that the competitive use of such credits is very costly and merely leads to bids and counter-bids by developed country export agencies which cancel each other out. Of course, this may have some benefits to the

Third World, in so far as the underdeveloped countries may gain access to capital goods on very favourable terms. However, even if this is the case, it holds true only for the range of middle-income UDCs, which are credit-worthy in the first place, and the poorer, less credit-worthy UDCs may see a reduction in aid resources as concessional finance is increasingly diverted into mixed credits. Moreover, despite the existence of interest rate subsidies and a grant-like component in export credit programmes, the overall effect may involve an increased flow of non-concessional finance to the Third World, and a consequent increase in debt burden and loan repayments.

The question of mixed credits raises other important issues, in particular those relating to the transfer of technology and the impact of new patterns of capital goods exports on issues of importance in the domestic area of Third World countries, e.g. indigenous technological capabilities, employment levels and patterns, income distribution, the appropriateness of technology, etc. For the growth of export credits and their increasing use in combination with aid has not simply resulted in the export to the Third World of 'more of the same'. Rather, the use of mixed credits has been accompanied by a shift in the technological composition of bilateral aid programmes, which arises directly out of the growing problems encountered by donor country interests in a period of recession. In order to analyse this further, it is necessary to return to a consideration of Britain's aid programme in order to evaluate the nature of the interests which come to be reflected in these changing patterns of allocation.

British Aid in the Fourth Phase: World Recession and the Growing Commercialization of Aid

While it is clear from the earlier analysis that British aid has long been dominated by commercial considerations, the growing problems in the world economy in recent years have resulted in an even greater emphasis on the creation of direct linkages between aid and trade. At the same time, there has occurred a shift in the sectoral composition of aid, away from the provision of support for traditional industries in a process of long-term decline (e.g. shipbuilding, textile machinery, etc.), towards the maintenance of a capacity in those strategic industrial sectors which are believed to be necessary as the basis for future industrial growth in Britain (e.g. civil engineering and large construction companies, the capital goods industries involved in the production of process plant equipment and heavy electrical plant, telecommunications apparatus, etc.).

The origins of this shift are to be found in the poor performance of British industry over the 1970s and the adoption of policies designed to establish even closer links between aid and trade. Over the period 1973–76 UK industrial production declined by 8 per cent, the rate of profit fell sharply, and the collapse of a number of secondary banks pointed to a growing financial instability. Britain's share of world trade declined still further, and had it not been for short-term capital inflows from OPEC countries, the UK balance of payments would have been in deficit to the tune of £3.6 billion in 1974 (Aaronovitch and Smith 1981: 83, 197–206). Unemployment grew to 7 per cent of the workforce by the end of 1977 and, of course, it was to increase still further as the crisis deepened. References to the 'de-industrialization' of the British economy began to be heard, and in an attempt to counter this decline, the Labour government of Britain adopted its industrial strategy in 1975. This policy involved the injection of large sums of public money and some notional planning in an attempt to revitalize and re-equip Britain's declining industrial sector (Aaronovitch and Smith 1981: 124–5).

In supporting the industrial strategy, the Ministry of Overseas Development had a particular role to play through the introduction of the Aid-Trade Provision (ATP). Under the original term of this provision,

> The government agreed to set aside 5% of the bilateral aid programme for development projects which are also of particular importance to our own exports and industry, and which will provide employment where it is needed in this country. The object is to give greater weight to these factors in proposals which, if looked at purely in terms of development would be worthwhile, but might otherwise be squeezed out of the programme by others of even higher priority (Overseas Development Paper No. 17, 1978).

In 1979, for example, this implied that some £31 million of aid money was made available in direct support of UK industry. However, in practice, the actual levels of support were much greater than this because the ATP often came to be used as the aid component of a package of mixed credits which, as will be seen later, gave a multiplier of four or five.

In many ways, however, the allocation of 5 per cent of aid-funding to the ATP did little more than formalize Britain's existing commitment to the use of aid to increase the export opportunities of domestic industries, and it was the election of the Conservative government which was to elevate commercial and political criteria to even greater dominance in the allocation of aid.

Paradoxically though, the government began by reducing the size of the aid programme as a contribution to reduced UK public expenditure and lowered inflation. The aid budget was cut by 6 per cent in 1980, and further cuts totalling 16 per cent in real terms between 1981–84 were proposed (*The Sunday Times*, London, 17 February 1980; *The Times*, London, 6 February 1981). However, not even the government of Mrs. Thatcher could totally abandon its support for domestic private capital, especially in a world in which British industry — competitive or not — was being strongly challenged in overseas markets by the aid and export credit schemes of its competitors.[3] Accordingly, the Conservative government removed the 5 per cent upper limit of the ATP so that in theory the whole of the aid programme could be used to subsidize exports; in addition, an unallocated margin of 3 per cent of the total aid budget — traditionally held back in order to assist with disasters and unforeseen emergencies — was increased, and was also to be made available for programmes or projects funded for commercial or political reasons. It has also become clear that in recent years the normal aid programme is increasingly being evaluated and disbursed on ATP-like criteria (*The Times*, London, 12 November 1981; Elliot, 1982: 14). Thus, by 1981–82, while the ATP formally accounted for some £53 million (8 per cent) of net bilateral aid, at the same time there was an actual commitment of some £198 million aid funds to projects based on ATP criteria. This constituted just under 30 per cent of Britain's bilateral aid programme totalling £674 million in that year. More to the point, direct expenditures under the ATP represented only a fraction of the value of orders generated. The British government's own data reveals that over the period 1978–82, ATP expenditures of £229 million generated export orders valued at £1,143 million, giving a multiplier of five (*COI Survey*, January 1984: 12–13). These returns would have been further increased by recent attempts on the part of the British government to exert greater control over its contribution to the multilateral agencies, thereby making this component of the UK aid programme as responsive to the needs of British industry as the bilateral programme (*Sunday Times*, London, 1 November 1981 and 20 March 1983).

Even these measures appeared to be unable to cope with the general problems of British industry in the 1970s and 1980s, or with the particular problems associated with those strategic sectors which, in the rest of Europe, represented a potential growth point, i.e. the process plant contractors, the producers of heavy electrical equipment, the civil engineering and construction sectors, etc.[4] As a result, British aid increasingly came to be focused on a narrow range of capital goods, as the UK government attempted to assist important components of the industrial base in a period of recession and intense competition.

The Commercialization of Aid and the Transfer of Technology

The continued existence and growth of the process plant and related sectors was of obvious significance if Britain was to arrest the process of industrial decline. For example, the Davy Corporation was considered to be so strategically important, especially in the extent to which it generated orders for hundreds of sub-contractors, that in 1981 the 'market-oriented' government of Mrs. Thatcher intervened to block a take-over bid by the US-owned Enserch Corporation (*Guardian*, London, 17 April 1984, 21 January 1984 and 1 August 1984).

As a consequence, these companies and the industrial sectors in which they operate have clearly been singled out for special treatment under Britain's aid programme. That this should be so is hardly surprising. The process plant sector alone accounted for around 100,000 jobs in Britain's manufacturing sector, and in 1983 had a turnover of £3 billion (*Guardian*, London, 17 April 1984). At the same time though, it was clear that this sector had long displayed considerable weakness, made worse by the recession of the 1970s. In 1976, for example, the director of the Process Plant Association was warning that

10,000–15,000 jobs in the heavy electrical equipment industry would disappear if no new orders for power stations were forthcoming. At the same time, it was noted that one company, Babcock and Wilcox, had already informed 10,000 of its employees in its power and process engineering group that redundancies were possible (*The Times*, London, 17 August 1976).

In an attempt to generate export orders for this sector, the British government (among other things) established the Overseas Projects Board (OPB), an offshoot of the British Overseas Trade Board, with a membership largely drawn from the process plant and civil engineering sectors. The aim of the OPB was to advise an export policy for large-scale projects in the process plant sector, and to offer financial assistance and support to this sector (*The Times*, London, 19 July 1978).

Despite such measures, and despite the introduction of subsidized export credits which were so vital to this sector, the threat to the employment of labour and capital remained and even worsened. By the early 1980s, the situation was critical; work-in-hand in the process plant sector fell from £4 billion in 1980 to £2.9 billion in 1982, while new orders fell by over three-quarters, from £2.6 billion to £629 million. Even with the large volume of aid and cheap export credits made available between 1981–84, the process plant sector still experienced a 50 per cent fall in the value of export orders over the period, while the UK share of overseas markets fell from 8 per cent in 1980 to 6.8 per cent in 1981. Perhaps of greater significance was the fact that UK-manufactured content for these overseas projects fell from 72 per cent of the total in 1980 to 27 per cent in 1982 (*Guardian*, London, 21 January 1984; 17 April 1984; 1 August 1984, 5 September 1984, 26 September 1984, 7 November 1984 and 5 January 1985).

Under such circumstances, the UK government was bound to accord this sector a high priority, and it has been calculated that in 1981–82 alone, the combination of aid under the ATP and cheap export credits represented a direct subsidy of £640 million to exports from this and related sectors, with three-quarters of this sum being received by only a dozen or so larger companies (NEDO 1981: 4; Byatt 1984; *Financial Times*, London, 1 February 1984).

However, under the conditions prevailing at the time it was frequently alleged that the scale of the financial commitment provided by the UK government was still not enough to ensure that British manufacturers were able to compete fairly with their industrial competitors (*The Times*, London, 6 September 1983, 18 April 1985 and 26 July 1985). Consequently, the 1980s have witnessed numerous changes in policy as the government attempted to come to terms with rapidly changing circumstances. For example, in response to claims by British manufacturers that they were hampered in their negotiations on overseas contracts by long delays in arranging a package of mixed credits, and by the fact that they could respond to events and claim such support only by producing evidence of the concessions made by foreign competitors, the Treasury laid down new guidelines on the operation of the ATP which allows companies to offer financing deals involving the provision of mixed credits at the initial stage of a contract, when the competitive bidding starts. In total, some 45 per cent of the ATP fund was made available for this purpose. (*Guardian*, London, 2 November 1984 and 7 November 1984). In addition, in response to claims of unfair competition which resulted in British companies losing a number of major contracts, e.g. the construction of a bridge over the Bosporus and the re-equipping and re-construction of Bangkok's bus services, the government announced an increase in the amount allocated under the ATP, from £66 million in 1985–86, to around £86 million by 1988–89. This, it was hoped, would double the volume of business generated by the ATP. At the same time, the government introduced measures which allowed ATP funds to be used as a means of softening the terms upon which commercial bank loans could be made available to credit-worthy Third World countries (*The Times*, London, 13 November 1985).

Such changes have also had major implications for the exercise of control over aid funds. For example, it should be noted that under the ATP, the initiative for the allocation of aid now comes not from the recipient country, or the Overseas Development Administration (ODA), but rather from an industrial or commercial concern which believes it needs assistance to win an export order. Moreover, the approach by a firm is made not to the ODA, but to the Department of Trade and Industry. As a consequence, the Overseas Development Administration is effectively by-passed, since the DTI not only acts as the first point of contact for industries seeking aid funds to support overseas projects, but is often the final arbiter of whether or not such funds are made available. In this context, very little concern is shown for 'developmental' criteria or the appropriateness of any particular project. Indeed, with the new guidelines announced in October 1984 and November 1985, it seems to be the case that a large part of the process of selection and evaluation of projects has simply been handed over to private capital, as a means of underwriting their overseas marketing operations.[5]

If one looks at the projects funded under the ATP from 1979 to 1983, it can be seen that the sums disbursed to the process plant and heavy electrical equipment sectors totalled some £133 million, and accounted for some 57 per cent of total ATP funding allocated in this period. If a multiplier of four-to-five is assumed, this implied that the ATP generated orders to these sectors ranging from £523 million to £665 million. In order to understand fully the significance of such data for both the donor and the recipient country, it will be useful to analyse in a more detailed way the particular circumstances surrounding the industries selected for favoured treatment, in so far as they received financial assistance not only under the ATP but also from the normal aid programme. What follows therefore, is a series of brief case-studies of a small number of UK industries in the process plant sector which have benefited under recent aid policies. Following on from this, the study will conclude with some evaluation of the impact of these patterns of aid allocation on the recipient countries.

British Aid and the Process Plant Industry

As mentioned earlier, for the purposes of this study the process plant sector is defined in such a way as to encompass technologies involved in steel production, chemical production and the electrical power industry. Each of these sectors will be considered briefly, beginning with steel plant equipment.

For many years since the mid-1950s, steel plants and steel plant equipment have been included in Britain's aid programme, in large part as a means of resolving the problems faced by the steel plant producing industry in Great Britain (Burch 1987). As Lord Brown, sometime Minister of State at the (then) Board of Trade stated in 1969, the steel plant industry experienced major variations in demand, and aid used to export its products to the Third World could be of 'substantial assistance' if used to 'prevent the cyclical up-and-down in the . . . industry' (H. of C., 285, 1970: 85). The reasoning behind this lay not only in the need to export in order to maintain foreign exchange earnings, but also to gain orders for the purpose of keeping existing capacity viable:

> One could use [aid] on steel plant where, if one does not keep this industry on the move, we shall find ourselves in a few years' time importing excessive amounts of steel plant because our own steel plant makers have suffered a decline (H. of C., 285, 1970: 85).

However, with the deepening of the world depression from 1978, there was a significant increase in the level of aid allocated to the provision of steel making plant in the periphery, as local steel producers in the centre scaled down existing capacity and investment plans in response to reduced demand (*Economic and Political Weekly*, Bombay, 5 December 1981, p.1985). The British steel industry was especially hard hit by the world recession, and the British Steel Corporation was both forced to reduce drastically its planned growth and significantly reduce its existing capacity. The combination of reduced home demand for steel plant and intensified competition in export markets had a significant impact on the operations of the large process plant and engineering concerns like the Davy Corporation, Babcock International, Balfour Beatty and so forth, which, as noted earlier, experienced a drastic fall in orders between 1980 and 1982. As a consequence, from 1981 a substantial volume of UK aid was committed to the export of steel plant and supporting inputs, mostly under the ATP.

One such project was the Nador steel mill in Morocco, where the UK made available a sum of £13.5 million in 1981–82. A much larger project, also agreed in 1981 but subsequently cancelled by the Indian government, was the Paradip plant in the state of Orissa. Construction of Paradip was originally proposed by the Steel Minister in the Janata government, but was subsequently cancelled as a result of funding difficulties. It was then revived as a project to be funded by West Germany and carried out by the firm of Mannesman-Demag. However, the Davy Corporation ultimately secured the order for the £1,250 million project, with an offer of £660 million in cheap credits and £150 million grant aid under the ATP. This made Paradip the largest aid project every supported by the UK government up to that time, and one which held out the prospect of 28,000 man-years of employment.[6]

The Davy Corporation undoubtedly found compensation for the loss of the Paradip order in the fact that it won a £330 million order for the Sicartsa steel plate mill in Mexico in 1981, against strong competition from French and Japanese interests. This order promised 27,000 man-years of employment and it was a British offer of £35 million in the form of grant aid under the ATP, combined with £183 million cheap credits, which tipped the balance in favour of the British company (Elliot 1982: 15; *The Times*, London, 12 November 1981; *Sunday Times*, London, 1 November 1981).

These cases clearly demonstrate the extent to which commercial criteria have been applied in recent years to help the ailing steel plant sector in Britain. Paradoxically though, it has been suggested that the support which UK producers of steel plant have received under aid also partly explains the problems that producers of steel and steel products have begun to experience in the markets of the periphery. In other words, the creation of a steel-producing capacity in the peripheral areas has been a factor leading to a reduction in demand for finished and semi-finished steel from the centre, and may even threaten markets in third countries or domestic markets. For example, the Paradip project contained a 'buy-back' clause which involved British purchases of steel equal to about 10 per cent of the average annual output of the UK steel industry in the 1980s. Although Paradip was cancelled, it represented the kind of policy preference being considered in official circles, which in turn, partly explains why the UK steel industry itself might be experiencing some reduction in home and overseas demand (*The Times*, London, 7 October 1982).

Another component of Britain's process plant contracting industry is concerned with the construction and equipping of chemical plants. In the context of Britain's aid programme, this sector has largely been concerned with the construction of fertilizer factories, or the provision of the large process technologies required for the manufacture of fertilizers. So far, little of the aid to this sector has come under the ATP; rather, orders for fertilizer plants and equipment have been placed as part of the 'normal' aid programme which is, of course, increasingly subjected to ATP-like criteria.

There is no doubt though that aid to this sector has been generated by the problems it has experienced over a number of years. For example, towards the end of the 1960s there emerged a substantial over-capacity in fertilizer production, which led to a substantial fall in new orders for plant and equipment; between 1966 and 1967 the value of work-in-hand on fertilizer factories in the UK fell from £43 million to £21 million, and was followed by a similar fall in the subsequent year. This pattern was repeated elsewhere in the industrialized countries, and the construction of fertilizer plants there had virtually come to a halt by 1970 (*Wall Street Journal* (Eastern Edition) 19 February 1974; *Forbes* 1 March 1974; Burch 1987).

Partly as a consequence of this, the UK chemical plant sector experienced a number of major problems. In 1967, for the first time since World War II, the home market witnessed a decline in the value of projects in hand, and by the first quarter of 1968, the British Chemical Plant Manufacturers Association was reporting some 30 per cent surplus capacity in process plant firms in the industry (*The Times*, London, 29 September 1967; *Chemical Age*, 23 March 1968). Virtually the only growth areas in fertilizer technology lay in those 'oil-rich developing countries' which were close to sources of feedstock, and in those countries of the periphery where the Green Revolution was having a significant impact. The combination of all these factors made the process plant and chemical engineering sectors prime candidates for assistance under Britain's official aid programme. This position was reinforced further as a consequence of the policies of the Labour government under Harold Wilson, which was seeking to 'rationalize' and expand this sector as one of the leading components in its programme for the 'technological revolution' which was to transform British industry and society. The Ministry of Technology, the Industrial Reorganization Corporation (IRC), and the National Economic Development Office all closely monitored the industry and sought to promote it through a combination of mergers, indicative planning and the expansion of markets at home and overseas (Ministry of Technology, 1969; NEDC, 1966; NEDC, 1969; NEDC, 1971).

All of these factors resulted in the inclusion of much fertilizer-producing technology in the aid programme, beginning in 1969, when the UK government initiated its policy of offering support for fertilizer factories, and let it be known that it would consider a request from India for a fertilizer project (*Chemical Age*, 29 August 1969). This initial proposal was to emerge as the Mangalore Chemicals and Fertilizers Factory, to be built at Mangalore by the firm of Humphreys and Glasgow. Subsequently, the UK government announced its formal decision to concentrate its aid to India on the provision of fertilizer-production technology and two or three other sectors. As a consequence, in the ten years between 1971–81, at least £173.75 million UK aid in the form of capital goods was allocated for fertilizer production. This represented something in the region of 13 per cent of all known aid over the period, although it is likely that the real benefits to UK producers were greater than suggested by the level of aid made available, since a multiplier of three–four was known to apply to many projects at this time.

These benefits were not inconsiderable either, with the British firm of Humphreys and Glasgow — one of the leading chemical plant construction companies in the UK — receiving a fairly large proportion of the total volume of aid sales. Undoubtedly these sales were related to the increasingly weak and uncompetitive position of the company; it can be no coincidence that the offer of aid to construct the Mangalore project was made at a time when the company was experiencing the most serious decline in the value of work-in-hand of all the major producers. The value of current orders with the company fell from £125 million in 1967, to only £84 million in 1968 and £42 million in 1970, with both home and overseas markets contracting. The subsequent improvement in the company's performance was almost entirely attributable to orders received under the aid programme. These orders — the one for Mangalore and the other for the Indian Farmers Fertilizer Co-operative (IFFCO) at Kandla/Kalol — were of immense importance to the company. For example, the total value of work-in-hand (including the two orders for India as above) stood at £94 million in 1973, of which £77 million was for export. The two Indian orders for fertilizer plants thus accounted for at least 50 per cent of all export contracts outstanding, and some 40 per cent of total work-in-hand. And if calculations as to the value of aid are made on the basis of the aid component only, rather than the total value of the orders, it still meant that over 22 per cent of the company's export orders and 18 per cent of the total value of work-in-hand were accounted for by official aid expenditures (Burch, 1987).

In common with other components of the process plant sector, the early 1970s saw some upturn in the demand for fertilizer production equipment. With plant orders running at a fairly high level, there was little need for the UK to make aid available for this purpose, at least until the situation again began to change from 1976 when, under the full impact of increased prices for oil, the former four-year cycle of growth and stagnation in investment patterns gave way to an eighteen-month cycle (Reuben and Burstall 1973). This was followed by an extended period of decline after the intensification of the world recession from 1978, during which time the position of the world and the UK fertilizer industry worsened. At the same time, the problems of the process plant manufacturers supplying inputs to the industry intensified, and increasingly official aid has been utilized in support of the large capital goods suppliers which underpin the industry. In 1981, for example, some £90 million aid was allocated to the purchase of such equipment for only two fertilizer projects when orders for steam generation equipment for the Hazira and Thal Vaishet plants were placed with Foster-Wheeler. These orders represented some 19 per cent of the international groups total work-in-hand, and would therefore have represented an even longer proportion of the order book of the company's UK subsidiary (Burch 1987).

Clearly, UK has been of critical importance in maintaining the viability and profitability of those companies located in the chemical plant sector of the process contracting industries. The same is true of the heavy electrical plant sector, which has also been the recipient of a large volume of aid orders over much of the post-war period, and especially over the period 1976–85. The reasons for this later commitment are clear. In the domestic market, orders for heavy electrical equipment declined in the 1970s, since the demand for electricity had not caught up with the high levels of capacity installed in the 1960s. Then, in 1976, the British government's central policy review staff reported that future home orders for electrical power plant would be entirely for the replacement of existing capacity, with no scope for expansion of output (Surrey *et al.* in

Pavitt 1980: 238). In export markets, the picture was little better. British producers maintained around a 9 per cent share of world exports in the 1970s, but were not able to regain export markets lost in the 1950s and 1960s. At the same time, British producers were effectively precluded from expanding in other major markets, such as North and South America and the European Community, as a direct result of the successful cartel operations of the International Electrical Association in the 1970s.[7] This left mainly the peripheral countries of the Commonwealth, and traditional markets such as Australia, Hong Kong, etc., as potential areas of expansion in a period when a considerable growth of markets was needed to avoid the critical situation which the industry faced from the late 1960s (*The Economist*, London, 27 January 1968). As a result of the limited ability to penetrate new export markets, and in view of the fact that in the future the home market was to consist largely of replacements, the British government was confronted with the probability that only about one-third of the capacity of the heavy electrical equipment industry would be utilized throughout the 1980s. As noted earlier, by the mid 1970s the Process Plant Association, representing the major manufacturers of heavy electrical equipment, were predicting that between 10,000–15,000 jobs would be lost unless new orders for power stations were forthcoming (*The Times*, London, 17 August 1976). The 1976 report of the Central Policy Review Staff was even more pessimistic in its estimation that some 33,000 jobs were at risk in the UK power plant industry, mostly concentrated in already-depressed areas, and that collapse of the industry would have involved very heavy social and economic costs to any government.

Given these problems, it is not surprising that this sector has been one of the main beneficiaries of orders placed under the UK aid programme, especially after the crisis of 1976 and the introduction of the ATP in 1978. In the period 1976–84, aid orders for heavy electrical equipment totalled nearly £110 million, and it is significant that aid sales increased markedly from 1976, following the release of the reports by the Central Policy Review Staff and the Process Plant Association, and when the major world recession became firmly established. Certainly from this time the pursuit of British interests in the formulation of the aid programme was becoming more openly acknowledged. As noted earlier, it was in 1973 that Britain specified the three or four areas in which it favoured offering assistance to India, the largest single recipient of UK bilateral aid. Power generation was one such area, and policy was formalized in a series of power sector agreements from 1976 (*Asian Recorder*, 20–26 August 1975). Subsequently, the introduction of the ATP resulted in some very large orders for British companies. For example in 1981, as part of a much wider aid package, Britain approved financial assistance to the £77 million Santa Cruz power project in Brazil, involving £13 million as an interest-free loan, and £55 million in export credits. The main contractor for this was the Davy Corporation — Britain's largest international process contractor and recipient of orders for other major aid projects — with much equipment to be supplied by Northern Engineering Industries (NEI), a subsidiary of NEI Parsons (*The Sunday Times*, London, 1 November 1981; *The Times*, London, 12 November 1981; *COI Survey*, November 1981: 336–7 and January 1983: 26–7).

This project was cancelled in March 1982, although in November of that year agreement was reached with Brazil on a replacement. Britain undertook to provide £24 million grant aid under the ATP for the provision of equipment for a coal-fired power station at Jacui. An unspecified financial package was also agreed, consisting of a loan covered by export credit guarantees, and a Eurodollar loan provided by private banks led by Lloyds Bank International. The finance was to be allocated to the purchase of a 350-megawatt steam turbine and a coal-fire boiler, and ancillary equipment and services, supplied by Klockner (UK), NEI Parsons and NEI International Combustion.

A similar ATP-funded project was for a coal-fired power station in Bihar state in India, to be built in association with a coal-mining project at Singrauli (Elliot, 1982: 15–16; *The Times*, London, 30 March 1982, 1 April 1982 and 11 November 1982; *Asian Recorder*, 28 May–3 June 1982). The major contractor for this project was again NEI Parsons, in association with General Electric and Babcock and Wilcox who were to supply turbines, boilers and coal-handling equipment. The total value of the contract was estimated at some £550 million. Of this sum, £65 million was to be made available under the ATP, while £75 million aid repayments due on past loans to India were 'forgiven' and converted into financing for local suppliers to the project. The UK also agreed to waive a rule of the IDA, the World Bank soft loan affiliate which was a major aid donor to India, which would have limited India's ability to draw on British funds committed to IDA. The IDA was subsequently empowered to make available £370 million of Britain's £555 million contribution.

Discussions on this project began in earnest in March 1982. At some point in the processes of negotiation, the location of this plant appears to have shifted from Singrauli in Bihar State, to Rihand in Uttar Pradesh. There were other changes made as well, and in the form finally agreed in late 1982, Britain made available a grant of £110 million, made up of £33 million bilateral aid, £17 million under the ATP, and £60 million for local costs. A further £7 million was provided from technical co-operation funds for consultancy services in design, construction, maintenance, operation and training, and the ECGD gave its backing to a loan of £344 million arranged by Standard Chartered Bank.[8] On the signing of the agreement, the Rihand station became the largest aid project financed by Britain. However, even this was to be surpassed in 1984 when the UK government allocated £131.4 million towards the cost of a commercial power station designed to supply the needs of the Bharat Aluminium Company (Balco) at Korba, in Madhya Pradesh (*Financial Times*, London, 23 July 1984 and 31 July 1984; *Guardian*, London, 31 July 1984). The main part of the agreement involved the allocation of £94 million towards the cost of a coal-fired power station supplied by GEC, and included four 67.5 megawatt turbine generating sets provided by Babcock Power. A further £37.4 million was allocated to Balco for local costs, while a loan of £25 million arranged by Lazard Brothers was to be guaranteed by the ECGD.

Clearly, British aid under the ATP has contrived to accord a very high priority to the power sector and supporting industries, and has been openly justified by reference to the number of jobs created in Britain (*The Times*, London, 2 March 1983; *Guardian*, London, 31 July 1984). In addition though, many orders for heavy electrical equipment or complete power stations have also been agreed outside of the ATP. For example, beginning in 1979, Britain allocated £100 million towards the Victoria dam and electric power project in Sri Lanka, part of the larger World Bank project to divert the waters of the Mahaweli River in order to irrigate large extents of land in the northern dry zone. Also in 1981–82, £70 million UK bilateral aid was allocated towards the provision of plant and equipment for a World Bank project to meet increased power requirements around Khartoum; in addition, £3.5 million was allocated to electric power schemes in Jordan, £4.5 million for the provision of generating equipment for small-scale hydro-electric plants in the Philippines, and a sum of £22.5 million to Mozambique, mostly used to fund power stations at Quelimane and Pemba (Burch 1987).

The benefits to the industry from these orders are indisputable, even if they are difficult to quantify precisely. In terms of employment, it can be suggested that in 1979, when aid from this sector stood at nearly £20 million, or about 8 per cent of known aid in that year, at least 2,500–3,000 jobs were directly generated by aid sales.[9] However, these are minimum figures for, as noted elsewhere, there is a multiplier of about four or five operating where the ATP is concerned, and the extent of job-creation and capacity-utilization resulting from aid is much greater than the basic data suggests. In addition, overseas sales have grown as a percentage of total sales as the home market reached saturation point, and aid was a significant factor in this expansion of export markets. This was especially important given the high levels of surplus capacity in the heavy electrical equipment sector in recent years, when each and every order covered some part of the large fixed costs carried by the industry. For these reasons, the producers of heavy electrical equipment have always been strong supporters of aid programmes (Cilingiroglu 1969: 23). Moreover, because of their strategic importance to the industrial economies, manufacturers in this sector have always been able to mobilize the political resources needed to ensure that the aid programmes from which they benefited have been maintained.[10]

Implications for Underdeveloped Countries

The material presented in the preceding section suggests that there has been, and continues to be, a significant shift of emphasis in the content of Britain's bilateral aid programme towards a concentration of aid resources on a fairly narrow range of highly capital-intensive technologies related to the process plant, heavy electrical equipment and civil engineering and construction sectors. Of course, it must be borne in mind that such a shift of emphasis is not confined to British aid; although differences in industrial structure imply some variations between different countries in the technological composition of national aid programmes. Nevertheless, much of the intense competition for orders in recent years, which has given rise to the use of 'mixed credits', has clearly emerged because of the simple fact that many industrialized countries do share similarities in their industrial base, and are fighting for limited orders for ˙ these industries. So Britain's experience suggests that most aid donors are moving in a similar direction and are making available subsidized sources of finance in order to stimulate exports from their own heavy capital goods sectors.

It is arguable that certain aspects of this situation will work to the benefit of those underdeveloped countries which, as noted earlier, will be able to get access to highly-subsidized capital goods more cheaply than might otherwise be the case. Against this, however, must be placed a number of other considerations. For example, the number of countries which benefit from these arrangements may be limited, and will be confined to those countries offering the greatest prospects for future export markets for process plant contractors. However, these countries would not necessarily be the same as those aid priority areas selected on the basis of the need to assist the poorest of the poor countries (*The Times*, London, 26 July 1985). Equally, these poorer countries will probably lose out on access to concessional finance because of the tendency of donor countries to re-allocate aid towards mixed credits by reducing the sums made available as 'normal' aid, rather than making additional funds available to subsidize exports.

In addition to these issues though, there are the concerns relating to the impact of this shift of emphasis in bilateral aid programmes on technology policy. There seems little doubt, for example, that such a shift in the distribution of aid funds means that there is less finance available for more appropriate technologies or projects, e.g. local level programmes to bring clean water to rural areas, social forestry projects, appropriate energy sources, etc. At the same time, it is also the case that the provision of capital-intensive process plant and electrical power equipment under aid may undermine the existence of, or the capacity to develop, local technological capabilities, in particular capital goods industries. This certainly seems to be the case for India where, in recent years, foreign technology in steel production, fertilizer production and heavy electrical equipment manufacture has been imported to the apparent detriment of local capital goods industries.

In the case of fertilizer production, for example, India had built up a considerable indigenous technological capability in the design, construction and operation of manufacturing plants in the period since independence.[11] Most of this capability was to be found in the Projects and Development Division of the state-owned Fertilizer Corporation of India (FCI) which, building on the experience of the Sindri and other projects started in the early years of the

Colombo Plan had, by the 1960s, acquired considerable expertise in this area. In addition, local industry had also acquired the knowledge and technological capability to supply much of the hardware to the new fertilizer project. Thus, by the 1970s, the Projects and Development Division was successfully able to construct four modern ammonia fertilizer plants, each having an installed capacity of 900 tonnes per day (tpd). Subsequently, the planning of new units received a substantial boost with the discovery of natural gas at off-shore locations on the western coast. On the basis of the availability of this gas, it was decided to build four plants at two locations, Thal-Vaishet in Maharashha and Hazira in South Gujarat.

However, a decision was also made to scale-up the production capacity of the new plants to 1,350 tpd, ostensibly because they would be more efficient and would generate greater economies of scale. It was also decided that for the construction of these plants, a foreign engineering constructor, selected on the basis of global tendering, should be solely responsible for the design and construction of plants operating at this higher level of capacity. But it has been argued that these decisions were somewhat arbitrary; for example, not all the major foreign contractors had much more experience that the FCI in designing and constructing plants capable of operating at 1,350 tpd, and in any case, such a plant reflected a scale of operations which was not necessarily the most economic in Indian conditions. But more to the point, this specification effectively excluded the Projects and Development Division of the FCI from competing, and ensured that foreign technology was utilized.

That this was the case is somewhat surprising, although it can be explained by reference to the interplay of interest groups associated with fertilizer production in India. For in many ways the public sector FCI had been too successful, and by the 1970s it was reported that 'strong forces were at work to bring about [the FCI's] fragmentation'. According to some Indian observers, local and overseas interests found a profitable and growing market for chemical fertilizers slipping away from then because of the activities of the FCI in acquiring a high level of technical, engineering and operational capability. Certainly there can be no doubt that from the late 1970s the technological capability of the FCI was weakened, firstly by the actions of the conservative Janata government in 'hiving-off' the Planning and Development Division of the FCI from the state corporation, and establishing it as a separate entity. In this new role, the division has been largely excluded from participation in the design and planning for the major expansion of fertilizer production which has occurred with the exploitation of natural gas from the Bombay High fields (including the Thal-Vaishet and Hazira plants with which British aid and technology was associated). Following on from this, under the impact of 'liberal' economic policies adopted by Mrs. Gandhi, the rest of the FCI was broken-up and reorganized into smaller units, with the stated aim of developing a public sector fertilizer industry, integrated on the basis of geographical cohesion or the use of a common feedstock and technology, rather than one functionally organized as a state corporation at the national level.

The changes were, of course, accompanied by the growing crisis in the process plant and other industrial sectors in the developed countries, and an increased willingness on the part of western governments to make available mixed credits and other concessional flows for the export of fertilizer production equipment. The combination of these internal and external factors meant that in the massive expansion of fertilizer production from the 1970s Indian technology and expertise were systematically downgraded. Among other examples, it has been suggested that Indian firms have been excluded from lists of approved suppliers, or that their proven technological capabilities have been ignored, e.g. in the production of reformer tubes and water treatment plant. Instead, foreign technology has been approved for use in the most recent plans for the industry, involving also a major expansion of private local and foreign capital investment supported by the aid of the developed centres.

In short, indigenous technological capabilities in this sector of Indian industry are being sacrificed to a growing reliance upon foreign 'turn-key' projects, a development which reflects the dominance of that coalition of interests in both the donor and recipient country which benefits from this particular strategy. But this is true not only for the fertilizer sector in India. There have been suggestions that the power industry has also come to reflect such trends with predictable consequences for indigenous technological capabilities. For example, Indian companies engaged in the production of power plant and heavy electrical equipment have been excluded from participation in several major projects funded under tied aid, including the Rihand power station and Balco power station financed by Britain. In both of these cases the public sector Indian company, Bharat Heavy Electricals Ltd., tendered for the contracts, but was unable to compete with the UK tenders heavily subsidized by grant aid and cheap export credits. In the event, it was suggested that under these circumstances, foreign technology would eventually come to supplant India's indigenous capabilities in this area. Referring to the Rihand power station and other projects in 1982, observers have noted that Bharat Heavy Electricals Ltd. had a very lean order book which, in the absence of substantial participation in major projects, would not take the corporation beyond 1985–86. Such concerns had not been fully allayed by 1985, either (*Economic and Political Weekly*, Bombay, 21 August 1982, 30 October 1982 and 13 July 1985).

A lack of space precludes any discussion of those other sectors such as steel, where it is also suggested a growing reliance upon foreign technology has led to the displacement of local technological capabilities in the capital goods sector.[12] There seems to be little doubt, however, that this is happening in the Indian case. Certainly, if the industrialized countries, in association with those Third World interest groups who benefit from access to foreign rather than local technology, continue to participate in turn-key arrangements for the provision of heavy capital goods, then the future is uncertain for Indian industry. Such considerations might also apply, if on a reduced scale, in the case of countries such as Indonesia, Malaysia, Kenya, etc. If such is the case, then aid programmes will end up repeating once more what has happened in the past; destroying local capabilities in the interests of local and foreign interest groups.

Notes

1. In East Africa, the Kenya–Uganda Railways spent about £16.5 million on capital investment between 1921 and 1933, and thereafter was burdened with interest and capital repayments of well over £800,000 per year. By 1932, Tanganyika's railway debt was £5.8 million, and repayments amounted to £310,000 per annum.

   The implications of these continuing financial burdens can be seen when they are compared with government expenditures on other services. In 1934, Kenya spent £170,000 on education, £125,000 on agriculture and £198,000 on medical services. These sums combined are a little more than half the total repayments of capital and interest on railway loans. In 1935, Tanganyika spent £81,000 on education, £150,000 on agriculture and £195,000 on medical services which, when added together accounted for a little in excess of the total repayments of £310,000 made annually by Tanganyika on its debt of £5.8 million (Brett 1973).

2. A number of railway loans were raised by the colonial territories on the London capital markets at this time, but these were commercial transactions and involved no concessional element.

3. That British industry and exports have been adversely affected by the use of mixed credits in general, and those of France in particular, has been claimed on numerous occasions by UK aid officials and businessmen. For example, in July 1983, the Chief Secretary to the Treasury expressed official disapproval of French use of mixed credits to subsidize exports, despite the existence of the ATP in the British aid programme. At the same time, Mr. Peter Godwin, a director of Lazard Brothers and Chairman of the Tropical Africa Advisory Group (TAAG, a business group advising government on export policy and promotion in the region), claimed that Britain was losing a large amount of business, even in its former colonial territories, as a consequence of French policy on the use of mixed credits, and that Anglophone Africa was a priority area for France's growing aid programme. See *Sunday Times*, London, 17 July 1983, *The Times*, London, 6 September 1983. In addition, there were also reports that the Hawker Siddeley Company had lost over forty overseas contracts (mainly for railways) in four years, as a result of 'unfair' international competition. On this issue, see *Guardian*, London, 2 November 1984 and 7 November 1984. See also the reports on the loss of orders for a new bridge over the Bosporus and the re-equipping of the Bangkok bus system, in *Sunday Times*, London, 16 June 1985 and 23 June 1965, and *The Times*, London,

20 July 1985, 26 July 1985 and 18 October 1985.

4. Barna (1983). For our purposes, process plant and equipment not only covers those industries which employ mainly chemical processes to convert materials into finished or semi-finished goods (e.g. petroleum refining, chemicals, steel, non-ferrous metals, etc.), but also industries such as heavy electrical power plant which use characteristic process plant components, such as pressure vessels. The major British companies operating in this and other sectors favoured under the ATP include the Davy Corporation, Babcock and Wilcox, GEC, NEI Parsons, Costain, Balfour Beatty, Hawker-Siddely, Trafalgar House, etc.

5. The current situation regarding the development content of ATP funds has been discussed in various places. *The Times*, London, 12 November, 1981, has described it thus: 'Notionally, projects financed out of the Aid and Trade Provision have to pass a "test of minimum development soundness", but former aid officials describe the procedures involved as a farce. It is admitted by the Overseas Development Administration (ODA) that in the case of many ATP projects, it is not always possible to carry out anything like the normal appraisal in the time available. The essence of the fund, according to its advocates, is the money can be developed swiftly.' See also the *Sunday Times*, 1 November 1981, which reports that aid funds were committed to the construction of a major Indian steel plant by Davy International with *no* report from the Overseas Development Administration on the 'development value' of the project. This charge was originally made by the Opposition spokesman on overseas aid, in *Hansard*, 11 February 1982, col. 1131. As the world recession worsened in the 1980s, the pressures on the ODA to discount development criteria intensified to the point at which 'one senior Whitehall source accused ODA officials of being "besotted by abstruse questions about whether a project was developmentally suitable or not" '. On the conflict over control of aid funds between the ODA and the Department of Trade and Industry (the department responsible for exports), see also *The Times*, London, 18 April 1984.

6. All data on this project is drawn from *Economic and Political Weekly*, Bombay, 26 September 1981, 5 December 1981 and 29 May 1982; *Sunday Times*, London, 1 November 1981 and 16 May 1982; *Asian Recorder*, 22–28 October 1981, p. 16,280 and 16–22 July 1982, p. 16,697; see also Lall and Chopra (1981), p. 195; *Guardian*, London, 17 April 1984 and 1 August 1984.

7. The International Electrical Association was a cartel of producers of heavy electrical equipment, whose main purpose was the achievement of the highest possible prices for its members' products, and the retention of the largest possible market share. For a discussion and analysis, see Mirow and Maurer (1982).

8. It soon became apparent that this project, agreed in haste and, as usual under the ATP, without regular processes of tendering, was in considerable

difficulty. NEI had had no previous experience of building a turn-key project in such a difficult environment and by March 1984 the project was over four months behind schedule. This poor performance was also causing concern amongst Department of Trade and Industry officials, because of the possibility that it might have jeopardized the chances of other export orders for UK industry. See the *Guardian*, London, 24 March 1984 and 29 November 1984.

9. These rough estimates are based on data from May and Dobson (1979), which implied that, in general, there was one UK job directly generated by every £7,860 worth of exports (the ODA data suggested a slightly lower figure of one job for every £6,500 worth of exports).

10. A good example of this occurred with the release in Britain of a report by an official of the UK Treasury which, in an attempt to reduce public expenditure and also avoid intense competition between the industrialized countries in the race for diminishing markets, suggested the elimination of export subsidies and mixed credits (Byatt, 1984). Treasury ministers were very supportive of the Byatt Report, and for a while it seemed that the recommendations it made would be accepted as policy (*Guardian*, London, 21 January 1984; *Financial Times*, London, 18 September 1984). However, those sectors of British industry which would have been most affected by the abolition of mixed credits — the large process plant and construction contractors mentioned earlier — were able to resist the policies of the Treasury. For these companies, the proposals to eliminate export subsidies could not have come at a worse time, and they mobilized support in various ways, through the Overseas Projects Board, through the National Economic Development Office, and through appeals to senior politicians. As was seen earlier, the level of support for the mixed credits which benefit these companies has continued to expand, and will apparently continue at least as long as the current recession continues. For a detailed study of this case, see Burch (1987) and also *The Times*, London, 27 March 1985, 18 April 1985, 20 April 1985, 26 July 1985 and 24 September 1985.

11. This discussion is taken from detailed material contained in the *Economic and Political Weekly*, Bombay, 10 May 1980, 23 August 1980, 5 October 1980, 25 December 1982 and the Annual Number, August 1984.

12. Other cases from India covering a variety of industries and sectors have also been documented in recent years following the gradual liberalization of imports. See, for example, the case of steel production technology (*Economic and Political Weekly*, Bombay, 1 December 1984 and 23 February 1985), and semi-conductors (*Economic and Political Weekly*, Bombay, 27 April 1985). For a general discussion of some of the issues documented here, see *Economic and Political Weekly*, Bombay, 13 July 1985 and the special Review of Management Issue, August 1985.

Bibliography

Aaronovitch, S. and Smith R. *The Political Economy of British Capitalism*, McGraw-Hill, Maidenhead, 1981.

Barna, T. 'Process Plant Contracting: A Competitive New Industry', in Shepherd, G. *et al, Europe's Industries*, Pinter, London, 1983.

Beckerman, W. *et al. The British Economy in 1975*, Cambridge University Press, Cambridge, 1965.

Brett, E.A. *Colonialism and Underdevelopment in East Africa*, Heinemann, London, 1973.

Burch, D.F. *Overseas Aid and the Transfer of Technology*, Gower, Aldershot, 1987.

Byatt, I.C.R. 'Byatt Report on Subsidies to British Export Credits', *The World Economy*, 7 (2) pp. 163–78, 1984.

Cmd. 2387. *Report of the East African Commission*, HMSO, London, 1925.

Cilingiroglu, A. *Manufacture of Heavy Electrical Equipment in Developing Countries*, World Bank Staff Occasional Paper no.9, Johns Hopkins University Press, 1969.

Development Assistance Committee. *Development Cooperation*, Organisation for Economic Cooperation and Development, Paris, 1983.

Elliot, C. *et al. Real Aid*, Independent Group on British Aid, London, 1982.

Feinberg, R. *Subsidising Success: The Export-Import Bank in the US Economy*, Cambridge University Press, Cambridge, 1982.

House of Commons, 285 (1970), *Select Committee on Overseas Aid*, HMSO, London.

Hyson, C.D. and Strout, A.M. 'The Impact of Foreign Aid on US Exports', *Harvard Business Review*, 46 (1) pp. 33–71, 1968.

Independent Commission on International Development. *North-South: A Programme for Survival*, Pan Books, London, 1980.

Lall, K.B. and Chopra, H.S. *The EEC and The Third World*, Radiant, New Delhi, 1981.

Little, I.M.D. and Clifford, J.M. *International Aid*, Allen and Unwin, London, 1965.

May, R. and Dobson, N. 'The Impact of the Aid Programme on British Industry', *ODI Review*, no. 2, 1979.

Ministry of Technology. *Report* of the Process Plant Expert Committee, HMSO, London, 1969.

Mirow, K.R. and Maurer, H. *Webs of Power*, Houghton Mifflin, Boston, 1982.

NEDC. *Process Plant Working Party: Process Industries Investment Forecasts*, HMSO, London, 1966.

NEDC. *Process Plant Working Party: Process Industries Investment Performance*, HMSO, London, 1969.

NEDC. *Process Plant Working Party: A Survey of Manufacturing Capacity*, HMSO, London, 1971.

NEDC. *Memorandum by the Director-General: Overseas Capital Projects*, HMSO, London, 1981.

Overseas Development Paper no. 17. *The Industrial Strategy: The Contribution of the Ministry of Overseas Development*, HMSO, London, 1978.

Reuben, B. and Burstall, M.L. *The Chemical Economy*, Longman, London, 1973.

Surrey, A.J. *et al.* 'Heavy Electrical Plant', in Pavitt, K. (ed.), *Technical Innovation and British Economic Performance*, Macmillan, London, 1980.

White, J. *German Aid*, Overseas Development Institute, London, 1965.

13

Science and Development
Underdeveloping the Third World

Khor Kok Peng

It is important to view the crisis in modern science within the context of the general crisis in the modern world, a crisis felt particularly in the Third World where the majority of the victims of modern science live. The crisis is that science has not been able to help fulfil the simple survival needs of a very large proportion of people in the world; it has not been able to conserve natural resources to ensure the decent survival of the majority of human beings in the generations ahead. Instead, it has helped to deprive the poor of the Third World of their basic requirements, and it has contributed to the depletion of natural resources that are required for the long-term or even the medium-term survival of man. It has also developed new chemicals, materials and processes which cause immense harm to human health and the environment.

Modern science is an instrument that has been used to serve the functions to which its controllers have assigned it. We need to examine what it has been used for, and why, and what the consequences are. We live in an unequal world, and much of the analysis of the inequality has been economic in nature with the role of science brought in only indirectly and in a peripheral manner. Yet science and its applied form, technology, play a strategic part. I shall attempt to look at the way in which resources are used, and how this has not satisfied the basic needs of a large proportion of people, and the role of science and technology in this scheme.

In the process of production man takes resources from nature with the help of tools or technology, and transforms them into products of his design. This holds true for all social systems and in all historical eras. This simple scheme, on which all human life depends, is complicated because there are different combinations of man, tools and nature, which produce many types of products using a variety of technologies, and which have different effects on the condition of nature.

Firstly, many natural resources are finite and can thus be physically depleted. The second problem is that the technology used by man may be unsuitable for the long-term balance or even survival of nature and its resources: for instance, pollution of water can cause death of marine life; or the application of chemicals can make agricultural soil less naturally fertile; or the build-up of carbon dioxide in the atmosphere threatens to cause climatic changes which will affect agriculture. Thirdly, there is the problem of the destruction of the planet or large parts of it through the production of weapons. Even if there is no nuclear war, large-scale accidents occurring in nuclear or chemical plants cause death and injury (as in the cases of Bhopal and Chernobyl). Another factor to consider is that the technology used in the production process, and even the product itself, is often hazardous to human health. There are thus costs to weigh against the possible benefits of the products. Lastly, when we drop the abstraction 'man' who deals with nature, but instead analyse 'people' in the production process we find that these people are divided very unequally around the globe and within each nation. In an unequal social structure, there is the unequal ownership of and control over resources, which means nature (land, forests, minerals, water) and the tools or technology to extract or process natural resources. People who control more resources and technology are better able to determine the production process: what to produce, how to produce it, what resources to use up to produce, and what technology to use. Since wealth and income are thus unequally distributed, we also have different baskets of goods and services obtained by different people: the rich have a bigger basket with a wide range of things, the poor have a small basket with few simple things, and the poorest have no basket at all and in all likelihood will perish for it.

We thus have this spectacle, on the one hand, of the powerful development of technological capacity, so that the basic and human needs of every human being could be met if there were an appropriate arrangement of social and production systems; and, on the other hand, of more than half the world's population (and something like two-thirds the Third World's people) living in conditions where their basic and human needs are not met. This would be tragic in itself, for it manifests how the world's social system distributes resources and technological capacity in an irrationally skewed and unequal fashion thus rendering the majority of human beings unable to feed or house or educate themselves, but it is rendered even more catastrophic by the fact that the same technological capacity that has facilitated the irrational composition of products is also so powerful that it has enabled the destruction or depletion of a very high proportion of non-renewable resources in the world. Day by day, this gigantic technological capacity uses up more energy, extracts more minerals, chops down more forests, results in more loss of topsoil, and pollutes more water, more land mass, more air and even the stratosphere. At current rates of production, many critical resources will run out within a few decades.

There is a finite stock of world resources available and in the process of production a portion of that stock is used up each year. These resources include energy, minerals, forests, water, and so on. The stock of resources represents the world's natural assets or wealth. That portion which is used up represents part of the annual flow of income, where the resources are transformed by human labour and the tools or technology created by man to become products and services. Those products which are used directly are consumer goods and services (such as food, clothing, health services). Those products which are made not directly for themselves but for use in producing consumer goods more efficiently are termed capital goods, and actually become the tools or technology used in the next round of production.

Seen in this way, it is clear that the Gross National Product (GNP) over which all nation states are so obsessed is only an annual *flow* which is very much dependent on the available stock of natural resources and the human and physical technology available to extract and process those resources. Given the available stock of resources, the rate at which production takes place depends on the level of technology resources. The higher the technological level, the higher the rate of production and GNP. This is what you find in an economics textbook, and so far as it goes, it is correct. What the textbooks do not explain is that the stock of resources is not a given. Most resources are non-renewable. The more we use the more they are depleted; the more they are depleted, the less there are available for use in production in future. In other words, the higher the GNP at present, the lower it will be in future, when the effects of resource depletion are felt.

This most simple and elementary of facts is almost completely omitted in economics textbooks; it is seldom in the consciousness of the planners and politicians who plan our future and rule our lives, or of the scientists and technologists who have made possible the rapid depletion of resources through the development of technological capacity. And, most importantly, it does not cause any loss of sleep to the businessmen and entrepreneurs whose motivations for expansion and profits and growth underly the direction and development of technology and the depletion of resources. The hiding of this elementary fact is perhaps the greatest cover-up perpetuated by the education system all over the world.

The second big cover-up is that the rapid extraction and utilization of resources is carried out very unequally in terms of control and benefits, with 80 per cent of world resources being used up in the developed world and only 20 per cent in the Third World. This unequal distribution also determines the nature of goods to be produced. People whose basic needs are satisfied but who still have thousands of dollars to spend a month will use the money for fashion, luxuries and indulgence. To cater for this elite market, high-tech technologies are created to produce products such as video recorders, compact discs, computers, motor cars, and services such as high-tech medicine, tourism and even tax-evasion legal programmes. A large portion of developed world GNP is spent on such consumer goods and on producing capital goods or technologies to make these consumer goods. Meanwhile the Third World gets to use only 20 per cent of the resources. Since national incomes are also unequally distributed, a large portion of these resources are used for the same high-tech consumer products as are enjoyed in the developed world, and in importing capital-intensive technologies to produce these elite consumer goods. Thus, only a small portion of world resources flows towards the making of basic goods required for the survival of the poor majority in the Third World.

In this on-going process of resources depletion and irrational use of resources, the main impetus and dynamics are located in political economy, the socio-economic systems, which give rise to competition for growth between companies and between nations. But the role of science and technology is crucial. If the level of technology is low, then we may still have the same inequality, but the degree at which resources are depleted would be less. In reality, however, technological levels are increasing rapidly under pressure of competition between firms and countries (not only in the economic but also military spheres), and so the depletion of resources also increases rapidly. Moreover the ever increasing technological capacity of the developed world leaves the Third World even further behind, thus in itself widening the inequality gap between nations.

If the present situation is already a tragedy, the future will be even more so, for when the world's resources run out, the position of those at the bottom will be even more unimaginably worse. This is the greatest indictment against modern science: that it has facilitated the high-tech exhaustion of resources and helped perpetuate a production system producing luxury and superfluous goods and services, whilst a majority of humanity do not have their needs fulfilled. And worse, this poor majority often have their resources taken away from them, to make way for high-tech infrastructure or projects such as hydroelectric dams, industrial estates, highways and urbanization. It is not 'abstract modern science' to blame, but the whole modern social, economic, cultural and political system, which provides the competition, the profit motive, and the militaristic competitiveness, that forms the dynamic impetus for the use of scientific knowledge and technology in this warped and skewed fashion. Science is the instrument of domination and control. Unfortunately, as the Chernobyl, Bhopal and Rhine disasters have shown, this instrument of science may also have a life of its own, making it uncontrollable and unstoppable.

The degree of resource depletion is a controversial issue as it partly depends on the known and projected reserves of the various resources, and that changes frequently as new reserves are found. However there is sufficient evidence that depletion has taken place rapidly since the industrial revolution, and especially since the Second World War ended in 1945.

According to Richard Barnet (1980: 16): in the twenty-five years between 1945 and 1970, 'the industrial world used more petroleum and nonfuel minerals than had been consumed in all previous human history. The United States bent, burned or melted about 40 per cent of the world's nonrenewable materials in those years'. In the case of forests, half the developing countries' forest area was cleared between 1900 and 1965, and since then the process of deforestation has accelerated. Almost a fifth of the remaining tropical forests will be destroyed or degraded by the year 2,000. The excessive loss of topsoil from world cropland due to erosion is now 23 billion tons a year, or at the rate of 7 per cent of total topsoil per decade. In the case of oil, at the 1983 oil production level of 18 billion barrels proven oil reserves will last thirty-seven years and ultimately recoverable reserves will stretch production to 114 years (World-watch Institute 1984: 9). S.R. Eyre, using 1968 data on known reserves and rates of depletion, estimated that by the turn of the century, eleven of the fifteen most important metals will have been exhausted (including copper, zinc, tungsten, silver, tin, aluminium). Iron and chromium alone will probably last to the middle of the next century, (Eyre 1978: 78–80).

If these estimates are considered too pessimistic, let us take a more recent estimate by the Gaia Atlas of Planet Management. Using mid-1980s data, the Atlas estimates that at 1981 consumption levels, silver will be depleted in twenty-four years, cadmium in thirty-nine years, zinc forty-one years, tin forty years, copper sixty-five years, lead forty-eight years, and nickel seventy-five years. The Atlas also estimates that known oil reserves will be depleted in thirty years at present consumption rates; and the time span would increase by only another thirty years if one allows for as yet undiscovered oil resources (Myers 1985: 113).

In 1980, the nations of the North, with only a quarter of world population, earned 80 per cent of the Gross Global Product (GGP). In the South, three-quarters of the world population claimed only 20 per cent of world income. The inequality in income is a manifestation of the similar structure of inequality in usage of world resources. For instance, in the case of energy, the Brandt Report commented, 'consumption of energy per head in industrialized countries compared to middle-income and low-income countries is in the proportion of 100:10:1. One American uses as much commercial energy as 2 Germans or Australians, 3 Swiss or Japanese, 6 Yugoslavs, 9 Mexicans or Cubans, 16 Chinese, 19 Malaysians, 53 Indians or Indonesians, 109 Sri Lankans, 438 Malians or 1072 Nepalese. All the fuel used by the Third World for all purposes is only slightly more than the amount of gasoline the North burns to move its automobiles' (Brandt 1980: 162).

Since 1980 the world has become even more unequal. The Third World remains dependent on the developed world for trade, loans, investments and technology. In the past few years, increasing amounts of funds have drained from the South to the North. In 1985 alone, US$74 billion left the Third World on its debt account alone: it obtained only $41 billion in new loans but had to pay $114 billion in debt servicing. If we included the outflow of profits by transnational companies in the Third World, capital flight from the Third World and the capital deficit of Middle East exporters, the outflow of capital from the Third World in 1985 alone would be US$230–240 billion. If we also include the US$65 billion lost due to the fall in commodity prices (an *Economist* estimate), the Third World's loss would be US$300 billion in one year. In 1986 the situation was worse with the collapse of oil prices and the increased prices of other commodities. Total loss could be US$300–350 billion. Whatever aid is given is a mere drop in the ocean of what flows from South to North, and even this drop is tied to conditions.

The North's grip over modern science and technology has contributed to the exploitation of the Third World's economic weakness. The rich countries use their industrial and agricultural technologies to produce surplus goods which they are unable to use themselves (part of the problem of over-development, or over-accumulation). Then they dump the surplus cereals or other crops or materials on the world market, causing prices of Third World commodities to collapse, and thus further reducing incomes and living standards of the poor. Modern technology and information systems have also enabled transnational banks and companies to expand into developing countries, drawing them further into the world market. Third World countries then find that protective tariff barriers block the entry of their industrial goods and that the rich countries have developed new technology to their own advantage. For instance, they might have reduced their usage of the Third World's raw materials by finding substitutes and by using less materials per unit of product. As a result, export prices and earnings in the Third World fall drastically even though they have to foot out more funds to service foreign debts.

In the case of Malaysia, for example, commodity prices fell an average 20 per cent this year, resulting in a loss of some M$9 billion, whilst payments to service foreign debt removed another $8 billion from the economy. As a result the per capita GNP of Malaysians fell by 11 per cent this year, and 16 per cent compared with 1984, the sharpest decline in Malaysian economic history.

In the Third World, the nature of development follows that of the North, except that ours is a dependent form of development. Growth takes the form of depletion of resources for export to the North, and the use of surplus from exports and from foreign loans to build expensive infrastructure and to invest in capital-intensive technology which mainly benefits big firms or big farmers. The commercialized sectors, with superior financial and physical resources and technology, penetrate, invade and take over the traditional, viable sectors, thus dislocating a large portion of people from their livelihood and homes. For instance, small fishermen using ecologically sound production systems are displayed by big commercialized trawler boats which destroy the marine ecology by overfishing and the use of destructive gear. Or else food-crop farmers have their lands taken back by landowners or bought by either government or private companies to be converted into middle-class housing estates, or free-trade zones for industries, or for highways, etc.

The result is a progressive diminishing of the sector which we can call the 'people's economy': the small-scale unit using family labour and simple but effective technologies producing basic goods for the ordinary people. The people's economy includes small-scale fisheries, peasant food agriculture, home-based and community-based industries making mats and baskets, building of traditional houses, and the manufacture of producer goods such as nets and ploughs. In many cases, this people's economy harbours the secrets of people's science and technology nurtured and tested through the generations, such as different varieties of hardy rice seeds in different localities, or fishing technologies which do not disturb the marine ecology, or indigenous shelter systems which make use of local renewable resources and which are designed to take into account the local climate and cultural environment, or indigenous medical systems which are derived from local plant life.

In terms of many criteria, such as the provision of employment, community or producer control over technology and the production process, equity, and ecological soundness, the indigenous technologies of the Third World are superior to the types of modern technology which have invaded the Third World. Yet these indigenous technologies are being wiped out under the impact of the commercialized sectors and under the threat of the consumer culture which lures tastes away from local to western culture, fashion and products.

Thus, being sucked in a dependent fashion into the modern world system has been disastrous for Third World nations whose futures in terms of sustainable development would have demanded the rational use of their resources for the genuine development of their people. It is time therefore for a re-orientation of the concepts science, technology and development.

It is clear that the crisis in modern science is part of the crises in industrialism and the modern world system, and that it has also been the catalyst and facilitator of this modern system and general crisis. As such the reshaping of science and technology must go hand in hand with a radical change in the overall social and economic system, if humanity is to survive. There must be a radical reshaping of the international economic and financial order so that economic power, wealth and income is more equitably distributed, and so that the developed world will be forced to cut down on its irrationally high consumption levels. If this is done, the level of industrial technology will also be scaled down. There will be no need for the tremendous wastage of energy, raw materials and resources which now go towards production of superfluous goods simply to keep 'effective demand' pumping and the monstrous economic machine going. If appropriate technology is appropriate for the Third World, it is even more essential as a substitute for the environmentally and socially obsolete high-technology in the developed world. But it is almost impossible to hope that the developed world will do this voluntarily. It will have to be forced to do so, either by a new unity of the Third World in the spirit of OPEC in the 1970s and early 1980s, or by an economic or physical collapse of the system.

But it is in the Third World that the new ecologically sound future of the world can be born. In many parts of the Third World there are still large areas of ecologically sound economic and living systems, which have been lost in the developed world. We need to recognize and identify these areas and rediscover the technological and cultural wisdoms of our indigenous systems of agriculture, industry, shelter, water and sanitation, medicine and culture. We do not mean here the unquestioning acceptance of everything traditional in the over-romantic belief of a past golden age which has to be returned to in all aspects. For instance exploitative feudal or slave social systems also made life more difficult in the past. But many indigenous technologies, skills and processes which are appropriate for sustainable development and harmonious with nature and the community are still integral to life in the Third World. These indigenous scientific systems have to be accorded their proper recognition, encouraged and upgraded if necessary.

Third World governments and peoples have first to reject their obsession with modern technologies which absorb a bigger and bigger share of surplus and investment funds, in projects like giant hydrodams, nuclear plants and heavy industries which serve luxury needs. We must turn away from the obsession with modern gadgets and products which were created from the need of the developed world to mop up their excess capacity and their need to fill up effective demand.

We need to devise and fight for the adoption of appropriate, ecologically sound and socially equitable policies for the fulfilment of needs such as water, health, food, education and information. We need appropriate technologies for agriculture and industry, and even more important we need the correct prioritizing of what types of consumer products to produce. We can't accept appropriate technology producing inappropriate products. We need technologies and products which are safe to handle and use, durable, fulfil basic and human needs, and which do not degrade or deplete the natural environment and resources. And perhaps the most difficult aspect of the fight is the need to de-brainwash the people in the Third World from the cultural penetration of our societies, so that life-styles, personal motivations and status structures can be delinked from the system of industrialism, its advertising industry and creation of culture.

In this effort to rebuild a human society, the role of science and technology is crucial. Just as we need a new economic and social order, so too we need science and technology under human control to be the servants of constructing the new order. Science, including knowledge from modern science, can be used to devise the appropriate technologies which can serve the masses in such areas as water supply, sanitation, shelter, health, agricultural and industrial production. For instance, if we realize that the limited supply of clean portable water is too precious to waste on servicing a flush toilet system for a minority of households, then science can devise a method of rechanelling clean water to the homes of the masses for drinking, or a method of deriving clean water at local level to serve the needs of the whole community. What is important to realize is that there is not enough water in the world for everyone to have a flush toilet. Once in the political and social sphere we determine that water should serve the needs of the majority and not the toilet convenience of the minority, then science can be the instrument of putting the principle of equality into force.

Finally, whilst a new science for the masses cannot succeed unless there is an accompanying or preceding change in social structures, it is also true that a change in socio-economic structure alone is insufficient for developing a new sustainable order. Control and distribution of resources is a crucial determinant of social order but a change in this aspect alone is insufficient and could lead to similar problems without there being an understanding of the limits of resources and the environmental, health, ethical and cultural aspects of science and technology. Therefore there can be no meaningful reform of science without a change in society at large. There can also be no meaningful reform in social structures unless there is a change in the understanding of science and its proper application to serve the people and to be in harmony with nature.

Bibliography

Barnet, Richard. *The Lean Years*, London, 1980.

Brandt, Willy. *North–South, a programme for survival*, London, 1980.

Brown, Lester. *State of the World 1984*, Worldwatch Institute Report, New York, 1984.

Eyre, S.R. *The Real Wealth of Nations*, London, 1978.

Myers, Norman. *The Gaia Atlas of Planet Management*, London, 1985.

Part Three

Third World Possibilities
Critique and Direction

14

New Paradigm Thinking
We Have Been Here Before

Claude Alvares

WESTERN institutions of philosophy and epistemology have a set pattern of destroying traditional world-views. They have either ridiculed them and 'proved' that they are insignificant, as anthropology has tried to do with 'other cultures'; or in traditions which have proved to be more resilient, they have intellectually co-opted and colonized them. Many prophets of the new paradigm, wittingly or unwittingly, are trying to colonize the 'ancient wisdom of the East' in an effort to rescue their intellectual tradition from bankruptcy. The work of Fritjof Capra, the doyen of the new paradigm movement, illustrates how the methodology of co-option works in practice.

Fritjof Capra's first successful book, *The Tao of Physics* (Wildwood House, London, 1975; Flamingo, London, 1986), created an intense feeling of euphoria among us 'orientals'. It seemed as if we had unexpectedly won an astonishing prize on a lottery ticket we had invested in centuries ago. Theoretical physicists from the advanced West, Capra reported, returning bewildered from the unfamiliar, weird terrain of the quantum world, and groping for words to describe the new landscape, had found a gratifying way out by falling back on the concepts and images of the speculative texts of the Hindus and the Chinese Taoists. Intellectual elites in Asia murmured silent approval of this confirmation of hoary wisdom. The *shastras* were right after all: the mythic dance of Shiva mirrored the actual behaviour of the elementary particles' pathways through the void.

Rumblings, premonitions of some of these connections between modern science and eastern mysticism had appeared before. Einstein, for example, had been attracted to the Indian metaphysics of Jagdish Chandra Bose. Oppenheimer, Heisenberg and Schroedinger had all mumbled vague vendantisms from time to time to describe the awesome and audacious worlds they had created with their mathematical models. And at least one major Indian philosopher, Sri Aurobindo, had attempted a grand synthesis between evolutionary theory and Indian metaphysics. What Capra did was to put all the allusions together in a fairly readable volume. The results turned out to be mixed. For laymen, the book provided fresh legitimation for ancient ritual. Scientists, however, found the comparisons Capra made between ideas in modern science and eastern mysticism appalling. Nobel Laureate Abdus Salam suggested that such comparisons cheapened both science *and* metaphysics.

In *The Turning Point* (Wildwood House, London, 1982; Flamingo, London, 1985) Capra admitted: 'None of [the book's] elements is really original . . . the interconnectedness and interdependence between the numerous concepts represent the essence of my own contribution.' The closest predecessor to the new book was *The Greening of America* by Charles Reich. In fact, the parallels between *The Turning Point* and *The Greening of America* are striking. Charles Reich was a sociologist, whereas Capra is a theoretical physicist, but both promise new, idealistic futures, brought in by what seem to them inherently compelling trends and forces. In *The Greening* Reich prophesied a takeover of American society by the flower generation, or Consciousness III. In his new book, Capra predicts a gradual but inexorable erosion of a world dominated by reductionist science, and a resurgence of a new society (a new America, actually) based on a holistic attitude towards all being.

'Capra begins *The Turning Point* by writing the obituary of the classical-physicist's approach to nature, after having interred Descartes, Newton and Galileo. But the ghost of a mechanical world picture continues to prevail in biology, psychology, medicine and economics: in these sciences, Capra finds reductionism triumphant. His principal aim is to show in detail the dire, negative implications of policies based on such science.

Thus, we have a biology without reverence for life, psychology without a psyche, medicine militating against health, economics with little common sense, and populations desperate for relief from the tyrannies of modern science. Yet, Capra observes, if one looks at the scene carefully, one can see a new vision emerging that seems capable of transcending the reductionist swamp. As evidence of this, he draws attention to the new systems' view of life, mind and consciousness, holistic methods of health care, fascinating integrations of western and eastern psychotherapies, new paradigms in technology and economics. This vision is also profoundly ecological, incorporates the demands of feminism, and is spiritual in its core. It will lead to profound changes in the organization of society and politics. What is also significant, writes our physicist, is that such a holistic vision comes closest to the metaphysics of non-western cultures.

The point could have been made more briefly. Capra can be dull and repetitive, for after all, what is being offered are essentially digests of books written by other western authors, including the ever popular Teilhard de Chardin, R.D. Laing, the Simontons (cancerologists), Rene Dubois, Gregory Batesom, Carl

Jung, Hazel Henderson and Ivan Illich.

Capra's proposal that the new physics looks similar to non-western philosophies earned him an invitation from the India University Grants Commission to deliver the Sri Aurobindo Memorial Lectures in 1980. This resulted in the publication of *The New Vision of Reality* (Bharatiya Vidya Bhavan, Bombay, 1985) which comprises the three Sri Aurobindo Memorial Lectures. The book is actually a fairly useful summary of both Capra's earlier bestsellers, being brief, tightly written and cheap.

It is important to go back in history to a period before Capra's arrival, to that other great synthesizer of western science and eastern mysticism, Sri Aurobindo himself. In Aurobindo's time, evolution was the dominant focus of the scientific world. With great self-confidence rooted firmly in Indian thought, the Pondicherry sage attempted to incorporate evolutionary insights within an Indian metaphysics of knowledge and experience. Before him, Jagdish Chandra Bose had attempted his own brand of synthesizing activity and had failed.

It is important to emphasize here that for both Aurobindo and Bose, the integrity of Indian metaphysics was never in question. After them, however, the state-sponsored legitimacy that western science was able to acquire in the Indian subcontinent led to a rapid devaluation of important constituents of Indian tradition. It is surprising but true nonetheless, that Indian tradition remained honoured more abroad than at home. The arrogance and clout of science-believers provided a further excuse to dismiss Indian metaphysics, as Macaulay had done a century earlier, labelling it all superstitious, irrational and evil. Gandhi's was the first major effort to raise the dignity of indigenous thinking.

Capra's effort differs from Aurobindo's in this significant sense: his base is not in Indian tradition, but in western science. He does not reject western science, but continues to hold modern physics as a reasonably reliable theory of knowledge. But like Aurobindo, his effort is to relate Indian thought once more to a dominant obsession of his time — this time, sub-atomic physics. Is Indian thought dignified or degraded as a result of the exercise? Aurobindo's efforts to marry western science and Indian thinking produced a result that was unrecognizable as either modern evolutionary theory or as Indian tradition.

A theory of knowledge that can suit different empirical facts, relating to different periods in man's history, has no truth value. Certainly Aurobindo's grand synthesis has mercifully passed into oblivion. If one ignores his disciples, we find that he has become for most others merely on additional fossil in India's intellectual history: there are some formidable glimpses or insights in his philosophy that still dazzle the mind, but the grand theory lies in shambles. Whilst his effort did no harm to evolutionary theory, it brought Indian tradition into grave danger by surrendering its claims to permanence or timelessness in exchange for a dubious claim of being fashionable and relevant to a particular scientific theory at a particular period in time.

The irony of Capra's invitation to Bombay, however, lay in the fact that it was now the land of 'eternal wisdom' that was calling in a mind with little evolutionary experience in that department, in order to shore up its own, more solid edifice of thought. It should actually have been the other way around. Massachusetts Institute of Technology should have invited a Sanskrit pandit to discourse on a possible new philosophical background to the new physics, if what Capra had written was plausible. And who knows, if Capra had not written his books, or if he had not been a theoretical physicist, we might have continued to *regret* our heritage in metaphysics, and carried on seeing it as an impediment to developing a great science and modern economy.

But there are some compensations in the Capra episode. We must remember that science in the Third World is seen as having almost divine qualities, and that even crimes can be justified as long as they are done in the name of science. How ironical then that our distinguished Indian scientists should now find their western colleagues, whom they held in higher regard than themselves, foraging in what they scorned as a rubbish heap! Here was a western theoretical physicist claiming that the most appropriate descriptions for the cosmology of the sub-atomic world were those used by Nagarjuna and Aurobindo.

For if Capra believed what he wrote, our science dogmatists and propagandists would have had no alternative but to call him a renegade, since they continued to assume that modern science remained the highest, the most reliable, in fact, the sole epistemology. If *I* abuse Capra in this essay this is because I have the contrary opinion: it is just not proper to make Indian metaphysics squat with a seventeenth-century, ethnocentric methodology. The values of both are directly opposed: they do not cancel out, but stand as two fuming bulls in the ring.

What does Capra specifically claim? There are three major ideas he broaches. First, he says, the notion of modern physics that we cannot decompose the world into independently existing, final, indivisible units, is not alien to Indian or Chinese traditions, both of which teach that it is the relations between things that constitute the real identity of things. As we are often told, the Tao is but the interconnectedness of things. Capra quotes atomic physicist Henry Stapp; who says: 'An elementary particle is not an independently existing unanalysable entity. It is, in essence, a set of relationships that reach outward to other things.' Now compare this, Capra says, with what Nagarjuna has said: 'Things derive their being and nature by mutual dependence and are nothing in themselves.'

The second point concerns relativity theory. When mystics intuit reality, says Capra, they seem to cross over in the fourth dimensional reality of Einsteinian space–time. How do we know, however, that reality has only four dimensions, or that the experience of mystics can be limited to such a four dimensional reality? Capra does not elucidate. Abdus Salam has suggested that unification theory may eventually demand the assumption of a thirteen dimensional real world. Finally, a dimension is an analytical tool: do such mental constructs mean anything to mystics at all?

The third major point that Capra makes is concerning the dance of Shiva:

modern physics proposes the notion of a continuous and dynamic interplay of primal forces manifest in the creation and destruction of sub-atomic particles/ events. This is precisely what the mystic dance of Shiva, the simultaneous creator and destroyer of the world, symbolizes. 'We can say with considerable confidence,' concludes Capra, 'that the ancient wisdom of the East provides the most consistent background to our modern scientific theories.'

In his second lecture, Capra poses the 'systems' view of nature against the reductionist dogma patronized by modern science, at least until the recent revolution in physics. Such a systems approach, he observes, is closer to the organic attitudes of most eastern traditions towards nature, or reality.

On closer examination, however, one discovers that the dichotomy between the systems or holistic approaches on the one hand, and reductionism on the other, is false. A systems or a holistic approach is still an *approach of the mind*, the latter an imperfect instrument that can never functionally match the capacities of intuition, mysticism or nature. In non-western cultures, in fact, the mind is barred effectively and rightly from pretending to be the primary epistemological medium: it is considered second class, a status that well befits its instrumental nature. Unless this is recognized, fundamental errors are going to be made. A mystic distrusts reason, recoils from discrete phenomena, resents separation. More important, mysticism directly encounters the ultimate, a claim from which science must methodologically bar itself.

For Capra, holism on the one hand and reductionism on the other are different extremes of a spectrum. 'We can see,' he writes, 'reductionism and holism, analysis and synthesis, are complementary approaches.' But this is nonsense. Both reductionism and holism are the constructions of science. And thus when one claims that the systems approach is a better scientific approach, and that it is also very similar to the organic view of life of the eastern philosophers, one is still exercising reductionism, reducing mysticism now to an understanding articulated by an analysing mind. Whereas the so-called mystical, tribal or metaphysical qualities of eastern traditions have one feature in common: they are a-scientific or, better still, trans-science.

What Capra is proposing then in his 'complementary' solution to the crisis in modern science is a totalitarian hypothesis. On the one hand, what he thinks is a reasonably reliable interpretation of reality, fabricated by analysis, by scientific method. On the other is this other view that has always issued from eastern traditions, which seems to be in agreement with the scientific picture today. Capra is not providing merely a new view, but a final picture of the world.

This is a tempting extrapolation from science. Aurobindo fell for it, and in the process laid the grounds for the rapid obsolescence of his philosophy. It is fairly clear that in a few decades our present understanding of reality will be quite different from what it is now. When such an occasion is reached, science, which maintains its right to modify, update or transform itself without compromising its capacity for 'truth', will continue on its enterprise, while the eastern traditions will have to suffer from being considered obsolete.

Capra's enterprise should be refused legitimacy for this one reason alone. Indian tradition has always ignored the claims of the analytical mind to achieve integral images of reality. Capra is overruling that same tradition when he proposes that the results of analytical thinking in modern science now equal the direct intuition of Indian (or Chinese) philosophers. In this sense, his disservice to India is greater than he realizes. For in his analysis, one is offered a final onslaught on the permanent qualities of Indian metaphysics — by having them rooted to the parochial, idiosyncratic perceptions of our era. The cosmic dance of Shiva will be frozen, like any figure in bronze, its timeless, enchanting imagery considerably diminished.

The attempt to bring western science and eastern tradition in agreement is an attempt to *improve science*. The metaphysical bleakness of science encourages constant foraging in other traditions. In that sense, Capra is basically using science in a fresh phase of colonization: whenever science is caught in a dead end, it looks around for new terrain. It usually overpowers other epistemologies by incorporating them. Thus, the systems view is not against the reductionist view: it merely makes up for the latter's crudity or deficiencies. It does not seek to displace reductionist knowledge, but uses it. This is obvious from the fact that modern science has respected neither mysticism nor the insights of non-western cultures overtly. The crisis today is a crisis of science, which is desperately looking for a humane metaphysics, not a crisis of Indian tradition: the latter hardly requires certification by what is, to its eyes, a lesser epistemology.

As Capra undertakes to examine a basically western pathology from a western standpoint and method, he can provide only false solutions. The West either cannot understand or concede that there are realities that are trans-science and trans-technique. This is its fundamental tragedy: reason's hubris. 'It is possible,' says Capra, 'to use a "bootstraps" methodology in a systems' approach': integrate various systems studies into a coherent method, so that they are internally consistent and provide an approximate understanding of reality or processes. But the mind itself again is a part of the bootstrap: being so, it must obtrude on any effort to approximate the whole. This it does because of its total dependency on presuppositions or assumptions. The mind cannot work without assumptions. But all assumptions must always distort reality. The West's efforts to work without assumptions, from Descartes to the Phenomenologists, have all failed.

'The current crisis therefore,' writes Capra, 'is not just a crisis of individuals, governments or social institutions; it is a transition of planetary dimension. As individuals, as a society, as a civilization, and as a planetary ecosystem, we are reaching the turning point.' The turning point imagery is taken from Toynbee's theory of the rise and fall of civilizations. As cultures mature into civilizations, they invariably stiffen and become more and more inflexible, unable to take on fresh challenges to their hegemony. At the same time, qualifies Capra, from within the womb of stratified societies, creative minorities can arise to take over the direction of consciousness, while the petrified majorities collapse.

This again is a crude form of reductionism, presupposing once again the predictive powers of reason. The western mind must forever remain trapped within the determinism of its own method and *The Turning Point* is an indication of this rather than any refutation of it. False prophets can prescribe only false therapies.

15

A Project for Our Times

Susantha Goonatilake

The flow of history opens windows and possibilities. It gives opportunities, propels movements forward and sometimes makes certain structured events seem almost inevitable. This apparent logic of history has given way in certain periods to the flowering of intellectual thought, as for example in the period circa the sixth century BC in the Gangetic Valley or during the period of the Renaissance. A particular conjuncture of socio-economic circumstances, a particular questioning of the givens of both history and reality makes such episodes almost a necessity. If we look at history from this perspective, is there a project for our times and for our place? The place being the broad continent of Asia and the time, the late twentieth century. What characterizes our times?

The world system, having acquired a momentum towards global hegemony from around the sixteenth century, is now undergoing considerable readjustment. Initiated by mercantile capitalism, then industrial and finally financial capitalism, the world system drew a near total social, economic and cultural blanket over the globe. Before completion of this process however, major fissures began to occur.

After the First World War, the Soviet Union detached part of its economic, social and cultural system from the global system though not its scientific and technological subsystem. Since the Meiji Restoration, Japan — under the slogan *Wakon Yosai* — has undergone a transition from agrarian feudalism to industrial polity and economy which have been closely intertwined, without going through the full bourgeois democratic experience of the western European countries. The slogan *Wakon Yosai*, 'Japanese spirit, western civilization' largely meant the uncritical absorption of western science and technology into an inegalitarian social system which was undergoing a relatively smooth transition from an unequal feudal society to an unequal capitalist one.

The period since the Second World War has seen the detaching in a formal political sense of chunks of Asia and Africa from the world system, a process Latin America had undergone in the early decades of the nineteenth century. The degrees of separation and of linkage depended largely on the internal socio-political dynamics that gave rise to the detaching process as well as to the external global environment. These dynamics varied widely in the different contexts such as those of say China, South Asia, West Asia and Africa.

Yet, in all these cases of separation from the hegemonic structure, the rupture has not been complete. Links, conscious and unconscious, have continued. On the economic front, planned economies based their prices partly on their external trade, taking cues from prices set by the world capitalist market. Although the product structure that was produced by these economies varied from one particular internal social system to another, they were by and large governed by the technological processes of manufacture which existed in Europe and America at the time.

Technology and science were generally considered by planning authorities as asocial and neutral. In socialist countries such figures as Boris Hessen in the Soviet Union of the 1930s, or Mao during the cultural revolution of the 1960s, took into account the social and cultural underpinnings of science and technology. Yet, the given existing science and technology was pursued as a desirable means or as a desirable end or both. In the capitalist countries of the developing world that disengaged in a political and economic sense, western capitalist science and technology was generally absorbed without question. In India, for example, Nehru perceived the new scientific establishment as the new temples of India; and in Japan science and technology was absorbed as constituting western civilization. Allowances were sometimes made for the organizational and social environment of the recipient countries, to enable the technology to be assimilated more easily. Thus, in the case of Japan, feudally derived structures such as lifetime employment and lifetime commitment to the company existed in the larger companies, leading to a distinctive internal social organization.

By and large this transfer of science and technology occurred on the assumption that they were both acontextual and asocial. The transfer was rooted by a process almost reminiscent of reverse engineering. That is, the given package of science and technology was sometimes dismembered (albeit sometimes only at the conceptual level), analysed and re-absorbed into the local system. Re-absorption took place at various levels; at school and university through the syllabus, at the blue-print or factory work bench or at the research level, implicitly and often explicitly by research programmes modelled after those used in the West. Thus the new science and technology was bought unquestioned.

Although social sciences in the West did not constitute a unitary frame and often separate themselves into various ideological strands, they were also by and large absorbed uncritically. The social sciences, we should recall, were derived from Enlightenment-influenced nineteenth-century social theorizing and their twentieth-century offshoots, which were all real intellectual responses to the historical experiences of one area — namely Europe.

Yet, in the developing countries they were internalized, unbundled as virtual acontextual 'technology' transfer packages. The result was that in capitalist or mixed economy countries, the social thought which was by and large sympathetic to the capitalist system was absorbed, whilst in the socialist countries the Marxist orthodoxy was internalized although variations of interpretation were allowed. In larger countries in the periphery, and especially in those where exposure to non-orthodox European thought such as that of Marxism had occurred, the several strands were internalized by different intellectual groups. Legitimation of knowledge in the developed centre takes place by and large through debate and social interaction whereas legitimation in the periphery is largely by imitation. Knowledge in the centre therefore takes a creative form, whereas in the periphery the social system of science tends to suppress creativity. Furthermore, because knowledge is mapped one-to-one in the periphery from many intellectual schools of the centre, various schools of intellectual fashions and models are absorbed in successive waves. They later settle as fossilized layers, incorporated in different groups which do not often interact organically across the layers.

These are some of the negative restraining features that characterize researchers and scientists and scientific knowledge in the Third World. It should be noted that these are social characteristics borne out of the Third World situation and not individual failings of the scientists themselves. Unproductive individual scientists from the Third World, once set in a productive First World research milieu, often break through their restraints. (For more details of how this functions, see my *Aborted Discovery: Science and Creativity in the Third World*.)

The Historical Construction of Knowledge

If that is the social model of science and technology in a dependent context, what is the model of knowledge in an 'organic' context? Recent research on the production of science in the West indicates that the scientific and technological enterprise is constantly buffeted by social forces at a series of levels. In short, scientific and technological output is at least a partial product of social forces in the environment. These social forces include micro-level social and historical changes as has been demonstrated by Boris Hessen (1930) on the physics of Newton, and by Forman (1971) on quantum physics and Dickson (1979) on algebra. In addition, an intermediate level of social change also impinges on science through, say, the impact of national policies on funding of science and generally, economic, political and social interests at the national level. (For examples of these influences see Mackenzie and Barnes (1975, 1979) on the biometry/Mendelism controversy; Farley (1975) on spontaneous generation/biology controversy; Mackenzie (1978) on statistics and Scharfstein (1979) on nineteenth-century cerebral anatomy, Ezrahi (1971) on the political impact of science in the USA and Rose and Rose (1976) on Lysenkoism.)

Science is also intimately influenced by the micro social world of the scientific community itself: especially at the micro level of the particular scientific group that is working on a problem. For examples of such social influences on science see Collins (1975) on lasers, Pinch (1976) on quantum mechanics, Wynne (1975) on the J Phenomenon, which cases describe the impact of the micro social environment of the scientists themselves on the output of science.

The process of knowledge creation is thus always influenced by the social environment, and what we experience as knowledge has been conditioned by changes in the social climate at three levels, namely at micro, at intermediate and at macro level. The flow of the stream of knowledge is governed and controlled at the personal level of the scientific worker by a process of detailed (social) programming individual scientists have been subjected to, such as in the ontological and epistemological assumptions of the given discipline. Thus a body of knowledge is created which is demarcated and legitimated by the social environment as being relevant, interesting and scientific. At the same time a delegitimizing process sets aside other 'knowledge' which is categorized as unscientific and irrelevant.

From this social historical perspective of knowledge, scientific knowledge, as it grows, takes the form and structure of an evolutionary tree of knowledge, social forces in the environment buffeting the trees and shaping the particular directions the branches of the knowledge tree develop into. The development and growth of the broad knowledge tree allows for bifurcations of its trunk into branches and the latter into yet smaller ones so that different scientific disciplines and sub-disciplines emerge.

The social output of knowledge results from what is essentially a worm's eye view engrained in a given discipline, brought through the detailed programming of the scientists, in their subject matter. A scientist at a knowledge frontier has almost by definition already crawled through the existing socially constructed knowledge tunnel of his predecessors and has acquired the knowledge of his predecessors. To this extent he can perceive only that which is immediately 'put' in front of him as interesting, and worthy of attention by the social process. The essential nature of science therefore becomes to a great extent relativistic in regard to different social contexts, different historical experiences giving different wormholes.

Thus, if we were to picture physical reality as a large blackboard, and the branches and shoots of the knowledge tree as markings in white chalk on this blackboard, it becomes clear that the yet unmarked and unexplored parts occupy a considerably greater space than that covered by the chalk tracks. The socially structured knowledge tree has thus explored only certain partial aspects of physical reality, explorations that correspond to the particular historical unfoldings of the civilization within which the knowledge tree emerged.

Thus entirely different knowledge systems corresponding to different historical unfoldings in different civilizational settings become possible. This raises the possibility that in different historical situations and contexts sciences very different from the European tradition could emerge. Thus, an entirely new set of 'universal' but socially determined natural science laws are possible. It should be noted that the social changes that took place under the dominance of western power, and the sciences that these gave rise to, cannot be considered universal. True alternatives in science, at least as a social and historical possibility, exist for other different civilizational contexts too.

Our Modern Science and Technology Times

Outside the European periphery a dependency which continues to get its cues from the historically derived science and technology structures of Europe exists. This implies that historically speaking the knowledge structures of non-Europe have been marginalized, and its earlier searches for valid knowledge are not being incorporated in their current scientific and technical enterprises. Science and technology practitioners in the periphery are *par excellence* a class of historically marginal men — almost by definition. They live bifurcated lives, the earlier knowledge structure — even its valid areas of exploration — are laid aside, sometimes as an embarrassment, sometimes as an area to which ritual *pooias* are made about a past greatness.

However, the past of the civilizations about which we have evidence (that means only the cases of literate cultures), indicates that there was a virile, dynamic tradition of searching for knowledge. This has been shown, say, in the case of China, by the work of Joseph Needham and his co-workers, in the case of South Asia by a bourgeoning set of researchers over the last fifteen years (Rahman 1975, 1977; Alvares 1979; Dharmapal 1971; Chattopadhyaya 1976 among others), and in the case of Islamic science an emergent literature which puts into new perspective the Islamic European connection (see for example *Proceedings of the Islamic Science Conference 1983 Islambad*).

This literature indicates that the science in these civilizational areas was, for its times, flourishing. Often, what has been ascribed to the European tradition has been shown on closer examination to have been done elsewhere by others earlier. (Thus Harvey was not the first to discover the circulation of blood, but an Arabic scientist was; Paracelsus did not introduce the fourth element 'salt' and start the march towards modern chemistry, but a twelfth-century alchemist from Kerala did so teaching in Saudi Arabia.)

Further, the new bourgeoning literature on pre-colonial non-European science indicates that there were many areas of science and technology that could have been further developed, but were left unexplored and stagnant and then forgotten because of the later dominant impact of the western system. Such examples vary from effective plant-based medicines drawn from the Ayurvedic and Unani pharmacopia (for example Raufalfia), to theories of psychology (for instance see the similarities with, and the tributes to, Asian psychologies by Maslow and other Humanistic psychologists), to different forms of logic

(Buddhist four-valued logic and Jain seven-valued logic vs. the traditional Aristotelian kind).

The output of scientific knowledge in the developed world is increasing at a rapid rate. This increase is going to raise problems of limits to growth resulting from qualitative changes in science. De Solla Price has over the last three decades been documenting evidence showing that ever since the western scientific revolution in the seventeenth century, various indicators of science double every seven years or so. Thus, the number of scientists, the number of scientific papers, the number of scientific societies follow this trend. If these trends are extrapolated from today's figures, one notes that roughly within the first quarter of the twenty-first century, (that is, within the life time of many scientists and would-be scientists living today) certain strange features appear. Thus, the number of scientists becomes greater than the population of the earth, the total world science budget becomes greater than the total world gross product and finally the bulk of scientific publications would become more than the mass of the earth.

The electronic revolution would save mankind from the last fate by storing information digitally and it could even give lie to the first prediction by seeing that more and more artificial-intelligence based analyses of data combined with robotic laboratories (both of which are now on the cards) would lessen the need for humans. This would either mean that within the next few generations human intervention in the scientific enterprise becomes qualitatively less important or that the human scientific enterprise will have to change in a more creative direction.

A significant characteristic of the science of our times is a crisis in epistemology that has occurred in two core areas of the scientific enterprise. For nearly two generations now, an epistemological crisis of a fundamental nature has haunted quantum physics. Reality as depicted in quantum physics has drawn away from the everyday experience of humankind to one depicted through the lenses of statistical abstractions. Thereby an ontological and epistemological Pandora's box has been created leading to raging debates on the fundamental nature of quantum reality. Secondly, ever since Godel's Theorem was postulated nearly fifty years ago, a similar crisis has gripped mathematics. Certainty has vanished leaving a fundamental crisis (Kline 1981).

The traditional nineteenth-century derived science and technology which has spawned many of the industries is energy and pollution intensive and dehumanizing. Yet in many industrialized countries, the traditional nineteenth-century image of smog-ridden cities through which polluted rivers flowed has been erased by concerted action (London for instance is much less polluted today than in Dickensian times). On the other hand, pollution-based industries have been exported to the Third World and in the absence of strong countervailing forces and regulating mechanisms, ecological disasters haunt the region.

At a global level more serious possibilities such as a greenhouse effect due to accumulated carbon dioxide melting the ice caps and devastating vast regions of the earth exist. The current contraction of genetic diversity across the globe and the elimination of species has been estimated to be of the same order of magnitude as the major catastrophes that occurred in the geological past.

The social sciences born largely in late eighteenth century and nineteenth century Europe — the conservative variety as well as the Marxist alternatives — are in a state of crisis. The crisis exists at the level of the fundamental conceptualizing of social forces and of their dynamics. The grand old laws of the motion of society seem today to reveal only creaking machinery seen through partially opaque lenses. The crisis in Marxism has become very apparent too within the last decade. With the rapid changes now occurring in China, Hungary, Poland and other Eastern European countries, Marxist orthodoxy as an explanatory tool and guide to action is under strain. (For a recent macro perspective illustrating this crisis on thinking see Wallerstein's paper (1986) titled 'Should We Unthink the Nineteenth Century?' or if you want a more 'native' and an earlier version of the same perspective see Goonatilake 1975, 1982a, 1982b, 1983, 1984.)

Thus the world of science and technology is, in a fundamental epistemological sense, in a state of crisis. This is occurring at a time when great shifts in economic power are taking place globally and also at a time when intellectual self-awareness and self-confidence is rising in parts in Asia. This fundamental uncertainty combined with the rising confidence clearly calls for a major 'project' for Asian creativity in science, bold in scope, fuelled by conscientized Asian scientific workers, launched with confidence and with the ability to hunt for and seek its goals.

The Search for Creativity

If the social structure of science in the Third World fits our description, how can we provide the social conditions for fundamental creativity in the Third World?

One method is of detaching a country from the existing international base of science. This had been attempted in varying degrees — in China during the cultural revolution and in a much more extreme form during Pol Pot's regime in Cambodia. However, at the time of detachment, the intellectual system is that of an imitative one and therefore a real danger of preserving fossilized handed-down truths at the time of cut-off exists. Revolutionary situations allow for internal social arrangements that could possibly restructure science. However, the existing literature on developing science in non-European contexts, even in revolutionary regimes does not give adequate examples of creativity. In fact recent criticism and exposure of the cultural revolution indicates that a real alternative scientific community did not develop for a variety of reasons. As a reaction the Chinese scientific system seems to be attempting a wholehearted attempt towards relinking with the former centres.

Transcending the links could be another strategy for Third World researchers. Such transcendence at the social psychological level would be to detach a scientist from a given micro social structure of science. This routinely happens in the developed countries when new sub-disciplines are formed with new sub-groups and corresponding sub-cultures of science. Such detached scientists in a Third World context would operate outside the existing formal system and develop alternative scientific social linkages.

An attempt to break out creatively would also require psychological consideration. The present programming of scientists through the educational system in most Third World countries puts great emphasis on routine learning and mimicry. The intuitive and the aesthetic, which are vital for creativity, are often subsumed at the expense of the mechanical and the verbal.

I argued earlier that the different perceptions of science and its ruling paradigms in different disciplines give a particular tunnel vision of knowledge. If our efforts in creativity are to develop new modes of perceiving reality through science and new tunnels of knowledge, how does one escape out of the present structured field of knowledge onto new pathways of knowledge? Several writers have emphasized that multi- and inter-disciplinary studies may be one such mode, but these attempts have their limitations. Multi-disciplinarity is an attempt to bundle together, perhaps rather haphazardly, several disciplines. The constituent disciplines leave many areas of physical reality unexplored, thus multi-disciplinarity and inter-disciplinarity attempts at best to provide additional glimpses of physical reality largely through existing channels of perceptions with no fundamentally major breakthroughs or new structured knowledge.

We know that, unlike the assumptions of a conventional mechanistic view of science, cross-flows of metaphors had often occurred in creative situations. Thus, specially at paradigm breaks, the intellectual elements that nourish a discipline constitute a surprising collection. These elements include a priori orientations of scientists which also encompass mystical and 'extra-scientific' beliefs, modes, metaphors and orientations from other often unrelated disciplines and perspectives, as well as beliefs drawn from the social ideas of the day. (For example see Rattansi 1973; Debus 1973; Elzinga and Jamison 1981.) Displacement of ideas across disciplines has been recorded by many. The impact of Malthus on Darwin and Darwin on Social Darwinism, social ideas on organic chemistry (Slack 1972), thermodynamics (Brush 1967), and relativity (Feuer 1971) are a few. However, it should be noted that although science in its growth takes in external elements outside a discipline, it is structured socially and a social perspective governs implicitly the process of discovery.

Drifts of metaphors from one discipline to another, a very fruitful source of new ideas, should not be confused with the relative aridity of conventional multi-disciplinary and inter-disciplinary studies. Drifts of metaphors could also arise from one's own culture, even from its past. Thus scientists from Dolton, Schrodinger (1958) to Heisenberg (1973) have mined the western past for new breakthroughs in their discipline.

To give an example from the South Asian region, the past traditions have a wide variety of schools that describe the nature of matter. Some deny atomism, while the Charvaks, and Jains and the Buddhists subscribe to many varieties of atomism. Some allow for particles to exist, while others extend the atomic view to time, time being viewed as quanta. Here is a rich minefield of metaphors that could be used to build new views. Holton (1973) has pointed out that from Greek times the theories of science have been built up by a small collection of what he calls themes and anti-themes. Examples of these are complexity-simplicity, reductionism–holism and continuity–discontinuity. If these constitute some of the ultimate building blocks of existing theories of science, further themes and anti-themes or derivatives thereof could be found from other cultures. This would give a larger thematic mix which would provide for a greater variety of building blocks for new concepts and theoretical frameworks.

Searching for the past should not be confused with efforts by apologist writers from the East who attempt to show that the eastern is contemporary and modern. This defensive attitude is often confusing and contradictory. Thus a cosmology of the big bang, the steady and oscillating state, have been shown to exist in Indian writings. Similarly writers have sought to demonstrate that Buddhism is equivalent to logical positivism, idealism, that it is progressive and close to Marxism as well as reactionary. (Goonatilake 1984: 162). In contrast, western writers have at times used the East to provide metaphors for new developments. This is a more fruitful exercise and one that is not taken seriously by Asian scientists.

Transport of metaphors and formal knowledge is easier in applied knowledge systems where local variables of ecology, climate and art meet with science. A clear example would be architecture where new formal architectural disciplines to replace the dominant European twentieth-century tradition are waiting to be developed. Such disciplines would bring in local climatic situations, the socially determined local silent language of spatial arrangements and sitting and living modes, as well as local aesthetics that could be transferred into a new synthetic mode of knowledge combined with existing western knowledge in engineering.

In a similar vein of thinking one could suggest other possible attempts drawn from Asian roots even in such areas as quantum physics and artificial intelligence. Thus quantum physics has particular problems of conceptualization which defy everyday western conventions, possibly because the given mathematics of the western tradition is based on a two-valued Aristotelian type logic which does not fit into such quantum phenomena as 'a particle is there/is not there'. Here, some of the multi-valued logics developed in the Asian tradition — especially aspects of Buddhist logic — have the potential of formalizing themselves into a mathematics that would take into account the peculiarities of twentieth-century physics.

Possible Asian inputs into artificial intelligence could make use of the fact that Asia has a vast store of knowledge on 'thinking about thinking'. The argument goes as follows: existing efforts at artificial intelligence have been at least partly attempts to operationalize formal descriptions of human thinking (for example, the 1950's attempts at deductive logic in the form of programmes to prove geometry theorems and 1970's attempts at inductive logic, for example the Bacon Programme). Programmers who are knowledgeable about eastern thought could abstract and formalize 'thinking about thinking' into actual artificial intelligence software.

The social science that was born in the late eighteenth and nineteenth centuries used as independent variables such factors as technology, the existing and past social relations of Europe, on which to build the major social theories. Whether the theories were those of Adam Smith, Karl Marx, Max Weber or Emile Durkheim, the ghost of European experiences haunted the subject matter, the explanatory variables used, as well as the particular trajectories within which the flow of history was traced. Whether the dichotomies were *Gemein-schaft* or *Gesselschaft* or feudalism vs. capitalism or other categories, the European experience was considered universal. The need to rethink this ethnocentric social science is acutely felt today.

The social sciences were particularly governed by subjective perceptions when viewing non-European peoples. Subjects such as Indology, Sinology and Anthropology formalized the study of supposed inferiors. This was the necessary accompaniment to a self-confident, aggressive view of the world that was internalized in the European psyche. European man was using the rest of the world as raw material for his advance and these studies, which were ideological developments paralleling the European advance, explained the rest of the world in terms of the need to dominate.

New social sciences from an Asian perspective would necessarily have to look at and explain Europe from a fresh perspective. And just as European fomal studies of other cultures used a self-serving viewpoint, any study of Europe by other cultures should likewise incorporate *ipso facto* such a self-serving viewpoint. Such an area of study would rightly have to be called — paralleling the nomenclature of Indology and Sinology — a Europology, a self-serving subject for non-European peoples. I have used the word Europology here heuristically to refer not only to geographical Europe but also to its settler bastion and cultural outposts in North America, Australia and New Zealand.

At the moment Europe and European culture have symptons of self-doubt: parts of the periphery are intellectually waking. The economic world has been shaken by geo-political tremors. Such events as the Soviet and Chinese revolutions, the Middle East oil boom and the emergence of successful capitalist states in East Asia have made for certain shifts in centres of economic power. The centre of the capitalist world according to many accounts is now shifting towards the Pacific belt with the possibility of East Asia playing a dominant role in the future.

The topography of the world of knowledge before the last few centuries could be delineated as several hills of knowledge roughly corresponding to the regional civilizations of, say, West Asia, South Asia, East Asia and Europe. The last few centuries have seen the levelling of the other hills and from their debris the erection of a single one with its base in Europe. This however, is not a 'world' hill, it is a very particular hill, not a universal one. The topography that should now emerge is again one of several hills. The search for a truly universal hill and of a truly 'universal' global science can begin only after this re-emergence. The project for our times is to create the new hills, in our own backyard.

Bibliography

Alvares, Claude. *Home Faber: Technology and Culture in India and West 1500-1972*, Allied, India, 1979.

Brush, S.S. 'Thermodynamics and History' *The Graduate Journal*, 7, 1967.

Chattopadhyaya, D. *Science and Society in Ancient India*, Research India Publication, Calcutta, 1976.

Collins, H.M. 'The Seven Sexes: A Study in the Sociology of a Phenomenon or the Replication of Experiment in Physics', *Sociology* 9, 1975.

Debus, Allen G. 'The Medico-Chemical World of the Paracelsians' in *Changing Perspectives in the History of Science*, Mikulus Teich and Robert Young (eds.), Reidel, Holland, 1973.

Dharmapal. *Indian Science and Technology in the Eighteenth Century*, Impex India, New Delhi, 1971.

Dickson, C. 'Science and Political Hegemony in the 17th Century', *Radical Science Journal*, no. 8.

Elzinga, Aant and Jamison, Andrew. 'Cultural Components in the Scientific Attitude to Nature: Eastern and Western Modes?', Technology and Culture, Occasional Report Series no. 2, Paper no. 146, May 1981.

Ezrahi, Yavon. 'The Political Resources of American Science', *Science Studies*, 1 (2) 1971.

Farley, J. *The Spontaneous Generation Controversy from Descartes to Oparin,* Johns Hopkins University Press, Baltimore, 1977.

Feuer, L.S. 'The Social Roots of Einstein's Theory of Relativity', *Annals of Science*, 27, 1971.

Forman, Paul. 'Weimer Culture, Causality and Quantum Theory 1981-27: Adaptation by German Physicists and Mathematician to a Hostile Intellectual Environment', *Historical Studies in the PhysiSciences*, University of Pennsylvania Press, Philadelphia, 1971.

Gilbert G.N. 'The Development of Science and Scientific Knowledge. The Case of Radar Meteor Research' in O. Lemaino (ed.), *Perspectives on Emergence of Scientific Discipline*, Mouton, The Hague and Paris.

Goonatilake, Susantha. 'The Social Sciences, their Present State of Flux, and the Third World Problems', Paper read at Agrarian Research and Training Institute, Study Circle, Colombo, November, 1975.

———. 'Social Production of Technology and Technological Determinants of Social Systems', Proceedings of the International Seminar on Science,

Technology and Society in Developing Countries, Bombay, November, 1979.

———. *Crippled Minds: Exploration into Colonial Culture*, Vikas Publishing House, New Delhi, 1982(a).

———. *Colonial Culture and Endogenous Intellectual Creativity*, United Nations University, Tokyo, 1982(b).

———. 'Dependence of Third World Science and a New Fall of Constantinople', International Conference on Science in Islamic Polity — its Present, Past and Future, Islambad, 1983.

———. *Aborted Discovery: Science and Creativity in the Third World*, Zed Press, London, 1984.

Heisenberg, W. 'Traditions in Science', *Bulletin of the Atomic Scientists*, December 1973.

Hessen, Boris. 'The Social and Economic Roots of Newton's Principle', International Congress of the History of Science and Technology, London, 1930.

Holton, Gerald. *Thematic Origins of Scientific Thought: Kepler to Einstein*, Harvard University Press, Cambridge, Massachusetts, 1973.

Kline, Morris. *Mathematics: the Loss of Certainty*, Oxford University Press, Oxford, 1980.

Mackenzie, D. 'Statistical Theory and Social Interests: A Case Study', *Social Studies of Science* **8**, 35–83, 1978.

Mackenzie, D. and Barnes, S.B. 'Biometirican versus Mendeliant: A Controversy and its Explanation', *Kother Zeitschrift fur Soziologie*, special edition, **18**, 1975, pp. 165–96.

———. 'Scientific Judgement: The Biometry-Mendelism Controversy', in *Natural Order*, Barnes and Shapin (eds.), 1979.

Pinch, T.J. 'What does a proof do if it does not prove', in E. Mendelsohn, P. Weingart, R. Whitley (eds). *The Production of Scientific Knowledge*, Reidel, Dordrecht, Holland, 1976.

Rahman. *Bibiliography of Source Material on History of Science & Technology in Medieval India, an Introduction*, Council of Scientific and Industrial Research, Indian National Science Academy, New Delhi, 1975.

———. *Triveni: Science, Democracy and Socialism*, Indian Institute of Advanced Study, Simla, 1979.

Rao, M.S.A. 'Introduction', in *A Survey of Research in Sociology and Social Anthropology*, vol. 1 Indian Council of Social Science Research, Popular Praskashan, Bombay, 1974.

Rattansi, P.M. 'Some Evaluations of Reason in Sixteenth Century Natural Philosophy' in Mikulas Teich and Robert Young (ed.) *Changing Perspectives in the History of Science*, Reidal, Dordrecht, *Holland,* 1973.

Rose, Hilary and Rose, Stephen. *Science and Society*, Penguin, Harmondsworth, 1970.

Scharfstein, Ben Ami. *Philosophy East/Philosophy West*, Basil Blackwell, London, 1970.

Schrodinger, Erwin C. *Science, Theory and Man*, Dover Publications, New York, 1958.

Slack, J. 'Class Struggle among the Molecules', in *Countercourse*, T. Psteman (ed.), Penguin, Harmondsworth, 1972.

Wallerstein Immanuel. *'Should We Unthink the Nineteenth Century?'* State University of New York, Binghamton, 1986.

Wynne, B. 'C.G. Barkla and the J Phenomenon. A Case Study of the Treatment of Deviance in Physics', *Social Studies of Science*, **6**, 1976.

16

Islamic Science, Western Science
Common Heritage, Diverse Destinies

Seyyed Hossein Nasr

To understand all that separates traditional science in its world-view, methodology, goal and significance from modern science, a comparison between Islamic and western science is revealing. There are many different forms and schools of traditional science such as the Egyptian, Indian and Chinese,[1] but either they were cultivated in areas far removed from the stages of the development of western science or they preceded it by many centuries and, therefore are seen more as the historical background of western science or as distant developments rather than as a parallel tradition. In the case of Islamic science, which is one of the most important schools of traditional science both because of the wealth of its achievements and the survival of its teaching, there is the extraordinary phenomenon of the growth of a major scientific tradition which shared more or less with the West the common heritage of antiquity and a similar religious and philosophical universe, but which, in contrast to what occurred in the West during the Renaissance and the seventeenth century, remained faithful to the traditional point of view. Moreover, this tradition was itself influential in the rise of medieval science in the West, which still possessed a traditional character before the scientific revolution, and yet Islamic science did not share in any way those upheavals which transformed the science cultivated by Robert Grosseteste in his treatise *On Light* to the physics of Newton's *Principia*.

It is true that Islam inherited certain aspects of the scientific heritage of the Mediterranean world that were not known to the West, in addition to the sciences of India and ancient Persia which reached the Occident through Islamic science itself. Islam absorbed nearly the whole corpus of Aristotelian science including the works of the Alexandrian commentators, Platonian cosmology, most of the important scientific achievements of Alexandria and its satellites in Pergamon, and the more esoteric strands of Greek science associated with both Pythagoreanism and Hermeticism.[2] Muslim scientists, moreover, became acquainted with Sassanid astronomy and pharmacology and the Indian sciences, especially medicine, astronomy and mathematics. They also obtained knowledge of certain aspects of Babylonian science that had not been transmitted to the Greeks.

Not all of these strands of the sciences of antiquity reached the Christian West. Much of the Aristotelian heritage, Hermeticism and Pythagoreanism remained unknown in Europe until the second millenium of the Christian era. It might be argued that the heritage of western science and Islamic science were therefore not the same. But the fact remains that both were heirs to the sciences of the same world and their knowledge of the natural order, concept of law, causality and general cosmology drew from the same sources although each developed these inherited concepts differently. Moreover, even if this difference of early heritage is accepted, the West itself became heir to early Islamic science and, through this science, to the sources which Islam tapped and itself developed for several centuries before the translations made in Toledo in the eleventh century began to make Islamic science available in Latin. Even if one leaves aside the earlier history of science in the West which points to the much richer development of Islamic science from the eighth to the eleventh century than anything to be seen in Christian Europe about the same time, the fact remains that by the thirteenth century medieval European science was developing along lines parallel to and usually based upon Islamic science. These two traditions were much closer to each other than medieval Latin science and Chinese science or even Indian and Chinese science.

Since Christianity and Islam belong to the same family of religions and the philosophical schools of Islam soon came to find their counterparts in both western Judaism and Christianity, one might have expected science to develop in the Christian West along lines similar to those which one observes in traditional Islamic civilization.[3] This parallelism would seem to be especially dictated by similarity of methods, cosmological and philosophical ideas concerning matter, motion, etc. and the goal and end of the sciences of nature as a means of discovering the wisdom of God found in both Islamic and medieval science. The school of Chartres, Albertus Magnus, Robert Grosseteste, Roger Bacon, Raymond Lull and many others seemed to be cultivating sciences very similar in nature, method and scope to those of the Muslims from whom they had learned so much.

Yet, in the West by the fourteenth century nominalism was already gaining the upper hand in theological circles while Christian philosophy was gradually being eclipsed. While science in the West was still of a basically traditional character during what is called the Renaissance, philosophical ideas based on rationalism and humanism were becoming dominant and preparing the ground for that scientific revolution which was brought about by Descartes, Galileo and finally Newton. Between Robert Grosseteste and Newton, at least the Newton of the *Principia*, or Roger Bacon and Francis Bacon, a transformation took place in the meaning of science which was neither emulated nor repeated independently in the Islamic world. The modern astronomy and physics of a Galileo or Newton were based on an already secularized view of the cosmos, the reduction of nature to pure quantity which could then be treated mathematically and a complete separation between the knowing subject and the object to be known based on Cartesian dualism.[4] A new science was indeed born, one which discovered much in the realm of quantity, but at the expense of the traditional world-view and neglecting the spiritual dimension of nature — the bitter fruits of which are only now being full tasted.[5]

In contrast to these transformations in the West, in the Islamic world the sacred character of God's creation continued to dominate the intellectual horizons of man. The symbolic sciences of nature as expounded during previous centuries from the time of Jabir ibn Hayyan to that of Suhrawardi continued to be cultivated while mathematical and physical sciences continued to be studied in the midst of the symbolic sciences and in the light of the metaphysical and cosmological principles derived from Quranic revelation. 'On the philosophical level such figures as the contemporary of Descartes, Sadr al-Din Shirazi, added a significant new chapter to the Islamic philosophy of nature; in the sciences themselves innovation decreased but their application continued in domains such as architecture and the making of dyes. A civilization which created the Shah Mosque of Isfahan or the Taj Mahal in India cannot be simply dismissed as being of no significance in the realm of sciences and technology, nor can the existing traditional science be considered insignificant simply because it did not change and develop in the manner of western science. Muslims continued to create and to preserve glories of art and thought, of traditional technology and science within their own world-view based on the harmony of man and nature and awareness of the spiritual significance of nature in man's life, while the West was developing a science based on considering nature as a thing or an 'it' to be quantitatively studied, conquered, controlled, manipulated and finally, despite the opposition of many scientists, raped with such ferocity that the results now threaten human existence itself. The process continued until the applications of that science based on power rather than contemplative wisdom provided such military advantage to the West that it was able to colonize most of the Islamic world and finally destroy, if not completely, at least to a large extent the homogeneous Islamic civilization that had developed parallel to that of the West for so many centuries.

Since the Eurocentric conception of history was taken for granted even in the intellectually colonized East, the development of science in the West was considered for several centuries as the crowning achievement of the whole history of science of mankind. Modern science was considered as the only valid science of nature and the question of the parallel development of science in another civilization was rarely posed. It is only now that the horrors of modern war and the ecological disasters brought about by the application of modern science along with the unprecedented alienation of man from God, nature and himself have become manifest that one may even ask about parallel developments of science elswhere. One can at last ask not why Islam or China with their long and rich scientific traditions did not produce a Descartes or Galileo, but rather why Europe did. To understand the roots of the crisis of present day humanity, it is necessary to address this last question and especially to enquire into the factors which caused the destinies of science in the West and the Islamic world to become separated and for the two civilizations to part ways.

The factors that led to the development of science in the West in a manner that science cultivated in the schools of Chartres or Oxford in the thirteenth century, and that of Paris and Oxford four centuries later seem to belong to two different universes rather than to a single civilization. The factor most responsible for the difference between the destinies of science in Islam and the West is the eclipse of the sapiential aspect of Christianity toward the end of the Middle Ages in contrast to Islam where this tradition has continued to the present. The gnostic and sapiential mode of Christianity was flowering in the teachings of such figures as Dante and Meister Eckhardt when the paganism of antiquity intruded in the form of Renaissance humanism.

Every science of nature relies upon a world-view concerning the nature of reality. Medieval Christianity shared with Islam a world-view based at once upon revelation and a metaphysical knowledge drawn from the sapienial dimension of the tradition in question, although, as far as the metaphysical significance of nature was concerned, this knowledge was not fully integrated into the mainstream of Christian thought. Once this knowledge was eclipsed and for all practical purposes lost, there was no means whereby a science based on metaphysical principles could be cultivated or even understood.[7] Without such a knowledge the traditional sciences became opaque and even meaningless. Soon they ceased to satisfy man's needs for causality. A vacuum was created which men sought to fill by means of a rationalistic philosophy grounded outside of the Christian tradition and a science of a purely earthly nature but which was satisfactory from the point of view of rationalism and empiricism. Having lost the vision of heaven, men discovered a new earth whose finding they considered as ample compensation for the heaven they forgot so rapidly.[8] Without metaphysical knowledge, the traditional sciences could not survive. First, they reappeared as occult sciences shorn of their metaphysical significance and finally their residue survived as mere superstition in the eyes of those for whom any science pointing to metaphysical principles beyond themselves and to realms of reality beyond the physical could not but be superstition. All that had been considered the highest form of knowledge became subverted to mere conjecture and shorn of the dignity of being called science while all that was accepted as science was accepted as such under the condition that this form of knowledge had no relation to any knowledge of a higher order. No single factor was as significant in the parting of ways of the West not only from Islam but from all other traditions than the loss of gnosis or sapience and the ever increasing eclipse of the metaphysical dimension of Christianity from the thirteenth century onward.

A closely related factor is the rise of nominalism in the fourteenth century. By depriving intelligence of the possibility of knowing the Platonic archetypes or ideas of things and in fact denying the very meaning of universals as possessing reality beyond that of names, nominalism affected profoundly not only theology but also philosophy. Nominalism, by basing religious truth upon faith rather than upon both faith and knowledge had no small role in secularizing knowledge and preparing the ground for the rise of modern science. The destruction of medieval Christian philosophy based upon ontology could not but lead, after a period of uncertainty and groping, to that rationalistic philosophy associated with Cartesianism which served as the necessary basis for the seventeenth century scientific revolution.[9] Without the withering criticism of nominalism, medieval Christian philosophy and theology would not have relinquished their claim to the role of knowledge in discovering the nature of things in the light of higher principles, leaving them undefended before the onslaught of secularism, rationalism and empiricism which were, as a result, able to gain a remarkably easy victory.

The domination of nominalism combined with a tendency to substitute logic for philosophy was both the result of the loss of the symbolic science of nature and instrumental in the destruction of such a science which is always wed to metaphysics. Medieval European man still understood the language of symbolism which dominated his art and science as well as nearly every level of expression of his religion.[10] A medieval cathedral is an expression of a symbolic and sacred science of the cosmos and in turn enables man to gain access to the realities to which such a science leads, provided he still possesses that symbolist spirit which Western medieval man shared to a large extent with the rest of humanity.

In the late Middle Ages there already had appeared this rationalistic tendency which had lost sight of the symbolic content of the traditional sciences of nature. Although the life of the symbolic sciences of nature did not cease completely, as seen in Hermeticism and to a certain extent Pythagoreanism which continued to be cultivated in certain circles, the intellectual arena of western Europe became ever more occupied by thought which was impervious and even blind to the language of symbolism. This type of thought helped destroy further the influence of symbolic modes of thought while a type of mentality blind to the symbolic significance of nature as well as scripture only helped to strengthen nominalism and rationalism. Soon, symbols became reduced to signs and facts; both the book of nature and the book of revelation were reduced to their literal and external level of meaning. Parallel with the loss of the symbolic sciences of nature, there occurred a marked decrease of interest in sapiential commentaries upon sacred scripture. What remained was a literal and external interpretation of religion left face to face with a science of the literal or factual aspect of nature which could then obviously see nothing in nature but brute facts to be gathered empirically and understood only rationally within a science that could no longer have any relation with the existing religion except in confrontation or indifference. Science developed in a direction in which it could no longer concern itself with whatever those facts or laws established by it could possibly signify beyond themselves.[11] A Jacob Bohme could still cultivate a symbolic science of nature in the seventeenth century; but he was not at the centre of scientific activity and not even in the mainstream of the religious and theological thought of his day.

Parallel with this loss of a symbolic science of nature one can observe the rapid process of the desacralization of the cosmos in medieval Christian thought. Early Christianity, faced with the danger of naturalism in the Graeco-Roman world and seeking to prevent at all costs the danger of cosmolatry,[13] did not emphasize the spiritual significance of nature and even drew a rigid line between nature and super-nature. Yet, in the early Middle Ages the religious significance of nature was not forgotten, at least in the writings of such men as Erigena and Hugo of St Victor, while the traditional Christian cosmos continued to be populated by the angels and spirits. Already by the thirteenth century, however, the more dominant schools of Scholasticism began the philosophical and theological process of desacratizing the cosmos and thereby making it a suitable object of study for a purely quantitative science of nature.

The reception given to Ibn Sina in the Latin West in comparison with that given to Ibn Rushd is indicative of this trend. Avicennan cosmology emphasizes the significance of angels who carry out the command of God in the cosmos and who make possible its life and order. For Ibn Sina cosmology is inseparable from angelology.[14] For Ibn Rushd, however, the 'souls of the Spheres' which are identified with angelic substances are dispensed with in favour of the intelligences. The fact that the Latin West was influenced by Averroism much more than by Avicennism and that even those deeply influenced by Ibn Sina tried to brush aside the central role he accorded to angels in both his cosmology and epistemology points to an important tendency taking place at that time. This tendency concerns the refusal to accept the angels of the traditional Christian cosmos as being essential and necessary to the governance and functioning of the cosmic and natural order. The traditional cosmos became thus philosophically and theologically prepared to be treated as that great mechanical clock whose laws would be discovered by a Galileo and a Newton by means of a mechanical science to be born through the seventeenth century scientific revolution. The fascination with mechanical clocks was already present in Europe long before Galileo wrote his *Discorsi*. Likewise, the angels had ceased to be considered as being metaphysically necessary to the running of the cosmos long before the advent of the seventeenth century philosophers and scientists even if ordinary men continued to believe in them. It seems as if the vision of nature in the mind of European man had already gained a strong mechanical component before an actual science based on the mechanistic point of view developed. Moreover, this science in turn helped to generalize and expand the mechanistic philosophy to such an extent that by the eighteenth century it had become part and parcel of the world-view of European man, going beyond the confines of the sciences of nature to embrace the whole philosophical world-view of the mainstream of western thought.

The desacralization of nature and the cosmos was abetted by the practical quest for gaining power over nature. The traditional sciences of nature sought to lead man to wisdom and enable him to perfect his soul through the contemplation of divine wisdom in his handworks. Even in alchemy where there was an attempt to accelerate the natural processes of giving birth to gold and even to gain power over nature, the whole process was contained within the matrix of tradition and protected by the presence of the sacred.[14] The ultimate goal of the true alchemist was in fact to gain power and control over his own soul and not the external world, to transmute the base metal of his soul into the gold of sanctity and not simply to manipulate substances in order to gain wealth.[15]

Gradually with the rise of mercantilism and the rebellion of western man against the traditional Christian image of fallen man there grew the desire of not only exploring the world but also of dominating it. The age of exploration was also the age of exploitation, domination and exercise of power over nature. Western science since the Renaissance has become increasingly associated with power and control. The goal of science in the minds of many its practitioners has become the control and manipulation of nature and not its contemplation. Many notable scientists in the West were and remain to this day opposed to the wedding between science and power, but there is no doubt that one of the factors which caused the destinies of science in the Islamic world and the West to follow such diverse paths despite so many common factors is this relation of science to worldly power, a relation which remained totally alien to Islamic science despite the claims of certain modernist and so-called fundamentalist Muslims today.

If one were to ask what elements within western Christianity were responsible for this development of the sciences of nature, one could point to the type of theology which developed in the Occident. In order to avoid the danger of naturalism, Christianity as formulated by the Latin Fathers drew too strong a distinction between the supernatural and the natural orders, did not emphasize sufficiently the cosmic function of the 'Word become flesh' and did not consider as central the spiritual message of nature. Despite the songs dedicated to nature by early Irish monks and even the development of Christian Hermeticism which christianized a whole traditional science of nature, the mainstream of Catholic theology did not concern itself as much with nature as did Islam or even Judaism which preceded Christianity and which interacted with it much more than did Islam. The voice of a St Francis of Assisi singing the canticle of the Sun was not typical of the Christian spirituality of the Occident any more than were the cosmic visions of Hildegard of Bingen. The discovery of nature by Renaissance art and science appeared, therefore, almost as a 'revelation' outside the mainstream of the Christian tradition, while the whole realm of nature was soon surrendered to science to be dealt with irrespective of the religious and spiritual consequences of the development of a purely quantitative science. The abdication of religion from the realm of nature, especially after Galileo, did not appear as a great defeat for the religious world-view because the rule of religion over this realm had already been a half-hearted one since the integration of a complete theology and metaphysics of nature into the main current of Christian theological thought had never been fully achieved. As a result, despite a St Francis or an Albertus Magnus, who was at once a theologian and a scientist, and despite the later religious reactions of German mystics and the Romantic movement against the total dominance of a purely mechanical science of nature, the ground was left clear for such a science to develop without any constraint or opposition of a serious kind and to claim for itself complete monopoly of knowledge of the natural realm. Any spiritual view of nature was relegated to the category of nature mysticism, while what remained of the traditional sciences of nature in the West became reduced to the category of the occult or even superstition, to survive solely in the margin of European intellectual life.

Finally, it must be remembered that what distinguishes the destinies of science in the West and the Islamic world is not only the presence of the metaphysical and cosmological doctrines of a Suhrawardi, Ibn Arabi or Sadr al-Din Shirazi at the heart of the Islamic intellectual tradition and the eclipse of doctrines of such an order in the West and their being relegated to the periphery of the intellectual life of western man. One must also consider in a more inward sense the continuous presence of contemplatives of a sapiential nature and gnostics in Islam and their extremely small numbers and almost complete disappearance in the West during the modern period. The contemplative who is of a gnostic nature is the channel of grace of nature. He hears the invocation of nature in the solitude of high mountains and deserts, along the shore of the sea and in the heart of forests. The mind of such a sage is indeed a mirror which reflects the light shining in his heart. His speculation is a reflection of the knowledge of the heart upon the plane of the mind, according to the literal meaning of the term *speculum* which means nothing other than reflection in a mirror. From this heart a light is reflected upon the mirror of the mind which in turn provides a doctrine concerning nature that cannot but reflect in conceptual terms that intimacy and inner *Sympathia* which the contemplative gnostic possesses with the inner reality of nature. To know nature according to the norms of the traditional sciences is to gain a knowledge which is permanent, which satisfies the mind while nourishing the soul. It is also to gain a knowledge which no form of quantitative science can replace, a knowledge without which man cannot ultimately survive on earth but with the aid of which he can live in harmony with himself and with nature because he lives in harmony with that reality which is the origin of both himself and the natural order.

Notes and References

1. On the meaning of traditional science and its contrast with modern science see R. Guenon, *Crisis of the Modern World*, trans. M. Pallis and R. Nicholson, London, 1975, chapter 4.; S.H. Nasr, *Knowledge and the Sacred*, New York, 1981, especially chapters 4 and 6.

2. Much of this tradition reached the Muslims through the Harraneans who were also known as the Sabaeans in Islamic history. See J. Padersen, 'The Sabians', in *Volume of Oriental Studies presented to E.G Browne,* Cambridge' 1922, pp. 383–91; and E. Drower, *The Mandaeans of Iran and Iraq,* Oxford, 1937.

3. In traditional civilizations there has in fact never been a continuous 'development' of science which the modern world envisages as normal to civilization. Traditional civilizations display periods of interest and activity in the sciences of nature interrupted by eras during which the intellectual energies of that civilization have turned to other domains without there being any sign of decadence of an intellectual or artistic nature. On the contrary, except for Islam, most other traditional civilizations seem to have turned more to the so-called 'exact sciences' at the moment of their own decay and demise. Babylonian and Alexandrian science provide striking examples of this phenomenon.

4. T. Roszak, *Where the Wasteland Ends*, New York, 1972; S.H. Nasr, *Man and Nature*, London, 1976; and W. Smith, *Transcendence and the Cosmos*, 1984.

5. S.H. Nasr, *An Introduction to Islamic Cosmological Doctrines*, Thames and Hudson, London, 1978.

6. On the relation between science and metaphysics see F. Brunner, *Science et réalité*, Paris, 1955.

7. F. Schuon, *Light on the Ancient Worlds*, trans. Lord Northbourne, London, 1984, chapter 2, 'In the Wake of the Fall.'

8. E. Gilson, *The Unity of Philosophical Experience*, London, 1938.

9. Symbolic science is treated amply by contemporary traditional authors because of the central role it plays in understanding both the languages and content of traditional writings. See R. Guenon, *Symboles fondamentaux de la science sacré*, Paris, 1962.

10. A symbol is always a symbol of some reality beyond itself while a fact as seen scientifically cannot but be ontologically separated from higher realms of existence. To say symbol, in the traditional sense of the word, is to say beyond what is immediately perceived. According to traditional metaphysics, only the Absolute Reality is totally Itself. Everything else in the universe is a symbol of a reality beyond the ontological level in which the being in question is preceived as a particular being.

11. Romans, 1:20.

12. F. Schuon, *Light on the Ancient Worlds*, chapter 3, 'Dialogue Between Hellenists and Christians'.
13. On Avicennan angelology see H. Corbin, *Avicenna and the Visionary Recital*, trans. W. Trask, New York, 1960; and S.H. Nasr, *An Introduction to Islamic Cosmological Doctrines*, chapter 15.
14. On alchemy as gynaecology and the attempt to speed up the rhythm of nature in transmitting base metal into gold see M. Eliade, *The Forge and the Crucible*, trans. S. Corrin, New York, 1962.
15. T. Burckhardt, *Alchemy: Science of the Cosmos, Science of the Soul*, trans. W. Stoddart, Baltimore, 1971.

17

Islamic Science
Current Thinking and Future Directions
Munawar Ahmad Anees and Merryl Wyn Davies

In less than a decade Islamic science has appeared from nowhere and become an intellectual movement which is now gaining rapid momentum. For the vast majority of scholars and scientists who have gained their intellectual awareness in a world dominated by western science and technology the whole debate appears truly vexing. Science, they think, is an objective, value-free enterprise, it is universal and has nothing to do with cultures and values: so what is this talk of Islamic science? What has Islam, a religion, to do with science? A scientist may be a Muslim or a Buddhist, Chinese or American but neither religion nor nationality should affect their science which is judged by external, universal criteria.

The heroic image of science as the rational, objective, value-free universal system of knowledge has been subject to searching scrutiny in recent decades. The picture that emerges makes it clear that rather than being *the* science, *the* way of knowing and knowledge *per se* this heroic science is a very particular and characteristic product. The science in which all students are trained today is western science, and far from being an academic discipline it is an ideological tool. Its products, technological devices, are employed to infuse and inculcate a self-perpetuating system that is built on metaphysical assumptions and abstractions about the nature of natural phenomena as well as human beings. It is destructive and wasteful, unmindful of its consequential growth and brutal in its practice. It is exploitative, for it employs any means possible without a hint of either hindsight or foresight. It is not universally verifiable for its self-imposed empirical limitations put it outside the pale of a unified system of knowledge. It is subjective rather than objective since it is coloured with the social, cultural and historical values of the society by which it is manufactured and distributed.

Criticism of western science is no isolated activity for it has grown from within science itself. Even if the socio-cultural criticism of science and technology — discernible through the agenda of a host of alternative movements — is discounted, there is an ever increasing number of working scientists and philosophers who have been prompted to re-assess their own standing, while the ediface of heroic science is demolished before their eyes. Western science, along with its methodological apparati, is regarded as the paradigm of rationality. Therefore, *faith* in rational thought and practice alone is the *modus operandi* of western science, though the boundaries of rationalism have never been satisfactorily defined, and the proposition is self-contradictory. As a corollary of 'crippled rationalism' anything that does not fit into the narrow peep-hole of western science is labelled as non-science and unworthy of intellectual pursuit. It is assumed that scientific rationalism is an immutable value; that instrumentalism is a vehicle for scientific validation; and that nature is best observed and manipulated under reductive strategies rather than an argumentative approach. Western science is a closed system, an impositional regimen by which any discipline can achieve the accolade 'scientific', while official science is made a useful tool of the Establishment. Human cognition is foreclosed in terms of its potential and possibilities. It is also isolated from external stimuli that may otherwise provide valid answers to some of the most pressing problems faced by humankind. The values inherent in the conduct of western science appear to be an odd mixture of an arrogant corporality, elitism, meaninglessness, artificiality, alienation, stubbornness, wastefulness and irrelevance.

Critical assessment of western science clearly indicates that merely dismantling it is neither desirable nor beneficial. The inherent arrogance and isolationism of science, augmented by notions such as evolutionary reductionism, make any improvements practically impossible. Alternative movements have emphasized the social radicalization of issues, a search for different uses of science, but their criticism of science is as fragmented and reductive as official science itself. An analysis of the philosophy of the alternative movements indicates they have not yet formulated any clear and specific mode of intellectual thought compatible with the contemporary situation. In the quest for social responsibility in science they remain nervous and nebulous in the discussion of specific value parameters for science, trapped by the dominant paradigm. Their 'alternative' is therefore a political agenda rather than a search for new epistemological dimensions. They are perceived by the Establishment as political rivals, aspects of whose programme can be accommodated and co-opted within the conduct of official science and science policy.

Yet sanity demands that corrective acts must be undertaken immediately, measures that will generate a system of science and technology that is organically linked to all other creatures on earth, instead of destroying and torturing them in the cause of science; whose pursuits are governed by ethics and not greed; a science that is subject to normative guidelines and not totalitarian ideology. Such a new science would operate within an objective framework but be constrained by considerations of values and cultural and social needs. Such a science is not a utopian dream — it is the dictate of our time.

Islamic science has been offered as a different science, one that by definition of its origin and sources — Islam and its world-view — must answer the needs of our time. It is a function of the dominance of the western paradigm that the introduction of Islamic science should be surrounded by questions: what has religion got to do with science?; is it really a science?; can it effectively replace western science?; in what respects does it differ from contemporary science? The questions are posed in this way because of the existence of western science, nevertheless, they are a valid starting point for questioning current thinking about Islamic science, most especially whether the ranks of this growing movement demonstrate a consistent and thorough allegiance to the espistemological and methodological peculiarities that make Islamic science distinct from science as it is practised today.

Arguably the ablest attempt in the English language to introduce Islamic science has been that of Seyyed Hossein Nasr. Beginning with his doctoral thesis, *An Introduction to Islamic Cosmological Doctrines*, Nasr heralded a refreshingly novel image of Islamic science.[1] Having demolished the arguments for considering modern science as value free in *The Encounter of Man and Nature: The Spiritual Crisis of Modern Man*, he proceeded to demonstrate how the value structure of Islam led to the emergence of a distinctly Muslim science.[2] However, due to Nasr's deep attachment to Sufism, his equation of all religions as one, and his ready acceptance of Greek thought and hellenic imagery, he presented too utopian an image of Islamic science. This seriously limited his audience in the Muslim world and amongst interested westerners. Nevertheless, this body of work has brought the endeavours of Muslim scientists, their philosophy and methodology of operation to the forefront of modern knowledge.

Nasr's personal interests have led him to give undue significance to minor diversions in the historical pattern of Islamic science. He dwells on things Islamic science is not: alchemy, astrology, occult, gnosis, metaphysics, validated by obscure Persian manuscripts. In the classical literature on science and technology in Muslim lands there is insufficient historical evidence to support these claims. Muslim lands have remained a complex mosaic of cultural patterns, true to the unity in diversity emanating from Islam. There is a necessary distinction to be made between folk practices that linger and are enmeshed in Muslim activity, and the particular nature of the folk practices that preponderate in Iranian culture due to its Zoroastrian ancestry, and the mainstream of the thought of the *ummah*, the international community of Muslim civilization. Muslim scholars provided the right perspective on these practices: Imam al Ghazali, for example, instead of condoning them considered them 'blameworthy'. Furthermore, there is no evidence that the Muslim masters in science and technology either used or encouraged these practices by lay persons.

The case of gnosis and metaphysics is slightly different. In practice, gnostic pursuits, totally oblivious of corporal existence, have always been looked upon with suspicion in Muslim culture. This reaction may be attributed to the fact that the dictates of gnostic extremism contradict the integrative view of life in Islam. As to metaphysics, similar Muslim reactions are on record. Scholars like Ibn Rushd, Ibn Baja, Ibn Sina and Imam al Ghazali are known to have made investigative studies of Greek and other metaphysical systems and produced an able defence of metaphysical studies from a Muslim perspective. It is, therefore, erroneous to insist that Islamic science was shaped either through the proliferation of folk practices or by perusal of gnosis and metaphysics. The source of Islamic science is Islam itself, its cosmology and world-view.

A potent factor in the modern resuscitation of Islamic science is its history which has had to be laboriously reclaimed by contemporary Muslims. Western historians of science have regarded it as intellectually dishonest, as nothing more than a holding pattern on a conveyor belt that linked the scientific originality of the Greeks to the scientific originality of the West. It has been necessary for Muslim writers to redress the misconception of the intervening period, during which Islam prompted a spectacular upsurge of intellectual activity amongst Muslims, as devoid of scientific originality. It has also been their task to consider whether the delinking of scientific information from the cosmology, epistemology and value constraints within which it existed for Muslim scientists, and the assimilation of this information devoid of constraints within the particular context of western civilization, is not the very foundation of the nature of western science. The argument from history serves a dual function: it establishes the inappropriateness of western science for Muslim society while showing there has been a science attendant upon the attuned to the essential orientation of Muslim civilization, Islam.

The role of modern science in creating the predicament of the Muslim world is a dimension explored in two major works and a series of related essays in leading journals by Ziauddin Sardar. His first book, *Science, Technology and Development in the Muslim World*, did much to shape the emerging face of Islamic science.[3] Beginning with the explication of Islamic epistemology, Sardar examined the status of science and technology in the context of the Muslim world today and concluded that replicating western systems was counter-productive. Apart from socio-cultural dislocations brought about by unmindful transfer of technology, he claimed that western science was imposing an alien political ideology that exploited science and technology as tools of tyranny. Effectively an argument was being advanced that derived its strength from a moral and value-laden structure and sought to integrate science within a larger world view.

Sardar's second work, *The Future of Muslim Civilization*, was an attempt to present an agenda for the Muslim world on a civilizational scale, and to reassess its status in the contemporary world.[4] He rebuked the false pride of those who bolster the historic image of Muslim civilization to produce merely the inertia of nostalgia. The historic achievements were genuine but a study of them should underline rather than obscure the fact that Muslim civilization has been in decline for four centuries. This decline, which includes the disappearance of Islamic science in any operational form, has been induced as much by attitudes and actions of Muslims themselves as it has coincided with and been compounded by impositions from outside. This comprehensive survey of history and contemporary predicament forms the basis of an attempt to construct a vision of the future. The future begins with the eternally valid concepts and values of Islam and consists of putting into operation its holistic, unified and universal world-view by recovery of its distinctive epistemology and methodology and their application to the resolution of contemporary problems.

This and works by other emerging authors are beginning to create a cohesive picture of Islamic science. It is a science at once in consonance with the fundamental Islamic values and the many demands of contemporary situations. Above all it clarifies what religion has to do with science. The crux of the argument advanced by Muslim scientists is that the epistemology and methodology of a knowledge system must rest on enduring values and not transient factors. Islamic science as it is emerging is neither a re-orientation of western science nor its unmindful imitator, though both tendencies can be found in subtexts of the literature on Islamic science. The mainstream of argument is that Islamic science is an entity on its own, not defined in comparison with and amendment of an already existing science. Islamic science is an integral part of Islam as a complete way of life, the only framework within which it can be defined; it cannot be inculcated in isolation from the mainstream of the Islamic intellectual and moral landscape. Islamic science is a sub-species of Islam (and not of science) that generates a world-view within the overall framework of Islamic values.

It is an open-ended system without a built-in experimental bias. As Sardar has pointed out, in the past the particular type of methodological technique employed has varied according to the nature of the questions posed by Muslim scientists. For Islamic science a diversity of approaches is valid and necessary, reliance on just one methodology of verification is a reductive limitation. It does not lose its accountability however; it is characterized by self regulation. Scientific activity is regulated by an open debate on the issues related to means and ends with no secret, underground, ultra-sensitive research and development fad; its conduct is motivated not by financial catches, group loyalties or strategic interest, but by concern for the ultimate use of knowledge.

A shift towards Islamic science is not to be equated with re-inventing the wheel. There is no subtle attempt to undermine or sabotage the cumulative human labour in amassing wisdom that has generated tools for human betterment as well as for understanding natural phenomena. However, Islamic science is part of, and incorporates, a means of discernment that questions and evaluates what constitutes wisdom based on its own holistic definition of the ultimate end of human betterment. The shift to Islamic science must entail a radical re-orientation of the norms and values under which knowledge is pursued, utilized and assessed.

There has been a steady development of thought on how the ethico-moral guidelines of Islam can be incorporated into shaping the structure of Islamic science. It was in 1981 that, under the auspices of the International Federation of Institutes for Advanced Study (IFIAS), Sardar co-ordinated an international conference on a comparative study of science and values in Islamic and western societies. The Stockholm seminar identified at least ten Islamic concepts — namely *tawhid*, unity; *khilafah*, trusteeship; *ibadah*, worship; *ilm*, knowledge; *halal*, praiseworthy; *haram*, blameworthy; *adl*, social justice; *zulm*, tyranny; *istislah*, public interest; and *dhiya*, waste — which could be used to develop value parameters around science. 'When translated into values, this system of concepts embraces the nature of scientific enquiry in its totality: it integrates facts and values and institutionalizes a system of knowing that is based on accountability and social responsibility'.[5]

In his keynote address to the seminar on 'Quest for New Science' organized by the Centre for Studies on Science, Aligarh, India, Sardar elaborated further some of these concepts.[6] Based on a conceptual analysis of these Islamic values a contemporary model of Islamic science has been put forward: the concepts *tawhid, khilafah* and *ibadah* are the paradigms of Islamic science. These paradigms are the overall framework in which the advancement of knowledge, *ilm*, is pursued for the promotion of public interest and social justice. All knowledge and man's inherent ability for its quest are to be regarded as tokens of trusteeship. The concept of *ibadah* bridges the realms of the so-called sacred and profane. By discharging the obligation for the pursuit of knowledge, *ibadah* is performed. Clear guidelines for such a pursuit are provided by the values embodied in the concepts of *halal* and *haram*. The pursuit of knowledge leading to blameworthy ends is not permitted, while knowledge aimed at public interest is encouraged and rewarded.

The emergence of Islamic science has occurred in conjunction with the call for the Islamization of knowledge. It is necessary to consider the distinctions between these two movements to understand what Islamic science is not, as well as to see the way in which the Islamization approach is used by some who claim to be advancing Islamic science, but are creating confusion and following seductive blind alleys. The action plan for the Islamization of knowledge excludes natural sciences and concentrates on education and behavioural sciences alone.[7] The proponents of Islamization justify the exclusion of natural sciences on the premises of *exactness*, and hence they assume the integrity of their position to be unquestionable. It is of course erroneous to believe that behavioural sciences and education have remained unaffected by the dominant paradigms of western science. The acceptance of psychology and parapsychology as soft-core disciplines of science has been contingent upon their adoption of the official methodology, while acceptance of the superiority of official methodology is the foundation of the social sciences and has determined the course of their growth and content. In this respect these disciplines are no exception to the lopsided world-view of 'exact' sciences.

The first step towards Islamization proposed by the workplan is mastery of modern — that is western, disciplines — followed by a mastery of the Islamic legacy. It is assumed that mastery at these two levels would lead to the establishment of the social relevance of modern disciplines and open new avenues to synthesize modern knowledge and the Islamic legacy, a synthesis that would maintain the broad outlines of modern knowledge yet incorporate amended Islamic values while preserving respect for the historic and value-laden legacy of Muslim identity. This simplistic trilogy is misleading and fraught with dangers. It shows that the workplan accepts the epigenetic and contextual relevance of modern disciplines without question. It fails to recognize that the genesis of the so-called Islamic legacy took place in an essentially different epistemological and cultural milieu that is irreducible to contemporary disciplinary approaches. The Islamization plan lacks a genuine critical assessment of western epistemology as well as an appreciation of its contemporary social critiques. A surface critique of the social manifestations of secular materialism in the West is substituted for linking the appearance to its epistemological roots. Furthermore, it fails to demarcate a distinct Islamic theory of knowledge that may be exploited for the proposed synthesis; without such a theory Islamisation is nothing but a modern day alchemist's brew whose objectives must remain as illusory as those of the original alchemists.

Islamic knowledge, the concept of *ilm*, does not exist outside the boundaries of the Islamic world-view. Indeed to translate *ilm* simply as knowledge does violence to its import and meaning as a multi-dimensional, integrative concept.[8] Theories of knowledge and classifications of *ilm* are essential and characteristic products of the Islamic legacy and the history of Islamic science. The essence of these schema is that knowledge is a whole entity, an organic unity derived from the all-encompassing, all-pervasive concept of *tawhid*. Fragmentation of knowledge is therefore unacceptable, *ilm* must be pursued in its entirety and one kind of knowledge must not be advanced at the expense of others. The nature of *ilm* presumes an on-going interaction and integration of revelation and reason as necessary aspects of real knowledge. The world-view of Islamic science is neither axiomatically rationalist nor dogmatically metaphysical. On the contrary, it is due to the inherent balance of these two aspects of its world-view that the unnecessary and wasteful conflicts between 'religion' and 'science' have been and are alien to the spirit of Islamic civilization.

Revitalization of the concept of *ilm* is a cornerstone of the Islamic science movement. Prompted by the concept of *ilm* and the value parameters of Islam, an integrative epistemological approach is the foundation of Islamic science. In history polymathy was the characteristic response of Muslim scientists: it was not uncommon for a Muslim scholar to master religious law, yet also be an astronomer, a medical practitioner and a philosopher. It is often argued that the development of rigid, compartmentalized modern disciplines was necessitated by the sheer increase in the volume of information in a given discipline. This is only partly true. Islamic science certainly does not accept the argument of necessity at face-value. The contemporary emergence of cross-disciplinary and multi-disciplinary approaches that are breaking down disciplinary boundaries in the pursuit of knowledge, as well as being employed as more appropriate ways of practical problem solving, strongly indicates a belated recognition of the unity of knowledge by western science and society. Islamic science is not *islamized science*. Similarly the characterization of science that is based on given knowledge groups is rejected. There cannot be a Muslim endocrinology or a Muslim astrophysics. This is a crude fragmentation of a whole body of knowledge whose essential nature is its integration and unifying force.

The mainstream of the movement sees Islamic science as an entity on its own. But the Islamization of knowledge debate has had its impact. The limitations of the Islamization approach, its lack of attention to an Islamic theory of knowledge, holds out the prospect of an easy accommodation leading to a synthesis between western 'scientific' knowledge and Islamic sentiment and belief, the primrose path to mental inertia. Based upon uncritical acceptance of the epistemological status quo, the dominance of and dependence upon western science, it suggests things can change while essentially staying the same. It is a very short and seductive step from this easy synthesis based upon the status quo to the notion that no synthesis at all is needed, because in the realm of 'science' it already exists. This is the appeal of Bucaillism. The popular attraction of Maurice Bucaille's *The Bible, The Quran and Science*[9] is the way it feeds the apologist Muslim's self-esteem by purporting to give scientific proof of the modernity of Muslim belief by employing scientific data to uphold the truth of revelation. This is then taken as demonstration of the Muslim truism that there can be no conflict between science and religion, both of which are construed in an entirely western, not Islamic, sense. Sardar has explained how Bucaillism:

> elevates science to the realm of the sacred and makes Divine Revelation subject to the verification of western science. Apart from the fact that the Qur'an needs no justification from modern science, Bucaillism opens the Qur'an to the counter argument of Popper's criteria of refutation: would the Qur'an be proved false and written off, just as Bucaille writes off the Bible, if a particular fact mentioned in the Qur'an does not tally with it or if a particular fact mentioned in the Qur'an is refuted by modern science. And what if a particular theory, which is 'confirmed' by the Qur'an and is in vogue today is abandoned tomorrow for another theory that presents an opposite picture? Does that mean that the Qur'an is valid today but will not be valid tomorrow?[10]

Bucaillism manages to unite a mythic image of both science and religion. The reductive literalism applied to the Qur'an, the interpretation of its meaning as encompassing concrete 'scientific' matters, strips both Qur'an and religion of conceptual meaning, allegorical and metaphysical import. Science is accepted as the final word on *truth*, in complete ignorance of science as a socially constructed, limited system of knowing. It would be pleasing to report that Bucaillism is only an opiate of the scientifically illiterate; sadly this is not the case.

Nobel Laureate M. Abdus Salam warmly endorses Bucaillism as the foundation of his argument for Islamic science.[11] For Salam the matter is very simple, and simplistic: science is Islamic by virtue of its operation by Muslims. Yet as Muslims who believe in Islam, these Islamic scientists draw nothing, except a generalized and totally uncritical validation of the quest for knowledge from Islam to organize the science they do. There are two discrete world-views, no hierarchy, and they carry on doing the only science there is, which is nothing but western science. The only problem Salam can define is that Muslims do not do very much western science, as demonstrated by a series of statistics on education, numbers and distribution of scientists and spending on science in Muslim countries. Apart from misplaced faith in an unanalysed heroic science, he also maintains misplaced faith in what constitutes the identity of the Muslim.

Islam is not an innate function of the person of the Muslim, rather by virtue of the innate nature of his or her person the Muslim acknowledges allegiance to Islam as a world-view that must be activated and operationalized in all spheres of life. Salam's conception of Muslim identity makes Islam a passive state of being, with vague, unspecified and certainly unanalysed limits. It is in this sense that Salam's Bucaillism and the approach of Islamization of knowledge merge, wittingly and unwittingly. Without an Islamic theory of knowledge Islam cannot be fulfilled by the Muslim, it cannot be a constructive force in achieving human betterment, and thereby must deny itself. Islam is admitted by all to define the goal of human betterment. It does so by insisting on the integration of means and ends under the same value-framework that defines the substance of human betterment and relating these to particular social and cultural milieux. The value-framework of Islam is unequivocally universal; Islamic science therefore must thrive on values to perpetuate these immutable values. Islamic science cannot itself be a sub-species of Islam. The unified framework can be achieved and sustained only by the essential attribute of the Islamic world-view that Bucaillism and the Islamization of knowledge ignore — a critical, questioning outlook that includes self-criticism.

Sardar has pointed out that despite the increasingly cohesive definition of Islamic science that has emerged many Muslim scientists still hold onto two impossible notions at once.[12] Bucaillism and the Islamization of knowledge are merely particular labels representing a more widespread tendency of Muslim scientists to talk of Islamic science and yet understand science as objective, value-free and universal. The consequence of this confusion is that, in their desire for the betterment of Muslim society, they can come up with nothing more wide-ranging than appropriate intermediate technology. Sardar makes a further distinction between Islamic science and science in Islamic polity.

Science in Islamic polity is the chimera so many Muslim scientists think can happen automatically: Muslim society, by institutionalizing certain Islamic referents of unspecified and unanalysed character and implication, can ensure that when science is operated by Muslims it will produce the kind of products and enquiry that should exist in Muslim society. This not only obviates the need for revitalization of *ilm*, Islamic knowledge, but violates the integrative Islamic conception of society that unifies the accountability and social responsibility of both the individual — here the scientist — and the collective — society — through its communal institutions of policy-making.

Rightly, Sardar emphasizes that what makes Islamic science different are the different institutional arrangements required to determine and operate its science policy in society, where science and the society it serves are sub-species of Islam so that the epistemology and methodology of both is an expression and activation of the same value-framework. Islamic science requires the scientist to understand his or her role, as well as the doing of science, differently. From the difference of goals characteristic of Islamic science and society must come the devising of different means. Islamic scientists are not called upon to speculate upon some uniform methodology in abstraction but to participate actively in the determination of social priorities for the formation of science policies. These science policies will require both basic and problem solving research, the nature of the questions will call for originality in activating Islamic science methodology through a series of available and potential techniques: in this process of devising and doing Islamic science will bear its distinctive fruits.

A cohesive picture of Islamic science exists in the contemporary literature. It is possible to point to what Islamic science is and what Islamic science is not. Indeed a concise list of what Islamic science is not has already been drawn up:

1. *Islamized science*, for its epistemology and methodology are the products of the Islamic world-view that is irreducible to the parochial western world-view.
2. *Reductive*, because the absolute macroparadigm of *Tawhid* links all knowledge in an organic unity.
3. *Anachronistic*, because it is equipped with future consciousness that is mediated through the means and ends of science.
4. *Methodologically dominant*, since it allows an absolute free flowering of method within the universal norms of Islam.
5. *Fragmented*, for it promotes polymathy in contrast to narrow disciplinary specializations.
6. *Unjust*, because its epistemology and methodology stand for distributive justice within an exacting societal context.
7. *Parochial*, because the immutable values of Islamic science are the mirror images of the values of Islam.
8. *Socially irrelevant*, for it is 'subjectively objective' in thrashing out the social context of scientific work.

9. *Bucaillism*, since that is a logical fallacy.
10. *Cultish*, for it does not make an epistemic endorsement of occult, astrology, mysticism and the like.[13]

Confusion and timidity, a fearfulness to dare to be distinctive, also exists. The task for the future is to resolve the confusion and sweep away the timidity. It can be seen that the model for Islamic science emerging through the works of Muslim scientists is a viable option out of the present chaos caused by western science. This model is an all-inclusive approach to epistemological problems, methodology of science and its application for social problem solving. This science does not exist for self-perpetuation. The value-framework does not make it parochial but offers a universal platform for its expression.

Kirmani has complained that the epistemology and general methodology advocated for Islamic science is exemplary, what is lacking is a definite blueprint for the working scientist.[14] It would indeed be simple if Islamic scientists could be told what to do, routinized within an off-the-peg paradigm of operative Islamic science that demands no thoughtful exertion on their part. He confuses paradigm as methodology of relating epistemological bases to relevant questions with methodology as techniques of routine investigation for obtaining specific problem solutions. Even so, Kirmani touches upon a sensitive point. The success of the western paradigm of science is that it cocoons working scientists, provides them with objectives they neither shape nor have to question since 'outside' forces of politics, power and money take care of all that, and most of all envelops their activity within an heroic mythology to obscure what is actually happening. Today, Muslim scientists think they know what they are doing because they are told what to do, that is the problematique of western science. Tomorrow, Islamic science will require them to think for themselves, to uncover what they should be doing. The paradigm of Islamic science is before us, how it is made into a routine technique of specific problem solving is the future task of Muslim scientists. There can be no blueprint for what can exist and be defined only by the activity of scientists within their society. The future direction for Islamic science is clear, it is to work out solutions to present-day predicaments on the basis of accountability, social justice and responsibility, to exercise the universal option of its value-framework to bring about desirable global change.

Notes

1. Seyyed Hossein Nasr, *An Introduction to Islamic Cosmological Doctrines*, revised, Thames and Hudson, London, 1978.
2. Seyyed Hossein Nasr, *The Encounter of Man and Nature: The Spiritual Crisis of Modern Man*, Allen and Unwin, London, 1968; see also his *Three Muslim Sages*, Harvard, Cambridge, Mass, 1964; *Science and Civilization in Islam*, Harvard, Cambridge, Mass. 1968; and *Islamic Science, An*

Illustrated Study, World of Islam Festival Trust, London, 1976.

3. Ziauddin Sardar, *Science, Technology and Development in the Muslim World*, Croom Helm, London, 1977.

4. Ziauddin Sardar, *The Future of Muslim Civilization*, Mansell, London, 1987.

5. Ziauddin Sardar in the introduction to *The Touch of Midas, Science Values and Environment in Islam and the West*, Manchester University Press, 1984, containing the contributions to the Stockholm seminar.

6. See Ziauddin Sardar 'Arguments for Islamic Science' in Rais Ahmed and Syed Naseem Ahmed (eds.) *Quest for New Science*, Centre for Studies in Science, Aligarh, 1984.

7. Ismail al Faruqi, *Islamization of Knowledge, General Principles and Work-plan*, International Institute of Islamic Thought, Washington, D.C., 1982.

8. See Munawar Ahmad Anees, 'Revitalising Ilm' *Inquiry* 3(5) May 1986, pp. 40–44.

9. Maurice Bucaille, American Trust Publications, Indianapolis, 1978.

10. Ziauddin Sardar, 'Between Two Masters, The Qur'an and Science', *Inquiry*, 2(8) August 1985, p. 41.

11. M. Abdus Salam, 'Islam and Science' *MAAS Journal of Islamic Science*, 2(1) Jan 1986, pp. 21–47.

12. Ziauddin Sardar, 'Islamic Science or Science in Islamic Polity: What is the Difference?' *MAAS Journal of Islamic Science*, 1(1) Jan. 1985.

13. See Munawar Ahmad Anees, 'What Islamic Science Is Not', *MAAS Journal of Islamic Science*, 2(1) Jan. 1986, pp. 9–19.

14. M. Zaki Kirmani, 'Imitative-Innovative Assimilation' *MAAS Journal of Islamic Science*, 2(2) July 1986.

18

Logical and Methodological Foundations of Indian Science

M.D. Srinivas

There seems to be a generally prevalent opinion both among the scholars and the lay-educated, that the Indian tradition in sciences had no sound logical or methodological basis.[1] While we know that the western tradition in abstract or theoretical sciences is founded on the logic of Aristotle and the deductive and axiomatic method of theory construction as evidenced in Euclid's elements (both of which have been further refined in the course of the work of the last hundred years in logic and mathematics), we seem to have no clear idea of the foundational methodologies which were employed in the Indian scientific tradition. This, to a large extent, has hampered our understanding of the Indian tradition in sciences, especially as regards their foundations and as regards their links both amongst themselves as well as with Indian tradition in philosophy.

The traditional Indian view, as it appears from the popular saying *Kanadam Paniniyanca sarva sastropakarakam*, is that the *sastras* expounded by Kanada and Panini are the basis for all other *sastras*. Here the *sastras* expounded by Kanada refer to the entire corpus of *Nyaya-Vaisesika Darsanas*, that is, the 'physics and metaphysics' as expounded mainly in the *Vaisesika Darsana*, and the epistemology and logic as expounded chiefly in the *Nyaya Darsana*. The *sastra* of Panini is the entire science of language (*sabda sastra*). In the Indian view these appear to be the foundational disciplines whose mastery is a prerequisite for a serious study of all other *sastras*, meaning all sciences, theoretical as well as practical, natural as well as social and also philosophy. So in order to have a reasonable idea of the logical and methodological foundations of Indian sciences, we should have an in-depth understanding of the methodologies, theories and techniques developed in the *Nyaya* and *Vaisesika* works as also the works on *sabda sastra*.

In this article I shall attempt to present an outline of the Indian approach to

261

just one particular logical and methodological issue, namely, the question of how the Indian tradition handles various foundational problems which involved the use of what are generally known as 'formal methodologies' or 'formal techniques' in the western tradition.[2] The foundational disciplines of logic and mathematics in the western tradition are considered rigorous mainly because they are formulated in a content-independent, purely symbolic or 'formal' language and the aim of many a theoretical science in the western tradition is to attain standards of rigour comparable to that of logic or mathematics. Such attempts have been made repeatedly in the West in various domains of natural sciences, some social sciences and much more so, in linguistics, the science of language. I shall offer an outline of some of the methodologies and techniques developed in the Indian tradition of logic and linguistics and compare them with the formal methodologies and techniques developed in the western tradition.

Firstly I shall discuss the distinctive features of Indian logic as compared with the western tradition of formal logic and explain how the Indian logicians provide a logical analysis of every cognition in terms of a technical language and use it to explicate logical relations between cognitions. I shall also discuss how the Indian logicians achieve a precise and unambiguous formulation of universal statements in terms of their technical language, without having recourse to quantification over unspecified universal domains. Then I shall consider the Indian tradition in linguistics especially the grammatical treatise of Panini, *Astadhyayi*, as a model or a paradigm example of theory construction in India. I indicate the manner of systematic exposition as well as the techniques employed in *Astadhyayi*, which appear to be common to the entire corpus of classical *sastric* literature wherein the *sutra* technique of systematization has been employed. I also explain how the Paninian grammar serves not only as a 'generative device' for deriving all the correct forms of utterances but also as a 'parser' for arriving at a precise and unambiguous 'knowledge representation' (in terms of technical language) of any correct utterance of Sanskrit language. Further, it is this systematic analysis of the Sanskrit language, which seems to have enabled the Indian *Sastrakaras* to develop a precise and technical language, suited for logical discourse.

In fact, the basic feature that emerges from my discussion of the Indian approach is that the Indian tradition did not go in for the development of purely symbolic and content-independent formal languages, but achieved logical rigour and systematization by developing a precise and technical language of discourse founded on the ordinary Sanskrit language — a technical language which is so constructed as to easily reveal the logical structures which are not so transparent and often ambiguous in a natural language, but at the same time has a rich structure and interpretability which it inherits from the natural language Sanskrit from which it is constructed. The Indian approach is thus free from many of the philosophical and foundational problems faced by the formal methodologies developed in the western tradition. More importantly, it seems to provide us an alternative, logically rigorous and systematic foundational methodology for natural sciences and philosophy.

The Indian Approach to Formal Logic

To understand the basic, foundational differences between Indian logic and western logic, let us first note the essential features of logic in the western tradition, which are well captured in the following extract from the article on logic by a renowned mathematical logician in the Encyclopaedia Britannica.

> Logic is the systematic study of the structure of propositions and of the general conditions of valid inference by a method which abstracts from the content or matter of the propositions and deals only with their logical form. This distinction between form and matter is made whenever we distinguish between the logical soundness or validity of a piece of reasoning and the truth of the premises from which it proceeds and in this sense is familiar from everyday usage. However, a precise statement of the distinction must be made with reference to a particular language or system of notation, a formalized language, which shall avoid the inexactness and systematically misleading irregularities of structure and expression that are found in ordinary (colloquial or literary) English and other natural languages and shall follow or reproduce the logical form.[3]

In other words, the following appear to be the basic features of western logic: it deals with a study of propositions, especially their logical form as abstracted from their content or matter. It deals with general conditions of valid inference, wherein the truth or otherwise of the premise has no bearing on the logical soundness or validity of an inference. It seeks to achieve this by taking recourse to a symbolic language which apparently has nothing to do with natural languages. All this is understandable, since the main concern of western logic in its entire course of development has been one of systematizing patterns of mathematical reasoning, and that too in a tradition where mathematical objects have often been thought of as existing either in an independent ideal world or as a formal domain.

In what follows, I shall attempt to contrast the above features of western logic with the basic features of Indian logic. The main point of this contrast is that Indian logic does not purport to deal with ideal entities such as propositions, logical truth as distinguished from material truth, or with purely symbolic languages which apparently have nothing to do with natural languages. As is well known, a central concern of Indian logic as expounded mainly by *Nyaya Darsana* has been epistemology or the theory of knowledge. Thus the kind of logic which developed here is not in any sense confined to the limited objective of making arguments in mathematics rigorous and precise, but attends to the much larger issue of providing rigour to the various kinds of arguments encountered in natural sciences (including mathematics, which in Indian tradition has more the attributes of natural science than that of a collection of context-free abstract truths) and in philosophical or even natural discourse.

Further, inference in Indian logic is both deductive and inductive, formal as well as material. In essence, it is the method of scientific enquiry. In fact one of the main characteristics of Indian formal logic is that it is not formal at all, in the sense generally understood, as Indian logic refuses to totally detach form from content. It takes great care to exclude from logical discourse terms which have no referential content. It refuses to admit as a premise in an argument any statement which is known to be false. For instance the method of indirect proof (*reductio ad absurdum*) is not acceptable to most Indian schools of philosophy as a valid method for proving the existence of an entity whose existence is not demonstrable (even in principle) by other (direct) means of proof.[4] In fact, the Indian logicians grant *tarka* (roughly translatable as the method of indirect proof) the status of subsidiary means of verification only, helping us to argue for something which can be separately established (though often only in principle) by other (direct) means of knowledge.[5]

The most distinguishing feature of the non-formal approach of Indian logic is that it does not make any attempt to develop a purely symbolic and content independent or formal language as the vehicle of logical analysis. Instead what Indian logic (especially in its later phase of Navya nyaya, say starting with the work of Gangesa Upadhyaya (fourteenth century)) has developed is a technical language which, by its very design, is based on the natural language Sanskrit but avoids inexactness and misleading irregularities by various technical devices. Thus the Indian tradition in logic has sought to develop a technical language which, being based on the natural language Sanskrit, inherits a certain natural structure and interpretation, and a sensitivity to the context of enquiry. On the other hand the symbolic formal systems of western logic, though considerably influenced in their structure (say in quantification,) by the basic patterns discernible in European languages, are professedly purely symbolic, carrying no interpretation whatsoever — such interpretations are supposed to be supplied separately in the specific context of the particular field of enquiry employing the symbolic formal system.

Logical Analysis of Cognition (*Jnana*) in Indian Logic

It has become more and more clear from various recent investigations that Indian logic deals with entities and facts directly. It is a logic of *jnana* (variously translated as knowledge, cognition, awareness) as contrasted with the western logic of terms or sentences or propositions. While Indian thought does distinguish a sentence from its meaning, and also admits that sentences in different languages could have the same meaning (which are some of the arguments used in the West in favour of introduction of the notion of proposition), there appears to be a total disinclination amongst all Indian philosophers to posit or utilize ideal entities such as propositions in their investigations. On the other hand what Indian logic deals with are the *jnanas*. Though philologically the Sanskrit word *jnana* is supposed to be cognate with the English word knowledge, a more preferred translation of *jnana* appears to be cognition or awareness, as *jnana*, unlike knowledge, can be either *yathartha* (true) or *ayathartha* (false).

Further, *jnana* is of two types *savikalpa* (often translated as determinate or propositional but not a proposition) and *nirvikalpa* (indeterminate, non-relational or non-propositional).[6] But what it is important to realize is that even the *savikalpa* or propositional *jnana* is not to be identified with a sentence or proposition, as has been emphasized by a modern Indian philosopher.[7] 'The *jnana*, if it is not a *nirvikalpa* perception, is expressed in language; if it is *sabda*, it is essentially linguistic. But it is neither the sentence which expressed it, nor the meaning of the sentence that is the proposition; for there is in the (Indian) philosophies no such abstract entity, a sense as distinguished from reference, proposition as distinguished from fact.'

In what follows I will give a brief outline of Indian logical anaylsis of *jnana*, as brought out in some of the recent investigations.[8] The main point that emerges is that though *jnana* is a concrete occurrence in Indian philosophy (a *guna* or *kriya* of the *jiva* in some systems, a modification or *vrtti* of the inner senses, the *antahkarana* in some other systems of Indian philosophy), it does have a logical structure of its own, a structure that becomes evident after reflective analysis. This logical structure of a *jnana* is different from the structure of the sentence which expresses it in ordinary discourse. There always remain logical constituents in a *jnana* which are unexpressed in the usual sentential structure. For instance in the *jnana* usually expressed by the sentence *Ayam ghatah* (this [is] a pot), the feature that the pot is being comprehended as a pot, that is as qualified by potness (*ghatatva*), is not expressed in the sentential structure. Thus the logical structure of a *jnana* is what becomes evident after reflective analysis, and the sentential structure of ordinary discourse only provides a clue to eliciting this epistemic structure of a cognition.

According to Indian logic every cognition (*jnana*) has a contentness (*visayata*). For the case of a *savikalpaka jnana* this *visayata* is of three types: qualificandumness (*visesyata*), qualifierness (*prakarata* or *visesanata*) and relationness (*samsargata*). For instance, in the *jnana* expressed by *ghatavad-bhutalam* (earth is pot-possessing) the *prakara* is *ghata*, the pot (not the word *ghata* or pot), the *visesya* is *bhutala*, the earth (not the word *bhutala* or earth) and since the pot is cognized as being related to the earth by contact, the *samsarga* is *samyoga*, the relation of contact. Thus the *prakarata* of the *jnana* 'Ghatavad-bhutalam' lies in *ghata*, the *visesyata* in *bhutala* and *samsargata* in *samyoga*. Thus in Indian logic, any simple cognition can be represented in the form a-Rb where a denotes the *visesya*, b the *prakara* and R, the *samsarga*, or the relation by which a is related to b. This analysis of a simple cognition as given by the Indian logicians is much more general than the analysis of the traditional subject-predicate judgement in Aristotelian logic or that of an elementary proposition in modern logic (say in the system of first order predicate calculus), as the Indian logicians always incorporated a *samsarga* or relation which relates the predicate to the subject.

Identifying the *visesya, prakara* and *samarga* of a *jnana* is not sufficient to characterize the *jnana* fully. According to the *Naiyayika* one has clearly to specify the modes under which these ontological entities become evident in the *jnana*. For instance while observing a pot on the ground one may cognize it merely as a substance (*dravya*). Then the qualifier (*prakara*) of this *jnana*, which is still the ontological entity pot, is said to be *dravyatvavacchinna* (limited by substanceness) and not *ghatatvavacchinna* (limited by potness) which would have been the case had the pot been cognized as a pot. The Indian logician insists that the logical analysis of a *jnana* should reveal not only the ontological entities which constitute the *visesya, prakara* and *samsarga* of the *jnana*, but also the mode under which these entities present themselves, which are specified by the so-called limiters (*avacchedakas*) of the *visesyata, prakarata* and *samsargata*. The argument that is provided by Indian logicians in demanding that the *avacchedakas* should be specified in providing a complete logical characterization of a *jnana* is essentially the following. Each entity which is a *prakara* or *visesya* or *samasarga* of a *jnana* possesses innumerable attributes or characteristics. In the particular *jnana* any entity may present itself as a possessor of certain attributes or characteristics only, which will then constitute the limiters (*avacchedakas*) of the *prakarata* etc. (of the *jnana*) lying in the entity concerned.

The Naiyayika therefore sets up a technical language to characterize unambiguously the logical structure of a *jnana* which is often different from the way this *jnana* might get expressed in the language of ordinary discourse. For instance, the *jnana* that the earth is pot-possessing, which is ordinarily expressed by the sentence *ghatavad bhutalam*, would be expressed by the logician in the form: *Samyoga sambandhavacchinna ghatatvavacchinna ghatanishtha prakarata nirapita-bhutalat-vavacchinna bhutalanistha visesystasali jnanam*: a cognition whose *visesyata* present in *bhutala* (earth) which is limited by *bhutalatva* (earthness) and is described (*Onirupita*) by a *prakarata* present in *ghata* (pot) and limited by *ghatatva* (potness) and *samyoga sambandha* (relation of contact).

The Naiyayika's analysis of more complex cognitions can now be briefly summarized. Each cognition reveals various relations (*samsarga*) between various entities (*padarthas*). Thus a (complex) cognition has several constituent simple cognitions each of which relate some two *padarthas* (one of which will be the *prakara* and other the *visesya*) by a *samsarga*. The *visesyata* and *prakarata* present in any pair of *padarthas* are said to be described (*nirupita*) by each other. Thus the various entities (*padarthas*) revealed in a complex cognition have in general several *visesyatas* and *prakaratas* which are further characterized as being limited (*avacchinna*) by the various modes in which these entities present themselves. Further a detailed theory is worked out (with there being two dominant schools of opinion associated with the Navadwipa logicians of the seventeenth to eighteenth century Jagadisa Tarkalamkara and Gadadhara Bhattacharya) as to how the different *visesyatas* and *prakaratas* present in the same entity (*padartha*) are related to each other. In this way a detailed theory has been evolved by the Indian logicians to characterize unambiguously the logical structure of any complex *jnana* in a technical language. For instance the Naiyayika would characterize the cognition that the earth possesses a blue-pot, which is ordinarily expressed by the sentence *Nilaghatavad-bhutalam* as follows:

Tadatmya sambandhavacchinna — nilatvavacchinna — nilanishtha prakarata niruputa ghatatvavacchinna-ghatanishtha-visesyatvavacchinnasamyoga sambandhavacchinna ghatatvavachinna ghatanishtha prakarata nirupita bhutalatvavacchinna bhutala nishtha visesyatasali jnanam: a cognition whose *visesyata* present in *bhutala* is limited by *bhutalatva* and is described by *prakarata* present in *ghata* which *prakarata* is limited by *ghatatva* and *samyoga sam bandha* and by the *visesyatva* in *ghata* which in turn is limited by *ghatatva* and is described by *prakarata* present in *nila* (blue) and limited by *tadatmya samban-dha* (relation of essential identity) and *nilatva* (blueness).

I shall now consider the question as to how the above logical analysis worked out by the Indian logician does serve the purpose of providing a representation of a *jnana* which is free from the various ambiguities which arise in the sentences of ordinary discourse, and also makes explicit the logical structure of each *jnana* and its logical relations with other *jnanas*. To start with let us discuss how the Naiyayikas formulate a sophisticated form of the law of contradiction via their notion of the *pratibadhya* (contradicted) and *pratibandhaka* (contradictory) *jnanas*. For this purpose we need briefly to outline the theory of negation in Indian logic as enshrined in their notion of *abhava* (absence).

Negation (*Abhava*) in Indian Logic

Abhava is perhaps the most distinctive as well as the most important technical notion of Indian logic. Compared with the Indian doctrine of negation, the notion of negation in western logic is a rather naive or simplistic truth functional notion in which all the varieties of negation are reduced to the placing of 'not' or 'it' is not the case that before some proposition or proposition like expression. This later notion does not for instance allow a subject term to be negated in a sentence and in fact most cases of internal negation in a complex sentence seem to be entirely outside the purview of western formal logic.[9]

The essential features of the notion of *abhava* are summarized in the following extracts from a recent study.[10]

The concept of absence (*abhava*) plays a larger part in Navya-nyaya (new-Nyaya) literature than comparable concepts of negation play in non-Indian systems of logic. Its importance is apparent from a consideration of only one of its typical applications. Navya-nyaya, instead of using universal quantifiers like 'all' or 'every', is accustomed to express such propositions as 'all men are mortal' by using notions of absence and locus. Thus we have 'Humanity is "absent" from a locus in which there is absence of morality' (in place of 'all humans are mortal').

Absence was accepted as a separate category (*padartha*) in the earlier Nyaya-vaisesika school. The philosophers of that school tried always to construe properties or attributes (to use their own terms: *guna* quality; *Karma* movement; *samanya* generic property; *visesa* differentia), as separate entities over and above the substrate or loci, this is, the things that possess them. They also exhibited this tendency in their interpretation of negative cognitions or denials. Thus they conceived of absence as a property by a hypostasis of denial. The negative cognition 'there is no pot on the ground' or 'a pot is absent from the ground' was interpreted as 'there is an "absence of pot" on the ground'. It was then easy to construe such an absence as the object of negative cognitions — and hence as a separate entity. Moreover, cognitions like 'a cloth is not a pot' . . . were also treated and explained as 'a cloth has a mutual absence of pot, i.e., difference from pot'. And a mutual absence was regarded as merely another kind of absence.

In speaking of an absence, Nyaya asserts, we implicitly stand committed to the following concepts. Whenever we assert that an absence of an object 'a' (say a pot) occurs in some locus (say, the ground), it is implied that 'a' could have occurred in, or, more generally, could have been related to, that locus by some definite relation. Thus, in speaking of absence of 'a' we should always be prepared to specify this such-and-such relation, that is, we should be able to state by which relation, 'a' is said to be absent from the locus. (This relation should not be confused with the relation by which the absence itself, as an independent property, occurs in the locus. The latter relation is called a *svarupa* relation.) The first relation is described in the technical language of Navya-nyaya as the 'limiting or delimiting relation of the relational abstract, counter positiveness, involved in the instance of absence in question' (*pratiyo-gitavacchedakasambandha*). Thus, a pot usually occurs on a ground by *samyoga* or conjunction relation. When it is absent there, we say that a pot does not occur on the ground by conjunction or that pot is not conjoined to the ground. By this simple statement we actually imply, according to Nyaya, that there is an absence on the ground, an absence the counterpositive (*pratiyogin*) of which is a pot, and the delimiting relation of 'being the counter-positive' (i.e. counter-positiveness — *pratiyogata*) of which is conjunction. While giving the identity condition of an instance of absence, Nyaya demands that we should be able to specify this delimiting relation whenever necessary. The following inequality statements will indicate the importance of considering such a relation:

1. Absence of pot $\neq$ absence of cloth.

2. An absence of pot by the relation of conjunction ≠ an absence of pot by the relation of inherence.

Thus for the Indian logician, absence is always the absence of some definite property (*dharma*) in a locus (*dharmi*) and characterized by a relation — technically, either an occurrence exacting relation (*vrttiniyamaka sambandha*) or identity (*tadatmya*) by which the entity could have occurred in the locus, but is now cognized to be absent. Thus each *abhava* is characterized by its *pratiyogi* (the absentee or the entity absent, sometimes called the 'counter positive') as limited (i) by its *prativogitavacche-dakadharma* (the limiting attribute(s) limiting its counter-positiveness) as also (ii) by the *pratiyogitavacchedaka sambandha* (the limiting relation limiting its counter-positiveness). Thus in the cognition *ghatabhavad bhutalam* (The ground possesses pot-absence), the *pratoyogi* of *ghatakhava* (pot-absence) is *ghata* (pot) which is *ghatatvavavacchinna* and *samyoga — sambandhavacchinna*, as what is being denied is the occurrence of pot as characterized by potness in relation to contact with the ground.

Further, it is always stipulated in Indian logic that *abhava* of some property (*dharma* is meaningful only if that property is not a universal property (*Kevalanvayi dharma*, which occurs in all loci) or an empty property (*aprasiddha dharma*) which occurs nowhere.[11] Thus 'empty' or 'universal' terms cannot be negated in Indian logic, and many sophisticated techniques are developed in order that one does not have to employ such negations in logical discourse.

The sophistication of the Indian logicians' concept of *abhava* (as compared with the notion of negation in western logic) can be easily seen by the formulation of the law of contradiction in Indian logic. Instead of considering trivial truth functional or linguistic tautologies of the form either 'p' to 'not-p', the Indian logician formulates the notion of *pratibandhakatva* (contradictoriness) of one *jnana* (cognition) with respect to another. Further, this relation of *pratibandhakatva* can be ascertained only when the appropriate logical structures of each cognition are clearly set forth and can thus be stated precisely only in the technical language formulated by the Indian logician for this purpose. For instance, it would clearly not do to state that the cognitions *ghatavad bhutalam* (the ground possesses pot) and *ghatabhavavadbhutalam* (the ground possesses pot-absence) are contradictory, because in the first cognition the pot could be cognized to be present in the ground by the relation of contact (*samyoga*) while in the second the pot could be assumed to have been cognized as being absent in the ground by the relation of inherence (*samavaya*).[12] These two cognitions do not contradict each other at all and in fact they can both be valid. The law of contradiction can be correctly formulated only when the logical structure of both the cognitions are clearly set forth with all the *visesyata, prakarata* and *samsargatas* and their limiters (*avacchedakas*) being fully specified and it is seen from their logical structure that certainty (*niscayatva*) of one cognition prevents (*pratibadhnati*) the possibility of the other cognition arising (in the same person). Consider the case when for instance the cognition that the ground possesses pot (*ghatavad bhutalam*) actually has the logical structure: *samyoga sambandhavacchinna ghatatvavacchinna prakarata nirupita bhutalatvavacchinna visesyataka jnanam*. This cognition is prevented by the cognition that the ground possesses pot-absence (*ghatabhavavad bhutalam*) only if the latter has the logical structure: *Svarupasambandhavacchinna samyoga sambandhavacchinna ghatatvavacchinna pratiyogitaka abhavatvavacchinna prakaratanirupita bhutalatvavacchinna visesyataka jnanam*. This prevented-preventer (*pratibadhya-pratibandhaka*) relation between these two cognitions is formulated in the following form by the Indian logician: *Samyoga sambandhavacchinna — ghatatvavacchinna prakarata nirupita bhutalatvavacchinna visesyataka jnanatvavacchinnam prati svarupa sambandhavacchinna samyoga sambandhavacchinna ghatatvavacchinna pratiyogitaka abhavatvavachinna prakarata nirupita bhutalatvavacchinna visesyataka niscayatvena pratibandhakatam*.

In regard to the knowledge having its qualificandness limited by groundness and described by the qualifierness limited by potness and the relation of contact, the knowledge having its qualificandness limited by groundness and described by qualifierness limited by constant absenceness and the relation *svarupa* (absential self-linking relation) the counter-positiveness (*pratiyogita*) of which absence is limited by potness and the relation of contact is the contradictory definite knowledge, contradictoriness resident in it being limited by the property of *niscayatva* (definite knowledgeness).[13]

Quantification in Indian Logic

As another instance of the Indian approach of making the logical structure of a cognition clear and unambiguous by reformulating it in a technical language, we consider here the method developed in Indian logic for formulating universal statements, i.e. statements involving the so-called universal quantifier 'all'. Such statements arise in the basic scheme of inference considered in Indian logic where one concludes from the cognition 'the mountain is smokey' (*parvato dhumavan*) that 'the mountain is fiery' (*parvato vahniman*), whenever one happens to know that 'wherever there is smoke there is fire' (*yatra yatra dhumah tatra vahnih*). A careful formulation of this last statement which is said to express the knowledge of pervasion (*vyapti jnana*) of fire by smoke has been a major concern of Indian logicians, who have developed many of their sophisticated techniques mainly in the course of arriving at a precise formulation of *vyapti*.

According to the Indian logicians a statement such as 'all that possesses fire' is unsatisfactory as an expression of *vyapti jnana*. Firstly we have the problem that the statement as formulated above is beset with ambiguities (nowadays referred to as the 'confusion in binders' or 'ambiguity in the scope of quantifiers'). For instance there is a way of misinterpreting the above statement using the so called *calani nyaya* — by arguing that if all that possesses smoke possesses fire, what prevents mountain fire from occurring in kitchen where one sights smoke, or vice versa. The Greeks also discussed some of these ambiguities in formulating universal statements. In the western tradition some sort of a solution to this problem was arrived at only in the late nineteenth century via the method of quantification. In this procedure, the statement 'all that possesses smoke possesses fire' is rendered into the form 'for all x, if x possesses smoke then x possesses fire', before formalization.

The approach of the Indian logician is very different from the above method of quantification. The Naiyayika insists that the formulation of *vyapti jnana*, apart from being unambiguous, should be phrased in accordance with the way such a cognition actually arises. Hence an expression such as 'for all x, if x is smokey then x is fiery' involving a variable x, universally quantified over an (unspecified) universal domain, would be totally unacceptable to the Indian logicians.[14] What they do instead is to employ a technique which involves the use of two *abhavas* (use of two negatives) which are appropriately characterized by their *pratiyogitavacchedaka dharmas* and *sambandhas*. The steps involved may be briefly illustrated as follows:[15]

The statement 'all that possesses smoke possesses fire' can be converted into the form 'all that possesses fire-absence, possesses smoke-absence'. Here, fire-absence (*vahnyabhava*) should be precisely phrased as an absence which describes a counter-positiveness limited by fireness and the relation of contact (*samyoga sambandhavacchinna vahnit-vavacchinna pratiyogita nirupaka abhavah*). Now the statement that smoke is absent by relation of contact from every locus which possesses such a fire-absence is formulated in the following precise manner: 'Smokeness is not a limiter of occurrentness limited by relation of contact and described by locus of absence of fire which absence described a counter-positiveness limited by fireness and contact' (*Samyoga sambandha vacchinna vahnitvavacchinna pratiyogita nirupaka abhavadhikarana nirupita samyoga sambandhavacchinna vrittita-anavacchedakata dhumatve*).[16] In the above statement we may note that the 'locus of absence of fire' (*vahnya-bhavadhi-karana*) is not the locus of absence of this or that case of fire, but indeed of any absentee limited by fireness, as also by the relation of contact (*samyoga sambandhavacchinna vahnitvavacchinna pratiyogita nirupaka abhavadhikarana*). This is what Indian logic employs instead of notions such as 'all the loci of absence of fire' or 'every locus of absence of fire'. In the same way, the phrase that 'smokeness is not the limiter of an occurrentness limited by relation of contact and described by locus of . . .' (. . . *adhikarananirupita samyoga sambandhavacchinna vrittita anavacchedakata dhumatve*) serves to clearly and unambiguously set forth the fact that no case of smoke occurs in such a locus (of absence of fire) by relation of contact.

I shall now make a few brief remarks on the Indian logicians' way of formulating statements of *vyapti* such as 'all that possesses smoke possesses fire', as compared with the method of quantification employed in modern western logic. Firstly, the Indian formulation of *vyapti* always takes into account the relations by which fire and smoke occur in their loci. But even more important is the fact that the Indian logician completely avoids quantification over (unspecified) universal domains which is what is employed in modern western logic. The statement that 'all that possesses smoke possesses fire' is intended to say something only about the loci of smoke — that they have the property that they possess fire also. But the corresponding 'quantified' statement 'for all x, if x possesses smoke then x possesses fire', seems to be a statement as regards 'all x' where the variable 'x' ranges over some universal domain of 'individuals' (or other sort of entities in more sophisticated theories such as the theory of types). The Indian logicians' formulation of *vyapti* completely avoids this sort of universalization and strictly restricts its consideration to the loci of absence of fire (as in the above formulation, known as *purvapaksha vyapti*) or to the loci of smoke (in the more exact formulation known as *siddhantavyapti*, which formulation is also valid for statements involving the unnegatable *Kevalanvayi*, or universally present, properties).[17]

Another important feature of the Naiyayika method of formulating *vyapti* is that it does not employ quantification over some 'set' of individuals viewed in a purely 'extensional' sense. It does not talk of the 'set of all loci of absence of fire', but only of 'a locus which possesses an absence the counter-positive of which absence is limited by fireness and relation of contact'. In this sense, the Indian method of formulating universal statements does take into account the 'intensions' of all the properties concerned and not merely their 'extensions'. As one scholar, has noted:

> The Universal statements of Aristotelian or mathematical logic are quantified statements, that is, they are statements about all entities (individuals, classes or statements) of a given sort. On the other-hand, Navya-nyaya regularly expresses its universal statements and knowledges not by quantification but by means of abstract properties. A statement about causeness to pot differs in meaning from a statement about all causes of pots just as 'manness' differs in meaning from 'all men'.[18]

As explained by another scholar:

> The Nayayikas in their logical analysis use a language structure which is carefully framed so as to avoid explicit mention of quantification, class and class membership. Consequently their language structure shows a marked difference from that of the modern western logicians . . . Naiyayikas instead of class use properties, and in lieu of the relation of membership, they speak in terms of occurrence (*vrittitva*) and its reciprocal, possession, moreover, instead of quantification, the Naiyayikas use 'double negatives and abstract substantives' to accomplish the same result . . . Any noun substantive in

Sanskrit . . . may be treated as a *dharma* (property) occurring in some locus and also as a *dharmi* (a property-possessor) in which some dharma or property occurs.[19]

According to the same author, in western logic 'classes with the same members are identical . . . But a property or an attribute, in its non-extensional sense, cannot be held to be identical with another attribute even if they are present in all and only the same individuals. Properties are generally regarded by the Indian logicians as non-extensional, in as much as we see that they do not indentify two properties like *anityatva* (non-eternalness) and *kritakatva* (the property of being caused) although they occur in exactly the same things. In Udayana's system, however, such properties as are called *jati* (generic characters) are taken in extensional sense because Udayana identifies two *jati* properties if only they occur in the same individuals.'[20]

It should be added however, that according to Udayanacarya there are a whole lot of properties which cannot be considered as *jati* and are generally referred to as *Upadhi*. In fact Udayanacarya has provided a precise characterization of all those properties which cannot be considered as *jati* or generic characters. Another point that should be noted is that the Indian logicians do consider the notion of a collection of entities, especially in the context of their discussion of number and the paryapti relation. But they refuse to base their entire theory on notions such as 'class' or 'set' viewed in purely extensional terms, and in this respect the Indian logicians' approach (which does not seem to separate extensions from intensions) is very different from most of the approaches evolved in the western tradition of philosophy and foundations of logic and mathematics.

Astadhyayi: The Paradigm Example of Theory Construction in India

Just as the modern western systems of axiomatized formal theories find their paradigm example in the exposition of geometry in Euclid's *Elements*, the Indian method of theory construction finds its paradigm example in the Sanskrit grammar of Panini, the *Astadhyayi*. As one scholar has noted,

Historically speaking, Panini's method has occupied a place comparable to that held by Euclid's method in Western thought. Scientific developments have therefore taken different directions in India and the West . . . In India Panini's perfection and ingenuity have rarely been matched outside the realm of linguistics. In the West this corresponds to the belief that mathematics is the more perfect of the sciences.[21]

Astadhyayi as a Generative Device

Over the last two centuries, the Indian grammatical tradition (especially the *Astadhyayi* of Panini and other works of Paninian school) has proved to be a major fountainhead of ideas and techniques for the newly emerging discipline of linguistics both in the phase of historical and comparative linguistics in the nineteenth century and in the descriptivist and structuralist and generativist phases of the twentieth century. In spite of such intensive study and considerable borrowing over a long period of time, the basic methodology and the technical intricacies of Panini's grammar were very little understood till the advent and development of the modern theory of generative grammars in the last few decades.[22] As a scholar has noted recently:

> The algebraic formulation of Panini's rules was not appreciated by the first Western students; they regarded the work as abstruse or artificial. This criticism was evidently not shared by most Indian grammarians because several of them tried to outdo him in conciseness by 'trimming the last fat' from the great teacher's formulations . . . The Western Critique was muted and eventually turned into praise when modern schools of linguistics developed sophisticated notation systems of their own. Grammars that derive words and sentences from basic elements by a string of rules are obviously in greater need of symbolic code than paradigmatic or direct method practical grammars . . .[23]

It is a sad observation that we did not learn more from Panini than we did, that we recognized the value and the spirit of his 'artificial' and 'abstruse' formulations only when we had independently constructed comparable systems. The Indian new logic (*navya-nyaya*) had the same fate: only after western mathematicians had developed a formal logic of their own and after this knowledge had reached a few Indologists, did the attitude towards the *navya-nyaya* school change from ridicule to respect.

The major proponent of the present day generative and transformational grammars refers to Panini's grammar as 'a much earlier tradition' of generative grammar, though 'long forgotten with a few exceptions'.[24] For another modern expert, Panini's *Astadhyayi* is 'the most comprehensive generative grammar written so far'.[25] This feature of Panini's grammar is explained in the following quotations:

> To Panini . . . grammar is not understood as a body of learning resulting from linguistic analysis but as a device which enables us to derive correct Sanskrit words. The machinery consists of rules and technical elements, its inputs and word-elements, stems and suffices, its output are any correct Sanskrit words. Thus the *Astadhyayi* is a generative device in the literal sense of the word. Since it is also a system of rules which allows us to decide the

correctness of the words derived, and at the same time, provides them with a structural description, the *Astadhyayi* may be called a generative grammar.[26]

Sharma describes it thus:

> Panini's *Astadhyayi* . . . is a set of rules capable of formally deriving an infinite number of correct Sanskrit utterances together with their semantic interpretation . . . The entire grammar may be visualised as consisting of various domains. Each domain contains one or more interior domains. The domain[s] may like wise contain one or more interior domains. The first rule of a domain is called its governing rule. These rules assist one in scanning. Given an input string, one scans rules to determine which paths should be followed with domains. These paths are marked by interior domains, each one headed by a rule that specifies operational constraints and offers selection in accordance with the intent (a set of quasi-semantic notions related to what we know about what we say before we speak . . . (denoted by) the Sanskrit term *vivaksa*). Where choices are varied in operation and there are innumerous items to select from, an interior domain is further responsible for sub-branching in the path resulting in its division into interior domains.[27]

Though various attempts have been made to find parallels to notions such as 'deep structure' or even 'transformations' in the Paninian system, it is now becoming clear that, though it is operating with concepts and techniques of comparable sophistication, the Paninian system of linguistic description is very different from the various models which have been and are being developed in modern western linguistics.

In fact the differences between the Paninian approach and those of modern linguistic theories have to do with several methodological and foundational issues. For instance while the Paninian system is viewed as a generative device, the inputs to this device are not formal objects such as symbols and strings which are to be later mapped on to appropriate 'semantic' and 'phonological' representatives. Further, the *vivaksa* or the intent of the speaker seems to play a prominent role in the Paninian system and, as has been noted recently, 'Panini accounts for utterances and their components by means of a derivational system in which one begins with semantics and ends with utterances that are actually usable'.[28]

Technical Features of Astadhyayi

We now turn to the various technical aspects of the *Astadhyayi* which reveal some of the basic features of theory-construction in the Indian tradition. The technical terms of the theory (*samjna*), the metarules (*paribhasha*) which circumscribe how the rules (*sutras*) have to be used, the limitation of the general (*utsarga*) rules by special (*apavada*) rules, use of headings (*adhikarasutra*), the convention of recurrence (*anuvrtti*) whereby parts of rules are considered to recur in subsequent rules, the various conventions on rule-ordering and other decision procedures as also the various so called 'metalinguistic' devices such as the use of markers (*anybandhas*) and the use of different cases to indicate the context, input and change — all these and many other technical devices employed in *Astadhyayi*,[29] are now coming to be more and more recognized as the technical components of an intricate but tightly-knit logical system, as sophisticated as any conceivable formal system of modern logic, linguistics, mathematics or any other theoretical science. But there is one crucial feature in which the Paninian system (like perhaps all other theoretical systems con- structed in Indian tradition) differs from the modern formal systems. While it employs countless symbols, technical terms and innumerable 'metalinguistic' conventions and devices, the Paninian grammar is still a theoretical system formulated very much in the Sanskrit language, albeit of an extremely technical variety. It is not a formal system employing a purely symbolic language.

It is sometimes remarked that the language employed in Panini's *Astadhyayi* (sometimes referred to as Panini's metalanguage) differs from ordinary Sanskrit so 'strongly that one must speak of a particular artificial language'.[30] This is a misunderstanding in the sense that though the technical language of Panini's *Astadhyayi* abounds in technical terms and devices, and does differ considerably from ordinary Sanskrit found in non-technical literature, it is all the same only a technical or shastric version of Sanskrit — i.e. a technical language constructed on the foundation of ordinary Sanskrit. As has been noted recently, many a technical device of Panini is arrived at via 'an abstrac- tion and formalization of a feature of ordinary language'.[31]

The relation between the technical language employed by Panini and ordi- nary Sanskrit can be made clear by considering an example. We discuss the so- called 'metalinguistic' use of cases in Paninina *sutras*. For instance consider the rule *ikoyanaci* (*Sutra* 6.1.77 of *Astadhyayi*). Here *ik, yan* and *ac* are symbols for groups of sounds, but are at the same time treated as Sanskrit word-bases. The word-base *ik* occurs in the *sutra* with genitive ending (*ikah*), *yan* with nomi- native and *ac* with locative ending (*aci*). The *sutra* stipulates that the vowels *i,u,r,l* (denoted by *ik*) should be replaced by *y,u,r,l* (denoted by *yan*) before a vowel (*ac*). The information as to what should serve as input, output and context is 'metalinguistically' marked with various case endings taken by the Sanskrit word-bases *ik, yan* and *ac*. For instance *ik* is used with the genitive ending (*ikah*) to indicate that it is the substituend or input, as per the metarule (*paribhasa*) *Sasthi sthaneyoga* (*Sutra* 1.1.49). The main point is that while there are various possible meanings indicated by the genitive case ending, Panini uses the metarule 1.1.49 to delimit the meaning of the genitive case-ending to indi- cate (wherever the metarule applies) only the substituend or the input of a grammatical operation. As one scholar has explained:

The rule 1.1.49 *sasthi sthaneyoga* . . . assigns a metalinguistic value to the sixth triplet (*sasthi*) endings. As noted . . . (the *sutra sese*) 2.3.50, introduces genitive endings when there is to be denoted a non-verbal relation in general. There are of course many such relations, such as father-son, part-whole . . . etc. The rule 1.1.49 states a particular relation to the understood when the genitive is used: the relation of being a substituend.[32]

In other words, these 'metalinguistic' case conventions are not arbitrary or artificial — they most often serve only to fix one unique meaning where several interpretations are possible in the ordinary use of the language.

In this context the oft-quoted criterion of *laghava* employed by the Sanskrit grammarians should also be properly understood. This has often been interpreted as brevity and is sometimes seen as the sole *raison d'être* of Panini's exposition — meaning thereby that most of the techniques employed by Panini are mere arbitrary devices to achieve brevity in exposition. Further, the tendency of the Indian grammarians to achieve brevity is often linked with other speculations concerning learning in ancient India — such as possible shortage of writing materials,[33] or the possible necessities of a purely oral tradition placing heavy demands on memory,[34] etc. Now, it is of course true that the Indian grammarians did indeed rejoice (as the saying goes) at the saving of even half of a mora (*matra*) in their exposition.[35] But this saving of moras was not to be achieved by arbitrary devices. As has been noted recently,[36] 'hundreds of moras could have been saved by selecting the accusative instead of the genitive case as marking the input of a rule' — but that would have meant a drastic deviation from the ordinary usage of the accusative.

Thus a 'metalinguistic' device like the use of cases to indicate context, input and output in a grammatical operation, is not an arbitrarily chosen convention for achieving mere brevity, but is actually a technical device founded on the basic structures available in the ordinary Sanskrit language and which serves mainly to render to language unambiguous, more precise. This, we could perhaps assert, is true of all the technical devices employed in the Paninian grammar. For instance, it has recently been argued that the Paninian use of *Anuvrtti* is not an artificial device for merely achieving brevity, but in fact a systematic and technical use of 'real language, ellipsis'.[37] As regards the criterion of brevity itself, it has been remarked that the point is rather that the rules are strictly purged of all information that is predictable from other information provided in the system. What Panini constantly tries to eliminate is not moras, but redundancy.[38]

Apart from developing a technical or precision language system for the formulation of grammatical rules, Panini's *Astadhyayi* also reveals several sophisticated devices which delimit the nature and application of these rules. Most of these techniques appear to be common for the entire corpus of classical shastric literature wherein the *sutra* technique of systematization has been employed. Here again we should take note of the generally prevalent opinion that the *sutra* style is employed in the Indian tradition merely for the purpose of achieving brevity in exposition. While brevity is indeed a hallmark of the *sutra* technique of systematization, there are a whole lot of other equally or even more important criteria that a *sutra* should satisfy. For instance, the Vishnudharmottarapurna characterizes a *sutra* as being 'concise (employing minimum number of syllables), unambiguous, pithy, comprehensive, firm and blemishless'.[39]

Though the Paninian (or other) *sutras* are often translated as rules, they differ substantially from what are generally understood as rules in modern linguistic theory. According to one scholar:

> Rules in modern linguistics are treated as statements independent of one another. They are formulated in such a way that they seldom require any information from other rules. Panini's rules by contrast are interdependent. That is, for the application of a given rule one may at times have to retrieve many rules, which may be very distant with respect to their placement in the grammar. This is what the tradition calls *ekavakyata* or 'single context'. Secondly, when it comes to interpreting a rule in modern linguistics, we find that each hardly needs any help from the others. By contrast, a rule in Panini usually requires the carrying over of previous [or later] rules, or other element[s], for its correct interpretation. This makes Paninian rules interdependent in contrast with rules in modern linguistics . . . This interdependence in the interpretation and application of rules required Panini to arrange his rules into domains and subdomains'.[40]

There are indeed several technical aspects of the *sutra* method of systematization — such as the use of *paribhasa, adhikara, upadesa, asiddha, vipratisadha* etc. These are extensively employed in Panini's *Astadhyayi*, but are not defined explicitly in the text. As has been noted recently these and similar technical terms are 'metagrammatical in the sense that they refer not to concepts about which grammatical analysis must theorize, but to the basic equipment which one brings to the very task of grammatical analysis. It should be noted that many of these terms are common property of the *Sutra* technique as applied not only in grammar but also in ritual and elsewhere'.[41]

Lest the main achievement of Panini's *Astadhyayi* be lost amidst all this analysis of its methodology and technical sophistication, we should restate what *Astadhyayi* achieves in about 4,000 *sutras* — it provides a complete characterization of Sanskrit utterances (or more appropriately, a characterization more thorough than what has been possible for any other language so far) by devising a system of description which enables one to generate and analyse all possible meaningful utterances. It also provides the paradigm example of 'theory construction' in the Indian tradition.

Sabdabodha and 'Knowledge Representation'

We have already noted how the *Astadhyayi* serves as a generative device which enables us to derive correct Sanskrit utterances and at the same time provides us with a structural description of these utterances. We shall now discuss how the Paninian analysis of Sanskrit utterances enabled the Indian linguists (*sabdikas*) to provide a full-fledged semantic analysis of meaningful Sanskrit utterances and formulate the cognition generated by an utterance (*sabdabodha*) in an unambiguous manner in a technical language. In other words, the Indian tradition of linguistics (*sabdasastra*) has endeavoured to fully systematize both the generation of the form of an utterance (*sabda*) starting from the intention of the speaker (*vaktr vivaksha*) as well as the analysis of the cognition generated by such as utterance (*sabdabodha*) in any hearer (*srobr*) conversant with the Sanskrit language.

The semantic analysis of Sanskrit utterances is outlined in the great commentary Mahabhashya of Patanjali. A detailed exposition of the semantic theories of Indian linguists may be found in the *Vakyapadiya* of *Bhartrhari* (believed to be fifth century AD), which is in fact a treatise on *Vaiyakaranadarsana*, dealing with all aspects of the Indian philosophy of language. Since, *sabda pramana* (the utterance of a reliable person (*apta*) as a valid means of knowledge) was accepted by most schools (*Darsanas*) of Indian philosophy, the analysis of *sabdabodha* (cognition generated by an utterance) was a major subject of enquiry. The entire analysis was deeply influenced by the techniques developed by the Indian logicians of the *Navya-nyaya* school. During sixteenth to eighteenth century the technique of *sabdabodha* was more or less perfected. There were of course three schools of thought represented by the *Navya vaiyakaranas* (such as *Bhattoji*) Diskshita, Kaunda Bhatta, Nagesha Bhatta, etc.), *Navya-Naiyayayikas* (such as Raghunatha Sironani, Jagadisa Tarkalamkara, Gadadhara Bhattacharya, etc.) and Navya-Mimamsakas (such as Gaga Bhatta, Khandadeva Misra and others). All of them gave systematic procedures as to how the *sabdabodha* of any utterance may be formulated in a precise and unambiguous manner in a technical language (based on ordinary Sanskrit), with the only difference that each of them had different views on: (a) what the entities (*padarthas*) associated with the various words (*padas*) in an utterance are,[42] (b) what the relations between these entities as revealed by the utterance are and (c) what the chief qualifier (*mukhya visesya*) of the cognition generated by the utterance is.

The basic technique of *sabdabodha* is briefly summarized in the following extract from a recent study:

A sentence is composed of words whether their existence is considered real as in the case of the Logician (Naiyayika), the Mimamsaka and others, or mythical as in the case of the Grammarian (Vaiyakarana) . . . *Sabdabodha* is the cognition of the meaning of sentence. It has been defined as 'the cognition effected by the efficient instrumentality of the cognition of words' (*padjnanakaranakan jnanam*) . . . 'the cognition resulting from the recalling of things derived from words' (*padajanya padarthopasthiti janya bodhah*) . . . 'the knowledge referring to the relation between each of the substances recalled by the words in a sentence' (*Eka padarthe aparapadartha samsarga vishayakam jnanam*).

In order to have a clear idea of this theory the various stages of verbal cognition (*sabdabodha krama*) may be studied with advantage. While comprehending the meaning of any sentence, first of all, we cognize the word and then its (denotative) potentiality (*sakti*) and from both of these put together the recalling of meanings is effected and thus import is generated. For instance in the sentence . . . '(Caitra) worships Hari' ((*Chaitrah*) *Harim bhajati*) there is first of all, the cognition of the several words 'Hari', the (accusative) case affix '*Am*', the root 'worship' (*Bhaj*) and the verbal affix '*tip*'. Next their (denotative) potentialities are comprehended in the following way:- The word 'Hari' by virtue of its denotative capacity (*abhidhasakti*) denotes Hari, '*am*' the case affix denoted objectness (*Karmatva*), the root '*bhaj*' denotes activity favourable to love (*prityanukula vyapara*), 'tip' denotes activity (*Kriti*), of course, in addition to the meanings of number, tense, etc. This is the cognition of the potentiality of words, the second stage of verbal import (*sabdabodha*) . . . subsequently as there exists among these several words (or among their meanings) mental expectancy (*akanksha*), compatibility (*yogyata*) and juxtaposition (*sannidhi or asatti*) a totality of comprehension is produced in the form 'Caitra is the substratum of activity favourable to love which has Hari for its object' (*Harikarmaka prityamukula Kritiman Caitran*)'.[43]

To elucidate the technique of *sabdabodha* let us consider the same Naiyayika method of *sabdabodha* of the sentence *Chaitrah harim bhajati* in some detail. Here there are six 'words' — *chaitra, sup, Hari, am, bhaj, tip*. In the Naiyayika method of *Sabdabodha*, *Chaitra* refers to the individual *Chaitra* (*Chaitra vyakti*) as qualified by the genus chaitraness (*Chitratva*) and form (*jatyakriti visistah*). The same is true of the word *Hari*. The case affix *sup* refers to singular number (*Ekatva samkhya*) and *am* refers to objectness (*Karmatva*). The root *bhaj* refers to the activity favourable to love (*prityanukula vyapara*). The verbal affix (*akhyata*) 'tip' refer to effort (*Kriti*), singular number (*samkhya*) and present tense (*vartamanakala*). The Naiyayika theory of *sabdabodha* further specifies the various relations by which all the above entities (*padarthas*) are related to each other.

The Naiyayikas express the *sabdabodha* of the sentence *Chaitrah Harim bhajati* in the form: *Ekatva samaveta Haritva samaveta Harinirupita Karmatvasraya prityanukula vyaparanukula Vartamanakalikaya Kritih tasyasrayah ekatvasamaveta Chaitratva-samavetah Chaitrah*: Chaitra as qualified by singularity and Chaitraness (via the relation of inherence) is the substratum of effort which is favourable to activity favourable to love residing in the objectness described by *Hari* who is qualified by singularity and hariness (via the relation of inherence). The above is only a simplified form of the more refined (*pariskrta*) *sabdabodha* wherein one would state precisely the various qualificandness (*visesyata*) and qualifierness (*prakarata*) resident in all the above *padarthas* along with their limiters (*avacchedakas*) — both the limiting attributes (*avacchedaka dharmas*) as also the limiting relations (*avacchedaka sambandhas*) which later are nothing but the various 'syntactical relations' (*anvaya sambandhas*) that have been indicated between the various *padarthas* in the above simplified *sabdabodha*.

The Vaiyakarana and the Mimamsaka formulations of *sabdabodha* follow a similar scheme; but the various *padarthas* associated with different *padas* and their *anvaya sambandhas* are slightly different in each scheme. Further the chief qualifier (*mukhya visesya*), which was *Chaitra* in the above *Naiyayika* formulation, would be the activity (*vyapara*) part of the meaning attributed to the verb root (*dhatu*) *bhaj* in the case of the Vaiyakaranas and the activity (*bhavana*) part of the meaning attributed to the verb affix (*akhyata*) '*tip*' in the case of the Mimamsakas. Each of the three schools have come up with detailed arguments to show how their formulation of *sabdabodha* is not only fully consistent but also superior to the formulations given by the other schools, from various fundamental considerations.

Whether it be the Naiyayika formulation of *sabdabodha* or the Vaiyakarana or the Mimamsaka formulation, what is achieved is indeed very significant. All of them provide precise and unambiguous characterization of the cognition generated by any particular utterance of Sanskrit language. If the utterance has ambiguities (be they due to the presence of polysemious words (*nanartha-kasabdas*) or of pronouns (*sarvanama*) or due to the sentence structure, etc.) then procedures are outlined as to how the actual import that is intended to be conveyed (*vaktr vivaksa* or *tatparya*) is to be arrived at and the *sabdabodha* done accordingly. The *sabdabohda* itself is formulated in a technical language which is unambiguous and clearly presents the full content (*visayata*) of the cognition (the various *padarthas* and their *sambandhas* as manifested by the cognition) as well as its logical structure. Indeed, as has been noted recently, the technique of *sabdabodha* seems to be a full-fledged scheme for arriving at what has been called a 'knowledge representation' of every utterance in the natural language Sanskrit.[44] What is significant is that while most of the techniques of 'knowledge representation' which are currently being investigated (in connection with natural language processing by computers) are mostly *ad hoc* schemes usually applicable to a particular class of sentences etc., the technique of *sabdabodha* is a systematic procedure based on a fundamental analysis of the nature of linguistic utterances, and the cognition they generate, which at the same time can be applied to obtain a 'knowledge representation' of *all* conceivable utterances in the natural language Sanskrit.

The Technical Language of Indian Sastras vis-à-vis Formal Languages

In conclusion, the main point I wish to focus upon is the power and potentiality of the technical language that has been developed in the Indian tradition as the basic tool for logical analysis. This discussion of Indian logic has perhaps indicated how the Indian logicians, instead of seeking to develop content-independent and purely symbolic formal languages as in the West, have sought to develop a technical or precision language founded on the natural language Sanskrit which avoids all possible inexactness and ambiguities. By means of this procedure of *parishkara* (refinement) the Indian logicians achieve precision, and also bring out clearly the logical structure of a cognition, the structure of which has an unambiguous representation in their technical or *sastric* language. Thus the technical language developed by the Indian logicians is indeed one of their major achievements — a fact which was not realized by the modern scholarship on Indian logic till recently,[45] partly because many of the comparable techniques in western logic are perhaps less than a century old. It is now generally recognized that the technical language developed by the Indian logicians allows them to achieve much of what is supposed to be achieved via the symbolic or formal languages of modern mathematical logic. According to one scholar:

> Navya-nyaya [the modern school of Indian logic started by Gangesa Upadhyaya in fourteenth century] never invented the use of symbols. It invented instead a wonderfully complex system of clichés by which it expresses a great deal that we would never think of expressing without symbols.[46]

Another scholar describes it thus:

> The technical language of Navya-nyaya is not I suspect so much a language as the groping for a kind of picture of the universe of individuals in their relationships with one another . . . There seems to be a kind of continuity extending from vague, ambiguous, inaccurate ordinary languages through languages filled with technical terms, to clear unambiguous, accurate maps of the kind exemplified by the mathematical physicists' formulas . . . The Naiyayikas style, it may be conjectured, is not intended for the purpose of communicating more easily, any more than the mathematicians' is; it is intended rather to provide a simple accurate framework for the presentation of the world as it really is. In short, the Navya-nyaya aim is not so far away from the apparent aim of those contemporary philosophers of this day and age in the West, who wish by use of techniques of symbolic logic to find a simple and accurate way of setting forth the picture of the world presented by the natural sciences.[47]

It is necessary to emphasize that these estimates of the technical or precision language employed in Indian logic seem to miss altogether the basic methodological principles inherent in the Indian approach. It appears to us that Indian logicians (instead of landing up somewhere in the 'continuum extending from vague . . . ordinary languages . . . to clear . . . mathematical physicists' formulas') deliberately avoided the purely symbolic and content-independent formal languages, just as they avoided postulation or use of ideal entities such as proposition, sense as distinguished from reference, logical truth as distinguished from material truth, etc. In striving to provide a logical analysis of cognitions, the Indian philosophers did not confine their analysis to a study of sentences or their meanings. However, at the same time, Indian tradition does not start with any pronounced contempt for the ordinary or natural languages. While it surely recognizes the imperfections in the natural languages as vehicles for logical discourse, the attempt in Indian tradition has been mainly to evolve a technical or precision language which is constructed on the basis of the natural language, Sanskrit, and which is free of whatever ambiguities, inaccuracies, vagueness, that the natural language might have. This technical language is so constructed as to easily reveal the logical structures which are not so transparent and often ambiguous in a natural language, but at the same time has a rich structure and interpretability which it inherits from the natural language from which it is constructed.

Perhaps, to a large extent, it was the strong foundation laid by the Paninian analysis of Sanskrit language, which enabled the Indian scientists and philosophers to, firstly, achieve a knowledge representation of all natural language utterances in terms of a technical language (thereby systematizing also the use of the natural language itself); and secondly, systematically to refine the natural language itself into a technical language with a transparent logical structure which could serve as a suitable vehicle for all precise and technical discourse.

The Indian approach of converting the ordinary discourse by *pariskara* (refinement) into a technical discourse, suitable for systematization and logical analysis of knowledge, indeed appears to be in conformity with the larger philosophical and methodological principles which have governed Indian thought all through. Instead of looking for ideal, context-free, and purely symbolic or formal languages which have no relation with natural languages, as possible tools for attaining 'perfect' logical rigour, the Indian tradition sets out to refine systematically the natural language Sanskrit to free it of all known ambiguities and inaccuracies and arrive at a technical language which can reveal the logical structure of a cognition as accurately as possible. In this sense, the process of *pariskara* is an evolving and even context-dependent process depending on the demands of a particular problem and the kind of ambiguities needed to be resolved. Our *Sastrakaras* always leave the options open for further *pariskaras* to be done as and when subtler problems need to be tackled. This is how, for instance the technique of insertion of *paryapti* was developed during the sixteenth to nineteenth centuries.[48]

The above features of the Indian approach need to be clearly contrasted with what has been sought to be achieved by the purely, symbolic or formal language systems developed in the western tradition and to what extent they have been successful so far. We shall here merely quote a recent estimate[49]:

> Traditional propositional logic is limited by two factors. Only truth functional connective has been studied and among these only those that are relevant to mathematics have been studied systematically. Originally logic was conceived of as a tool to study the logical properties of natural language. By translating arguments in natural language into propositional calculus one hoped to obtain the arguments in a more perspicuous form, where it would be easier to see whether they were valid. However, the translation turned out to be difficult; natural language with its vagueness and ambiguity had to be transferred into a somewhat arbitrarily chosen unambiguous system of formal representation. Since such a system was considered a great advantage in other respects, logic became increasingly estranged from the study of natural language. We still have not discovered how best to study and formalize non-truth functional relations between sentences.[50]

What estimates such as the above reveal is that while the modern western formal logic might have some relevance for providing foundational rigour to arguments in modern mathematics, it has so far totally failed in explicating logical relations between sentences as used in ordinary language or in most scientific and philosophical argumentation. When it comes to the foundations of mathematices itself, it has now become common knowledge that the formal and logical approaches being developed from the turn of last century have helped little in rendering them secure.

Formal methods, whatever their philosophical short-comings, got wide acceptance in the western tradition as they professed to free the ordinary discourse of all vagueness and ambiguity and provide logical rigour. What the Indian tradition seems to show is that one need not sacrifice the richness or the content of natural languages in order to achieve clarity, precision or logical rigour. In fact, in developing a technical or precision language based on the natural language Sanskrit, the Indian *shastrakaras* seem to have evolved a very powerful tool for the formulation of scientific theories, a tool very different from the modern mathematical logic or the attendant formal systems, and which needs to be investigated in much greater detail for its power and potential. A clear comprehension of the basic methodologies as outlined in the *sastras* of Kanada and Panini will also help us in rediscovering the foundations of all Indian *sastras* and restore the vitality and creativity that they seem to have displayed all through history.

References and Notes

1. For instance, the following is a recent assessment of Indian logic as found in the latest edition (1973) of the Encyclopaedia Britannica (Article on 'Logic, History of' by Czeslaw Lejewski).

   > Compared with the logic of the ancient Greeks, Indian logic is not very impressive . . . The development of Indian logic was severely handicapped by the failure of its logicians to make use of variables. As a result, no logical principles could be stated directly. Finally in Indian thought logical topics were not always separated from metaphysical and epistemological topics (on the nature of being and knowledge respectively . . . Both in the West and in the East, the origin of logic is associated with an interest in the grammar of language and the methodology of argument and discussion, be it in the context of law, religion or philosophy. More is needed, however, for the development of logic. It appears that logic can thrive only in a culture that upholds the conviction that controversies should be settled by the force of reason rather than by the orthodoxy of a dogma or the tradition of a prejudice. This is why logic has made much greater progress in the West than in the East.

   As another example we may present the following evaluation of the Indian tradition in mathematics by a contemporary historian of mathematics (Morris Kline, *Mathematical Thought from Ancient to Modern Times*, Oxford, 1972; 190).

   > As our survey indicates, the Hindus were interested in and contributed to the arithmetical and computational aspects of mathematics rather than to the deductive patterns. Their name for mathematics was *ganita*, which means 'the science of calculation'. There is much good procedure and technical facility, but no evidence that they considered proof at all. They had rules, but apparently no logical scruples. Moreover, no general methods, or new view points were arrived at in any one of the mathematics.

2. The '*sastras* of Kanada and Panini' also contain a detailed exposition of various epistemological and philosophical issues which have to be comprehended for a clear understanding of the foundations of Indian sciences. I do not touch upon these issues in this article.
3. Alonzo Church, article on 'Logic' in the Encyclopaedia Britannica 14th Edition (1959).
4. It is also a characteristic of Indian mathematics that it eschews the method of indirect proof or *reductio ad absurdum*. For the generally 'constructivist' character of Indian mathematics, see for instance: Navjyoti

Singh, 'A comparative Study of Foundations of Mathematics in India, China and the Modern West', PPST Bulletin 9 (1985) pp. 53–73; and Chhatrapati Singh, 'The Philosophical Foundations of a General Theory of Numbers', paper presented at the NISTADS Conference, Delhi 1984.

5. The Indian logicians' attitude to the method of indirect proof is very clearly brought out in the following excerpt from a recent translation of a portion of Udayanacarya's *Atmatattvaviveka* (tenth century AD). The text is in the form of an argument between the Naiyayika ('proponent', who does not accept the method of indirect proof) and a Pauddha ('opponent' who is arguing for the method of indirect proof):

(Proponent) . . . There cannot be any means of knowledge to establish a non-entity (i.e. a fiction, *avasta*). If it could be established by some means of knowledge it ceases to be non-entity.

Opponent: If so then your talk about the non-entity becomes self-contradictory.

Proponent: Does this self-contradiction point out that there is a means of knowledge to establish the non-entity? Or (second question) does it reject the prohibitive statement that we should not talk about non-entity? Or (third question) does it imply that we must concede such statements (about non-entity) which are inauthenticated, i.e. not established by any means of knowledge? The first alternative is not tenable. Even a thousand of self-contradictions cannot conceivably show that (the non-entity like) . . . rabbit's horn . . . is amenable to (a means of knowledge, such as) perception and inference. If it could, what is the use of this silly fight over the nature of non-entities? The second alternative is acceptable to us, because we admit only valid means if knowledge.

Opponent: If the prohibitive statement is rejected, no statement with regard to non-entities will be possible.

Proponent: What else can we do but remain silent in regard to a matter where statement of any kind will be logically incongruent? Silence is better in such cases . . . You yourself may please consider as to who is the better of the two: One who is making statements about entities that cannot be established by any means of knowledge? Or, the other person who remains speechless (on such occasions)?

Opponent: But although you are a wise man, you have not remained silent yourself. You on the other hand have made a prohibitive statement with regard to our talk about non-entities.

Proponent: True, in order to avoid a self-contradictory object not established by any means of knowledge, you have conceded that one can make statements about the non-existent. Similarly, in order not to allow any statement about the non-entities in our discourse on the means of knowledge, we concede that a self-contradictory statement (prohibiting the use of non-entities) is possible, although it is not supported by any means of knowledge. If you treated both the cases in the same manner, we would not

have said anything about non-entities (We have made the above self-contradictory statement because you first raised the question). (cited from
B.K. Motilal: *Logic Language and Reality*, New Delhi 1985, pp. 103-4).

I consider the above passage to be a remarkably clear statement of the
Indian logicians' position that they would rather live with self-
contradiction than accept the existence of entities which are inaccessible to
any (direct) means of knowledge (as demanded by those who argue for the
validity of 'indirect proof' as a means of knowledge). The Indian logician
would however prefer to avoid these self-contradictions by refusing to
admit these *Aprasiddha* entities into his discourse altogether.

6. A *Savikalpa jnana* is the cognition of an object as qualified by some
   qualifier — for instance the cognition of a pot (*ghata*) as a pot, i.e. qualified by potness (*ghatatva visista*). On the other hand a *nirvikalpaka jnana* is
   merely the cognition of an object with no qualifier — for instance the
   cognition of a pot as mere 'some thing' (*Kincit*). More precisely a
   savikalpaka jnana can be defined as a cognition which penetrates the relation between a qualificand and a qualifier (*visesya visesanayoh sam-
   bandhavagahi jnanam*).
7. J.N. Mohanty: 'Indian Theories of Truth', *Philosophy East and West*, 30,
   (1980) p. 440.
8. See for instance Ref. 7 and the following: Sibajiban Bhattacharya: 'Some
   Features of Navya-Nyaya Logic', *Philosophy East and West*, 24 (1974);
   and D.C. Guha, *Navya-Nyaya System of Logic*, New Delhi, 1979.
9. See for instance J.L. Shaw, 'Negation, Some Indian theories', in *Studies in
   Indian Philosophy*, D. Malvania and N.J. Shah (eds.), Ahmedabad, 1981.
10. B.K. Matilal, *The Navya-Nyaya Doctrine of Negation*, Harvard, 1968,
    pp. 3, 4.
11. See for instance J.L. Shaw, 'The Nyaya on Cognition and Negation' *Jour-
    nal of Indian Philosophy*, 8 (1980) Further according to Shaw:

    [In Indian logic] what is negated is an object which is the second term of
    dyadic relation . . . Let us consider the form a-(Rb) . . . What can be
    negated is 'b' as the second term of the relation 'R'. To say 'that the counter
    positiveness resident in b is limited by the limiting relation R' is equivalent
    to saying that 'b is the second term of the relation R'. So what is negated is b
    as the second member of relation R . . . Nyaya theory of negation . . .
    cannot be said to be a term negation, or a sentence negation, or a proposi-
    tional function negation in the usual sense of these terms [in western logic].
12. According to Naiyayikas, *samavaya* (translated as inherence) is the relation
    which holds between qualities (*guna*) or action (*Karma*) and substances
    (*dravya*), between a Universal or genus (*jati*) and individuals (*vyakti*),
    between a whole entity (*avayavi*) and its parts (*avayava*) etc.
13. See D.C. Guha Ref. 8, p. 11. Note that *Svarupa sambandha* is the relation
    between an *abhava* (absence) and its *adhikarana* (locus).

14. Another reason why the above statement as formulated is not acceptable to the Naiyayika is that it does not take account of the relations that smoke and fire bear to their loci. The Naiyayika scheme of inference allows us to infer a cognition of the for p-R_ss (where p is the *paksha* the mountain, s the *sadhya* the fire and R_s is the relation by which s occurs in p), from the cognition p-R_hh (where h is the *hetu*, the smoke and R_h is the relation by which the h occurs in p) if one has the *vyaptijnana* that *sadhya* is pervaded by *hetu*. Even if one uses the quantified form (a là Western logic) of the Universal statement expressing the *vyapti jnana*, it will have to be phrased in the form 'for all x, if x-R_hh, then x-R_ss', where the quantified variable 'x' appears as the first member of the binary relations, Rh, Rs.

15. See for instance, D.H.H. Ingalls, *Materials for the Study of Navya-Nyaya Logic*, Harvard, 1951, pp. 59, 61.

16. In a more correct formulation, the 'occurrentness' is to be characterized as being 'limited by occurrentnessness' (*Vrittitatvavacchinna*). See for instance, Ref. 12 (p. 61) citing the *parishkara* of Bandit Sivadatta Misra.

17. See for instance Ref. 15. pp. 61, 62.

18. ref. 15. p. 50.

19. B.K. Matilal, *Logic, Language and Reality*, N. Delhi, 1985, pp. 167, 168.

20. Ref. 16, p. 130.

21. J.F. Staal, 'Euclid and Panini', *Philosophy East and West*, 15 (1965) pp. 114.

22. See for instance Refs. 6, 12 and the following: J.L. Shaw, 'Number: From the Nyaya to Frege — Russel', *Studia Logica*, 41 (1982) pp. 283–91; R.W. Perret, 'A Note on the Navya-Nyaya Account of Number', *Journal of Indian Philosophy*, 13 (1985) pp. 227–34; and Chhatrapati Singh: 'What is a Set?', Indian Law Institute Preprint (1985).

23. H. Scharfe, *Grammatical Literature*, History Indian Literature series, J. Gonda (ed.), Wiesbaden, 1977, pp. 112, 115. Scharfe also mentions E. Obermiller's attempts in the 1920s to write a Russian grammar in Paninian style.

24. N. Chomsky, 'Principles and Parameters in Syntatic Theory', in *Explanation in Linguistics*, N. Hornstein and D. Lightfoot (eds.), Longman, 1981, p. 82.

25. P. Kiparsky, *Panini as a Variationist*, MIT Press, Mass., 1979, p. 18.

26. S.D. Joshi and J.A.F. Rodbergen, *Patanjali's Vyakarna Mahabashaya*, Karakahnika, Poona, 1975, p. i.

27. R.N. Sharma, 'Referential Indices in Panini,' *Indo-Iranian Journal*, 2 (1975) pp. 31, 32.

28. G. Cardona, *Panini: A Survey of Research*, Mouton, 1976.

29. For a survey of modern scholarship on Panini upto 1974, see Cardona *op. cit.*

30. H. Scharfe, *Panini's Kunstsprache*, Wissen Schaft Zeit Martin Luther Universität, 1961 (Passage translated in Ref. 30, p. 201).

31. Ref. 25, p. 3.
32. G. Cardona, 'On Panini's Metalinguistic Use of Cases', in Charudeva Shastri Felicitation Volume, N. Delhi, 1974, p. 307.
33. This speculation has been credited to D.H.H. Ingalls; see Ref. 29 p. 316.
34. This view has been recently revived by R. Roscher, fifty years after it was suggested by B. Faddegen; see Ref. 29 pp. 204, 205.
35. *Ardhamatralaghavena Putrotsavan Manyante vaiyakaranan* (*Nagesa Bhatta in Paribhashendusekhara*).
36. Ref. 25.
37. Ref. 29 p. 205.
38. Ref. 25.
39. *Alpaksharam asandigdham saravat viswato mukham/Astobham ahavad-yanca sutram sutravido viduh.*
40. R.N. Sharma, 'How does Panini Derive Sentences', All India Conference of Linguists, Calcutta, 1979, p. 3.
41. Ref. 25, p. 218.
42. 'Words' or *Pada* when used in *sabdabodha* refer to smallest meaningful units or 'morphemes'.
43. V. Subba Rao. *The Philosophy of Sentence and its Parts*, Delhi, 1969, pp. 1–3. In the passage cited the author is using the Naiyayika techniques of *sabdabodha*.
44. R. Briggs, 'Knowledge Representation in Sanskrit and Artificial Intelligence', *A.I. Magazine*, Spring 1985, pp. 32–39. Briggs has shown the parallelism between the 'semantic Nets' technique of knowledge representation used in artificial intelligence and the *sabdabodha* technique of Vaiyakaranas by taking various examples. His conclusion is that 'Many versions of semantic nets have been proposed, some of which match the Indian system, better than the others do in terms of specific concepts and structure.'

19

Appropriate Technology
A Reassessment

Amulya Kumar N. Reddy

Over the past thirty-five to forty years, a large number of countries, particularly in Asia and Africa, achieved political independence, and set out on the path of economic transformation. A large number of industries were established: modernization included the introduction of western food, clothing and houses, hospitals, universities, cars, airlines, telephones, radio, television. At the same time, many of these countries sought to modernize their agriculture, particularly through the so-called Green Revolution based on high-yielding varieties and large inputs of fertilizer, pesticides, water. The Gross National Product (a measure of the amount of goods and services produced by the country) has shown impressive increases in most of the these countries.

This pattern of transformation was inspired by a simple belief, namely, all that a backward country needs to do to develop is to retrace the path followed by the industrialized countries and to adopt the goals which they pursued. In particular, the aim should be to maximize growth in the volume of goods and services, that is, to maximize the GNP. By implementing such a strategy, it was assumed that the acute poverty of the masses in developing countries would be eliminated by the benefits of growth trickling down to the poorest sections of society.

By now, sufficient time has elapsed to see the results of such a growth-and-trickle-down strategy. The main result has been the consolidation of small islands of urban splendour amidst vast oceans of rural misery, and the perpetuation and aggravation of what has been termed a 'dual society' — a small politically powerful elite (constituting a mere 10–15 per cent of the population and consisting of industrialists, landlords, bureaucrats, professionals and white-collar labour) living in conspicuous affluence amidst the abject poverty of the politically weak masses. Recent history also shows that, in most of these backward countries, the greater the industrialization on the pattern of the advanced countries, the greater the polarization into a dual society, and the wider the gap between the elite and the masses. It is not even clear that the percentage of people below the poverty line has decreased.

The sole beneficiaries of a dual society are the sections of society belonging to the elite. It is reasonable therefore to infer that it has deliberately sought to perpetuate such a society.

My submission, in this article, is that a major instrument of the elite has been technology, which has been liberally imported from the industrialized countries and sometimes naturalized by a process of imitation and adaptation in the well-known import substitution drives. To substantiate the submission, it is necessary to understand the interaction between technology and society.

The basic feature of this interaction is that the pattern of technology is shaped by, and in turn shapes, the society in which this technology is generated and sustained. More specifically, technology responds to social wants which are in turn modified and transformed by technology through a causal chain, or rather causal spiral. The relevant conclusions which emerge from the technology-society interaction scheme are the following:

1. All social wants are not necessarily responded to by the institutions responsible for the generation of technology, namely the educational, scientific and technological institutions. There is a process of filtering these wants, so that only some of them are transmitted as demands upon technological capability, and the rest are bypassed by these institutions. In other words, there are ignored wants which institutions do not seek to satisfy by research and development.

The filtering process is usually operated by decision-makers, firstly, in the bodies which control the research and development institutions (decision-makers in government, agencies and corporations), and secondly, in the institutions themselves. These decision-makers are either conscious agents of political, social and economic forces, or are unconsciously influenced by these very forces.

In untempered market economies, only wants which can be backed up by purchasing power become articulated as demands upon the research and development institutions, and the remaining wants are bypassed, however much they may correspond to the basic minimum needs of underprivileged people. Thus, like all commodities in these economies, technology too is a commodity catering to the demands of those who can purchase it, and ignoring those who cannot afford it.

2. The generation and dissemination of technology involves the so-called innovation chain which is the sequence of steps by which an idea or concept is converted into a product or process. This sequence of steps varies with the circumstances, but can often be schematically reprsented thus:

Formulation of research and development objective → idea → research and development → pilot plant trial → market survey → scale-up → production/ product engineering → plant fabrication → product or process.

3. It is essential to note that socio-economic constraints and environmental considerations (if any) enter the process in an incipient form even at the stage of formulation of the research objective and then loom over the chain at several stages. These constraints are in the form of preferences or guidelines or paradigms, for example, 'Seek economies of scale!'; 'Facilitate centralized, mass production!'; 'Save labour!'; 'Automate as much as possible!'; 'Don't worry as much about capital and energy (in the days before the energy crisis) as about productivity and growth!'; 'Treat pollution effluents or emissions as externalities!'. Above all, 'Modernize!' meaning 'westernize!'.
4. Thus, every technology that emerges from the innovation chain already has engrained into it the socio-economic objectives and environmental consider-ations which decision-makers and actors in the innovation chain introduced into the process of generating that technology. Further, at a previous stage in the spiral (cf. 1. above), the very decision to respond to a particular social want by generating the necessary technology is the result of a deliberate filtering process wielded by decision-makers.
5. The technology that emerges from the innovation chain will become an input, along with land, labour and capital, to establish an industry or agricul-ture or a service if and only if the aforesaid socio-economic and environmental constraints are satisfied. Thus, it is not only the technical efficiency of the technology, but also its consistency with the socio-economic values of the society, which determine whether a technology will be deployed and utilized.
6. Social wants are not static. The products and services that are produced create new social wants, and in this process the manipulation of wants through advertising, for example, plays a major role, and thus the spiral:

Social wants → products/services → new social wants → . . .

The interaction between society and technology on the lines described above has an important implication. Every pattern of technology is socially condi-tioned — it is a product of its times and circumstances and it bears the stamp of its origin and nurture. It is in this sense that technology can be considered to resemble genetic material which, given a favourable milieu, tries to replicate that society. The replication is neither automatic nor inevitable; it is successful only when a host of factors are favourable — hence, the argument is not tanta-mount to technological determinism. Further, it has been emphasized that technology itself is socially conditioned, therefore technology is not viewed as an autonomous factor and a motive force outside society. Of course, all this would be obvious to an archeologist who must proceed from the material products, the tools, artifacts and so on, to reconstruct a vanished society and its culture, or to a social anthropologist who cannot but consider technology-industry/agriculture-society interactions.

The Western Pattern of Technology

The cultural and intellectual dominance of the industrialized countries over the developing countries has been so overwhelming that any thought of development automatically conjures up a picture of the pattern of technology (which for convenience shall be referred to as western technology) that obtains in Western Europe and North America. This picture is only reinforced by the fact that the centrally planned economies of Eastern Europe and the Soviet Union, Japan, and the newly industrializing countries such as South Korea have also adopted virtually identical patterns.

A fundamental question, therefore, arises: is the western pattern of technology a unique, inevitable and unavoidable pattern which developing countries must necessarily follow? In exploring this question, it must first be realized, following from the model of technology-society interactions presented above, that, like all patterns of technology, the current western pattern of technology is also a product of specific historical conditions, namely the epoch of history corresponding to the past thirty to eighty years.

In this epoch, a set of industrialized countries were first able to control the politics of a number of colonies and after their independence, to dominate the economies of these ex-colonies. These relationships of dominance enabled the industrialized countries to commandeer, and/or enjoy from the developing countries, natural resources, including non-renewable minerals and fossil-fuel energy, at much lower prices than would have been the case if relationships of equality had prevailed. This is why the prices of raw materials from the Third World have not risen as sharply as the prices of manufactured goods from the industrialized countries. (The 1973 oil price hike and the conflicts at the United Nations Conference on Trade and Development and the North-South Conferences are all part of the drive to redress these historically-enforced inequalities.)

This situation also resulted in the accumulation of capital in the industrialized countries at rates which would not have otherwise been possible. At the same time, the industrialization of the industrialized countries has invariably taken place amidst shortages of labour.

All these factors — the easy availability and low prices of raw materials, energy and capital, and the scarcity of labour — have had an overwhelming influence on the pattern of western technology. This is because every technology is viable only within certain limits (upper or lower) of the prices of raw materials, energy, capital and labour, and if the prices of one or more of these inputs changes drastically, the validity of the technology may be undermined. The point has been dramatically demonstrated with the vast number of energy-intensive western technologies based on the cheap Middle East oil of the pre-1973 days, all of which are now undergoing thorough reassessment. Thus, the old (and still prevailing international economic order) has resulted in the capital intensive, energy-profligate, recklessness with regard to non-renewable natural resources, and the labour-saving character of the western pattern of technology.

The second crucial feature of the period of history which spawned the western pattern of technology is that the vast majority of the technological innovations underlying this pattern have emanated from the basic driving force of capitalism: the maximization of profit and accumulation. This intrinsic compulsion to minimize internal costs of enterprises and to disregard as externalities all effects on the social and natural environment of the enterprises has led to the three intrinsic tendencies of western technology: amplification of inequalities between and within countries; increase of alienation of men from each other and from their work and diminution of social participation and control; and degradation of the environment.

The intrinsic tendency of amplifying inequalities between and within countries results from the following three features. Firstly, western production technology has become increasingly capital-intensive, and therefore gravitates to areas and locations where that capital can be mustered and exploited, usually towards rich nations and away from poor nations, and towards the urban areas of developing countries at the expense of their villages. Secondly, the associated increase in energy intensiveness leads to increasing automation and decreasing dependence on labour, that is, in the absence of careful planning, to greater unemployment. This produces, in industrialized countries, relative poverty for the minority and, in developing countries, a potentially catastrophic accentuation of the gap between affluent elites and the poverty-stricken masses. Thirdly, having largely solved the minimum needs of the populations in industrialized countries, western product technology is increasingly oriented towards luxury goods for private consumption, and towards military applications. For, when there is inequality in the distribution of purchasing power, the resulting skewed demand structure drives such technology to respond more avidly to the luxury demands of the rich and to other non-essentials, and assign lower priority to the basic needs of the underprivileged.

The inherent tendency of increasing alienation and diminishing social participation and control is an inevitable result of various features. Western production technology has relentlessly pursued so-called economies of scale, mass production and automation. In doing so, it has generated a highly skewed pattern of demand for skills. Only the few are required to possess a high degree of intellectual training or manual skills, while the barest minimum of intelligence and dexterity is expected from the vast majority of the working force, which naturally becomes alienated. This trend is only aggravated by the deliberate organization of the labour process to increase profits, rather than to enrich the lives of workers.

However, training and skills lead to control over technology, and thereby to power — hence, western technology tends to concentrate power in the hands of the few and deprive the majority of control over their destinies. Push-button warfare is the ultimate example of the technology-power equation.

Furthermore, the virtually complete exclusion of craftsmanship and creativity from work in modern factories which are, in addition, dominated by machines, results in the alienation of men from their work. Alienation between man and man is increased because western product technology is specifically designed to respond to and evoke demands from those privileged with purchasing power, and therefore results in the proliferation of luxury goods for individual consumption and the generation of consumption-oriented lifestyles.

The disastrous impact of the western pattern of technology on the environment is a consequence of four main factors. Western industry's obsession with an ever-expanding scale of production results in an increasing perturbation of eco-systems (for example, the sources of pollution become more concentrated) till there is a real possibility of pushing them beyond the limits of stability. At the same time, the constant drive to manufacture products, which are ever-changing in appearance and form, but similar in function and content, is the cause of the rape and exhaustion of natural resources, the alarming degree of product obsolescence and the 'throwaway' philosophy.

Thirdly, western industry has generated risks to the biosphere of increasing gravity ranging from trivial and acceptable to remediable, avoidable and catastrophic, and has increased the probability of the occurrence of any given category of risk. Thus, human civilization and life itself have become threatened by technological 'progress', particularly in weapons.

Finally, the tendency of western technology to magnify inequality results in the very rich countries (and the rich groups within poor countries) damaging the environment through over-consumption and the very poor being able to ensure their survival only at the expense of their environment.

The Need for an Alternative Pattern of Technology

In considering the industrialization of developing countries, two fundamental issues must be raised: is it feasible for these countries to emulate the western pattern?; is it just and moral to do so?

The feasibility aspect is easily considered by noting that developing countries like India just cannot replicate the favourable environment which the industrialized countries enjoyed, particularly during the early stages of their industrialization. Very few of the developing countries have within their frontiers the range and quantity of raw materials necessary for the western pattern of technology. Barring the oil-producing countries, most of them are critically short of energy. Most of their agricultural systems have been distorted into production of commercial crops for the industrialized world resulting in frequent food deficits. They do not have captive external markets for manufactured goods, and in attempting the export drives being recommended to them, they find that their industrial technologies are either not competitive with those from the industrialized countries, or when they are, they are faced with a rising tide of protectionism. Developing countries often find that it is too costly to generate indigenous technologies that are competitive. Hence, they are forced to import western technology. And when they do so, they learn that technology exports from industrialized countries have become a new mechanism for reinforcing their dependency.

It is also clear that a desperate scarcity of raw materials would develop if all the developing countries were to attain the per capita consumption levels of the industrialized countries. For instance, the USA alone is consuming about 25 per cent of today's world oil production, averaging three tonnes per year per capita. In fact, the USA consumes as much oil for leisure (pleasure boats, etc.) as India does for all its requirements. If every Indian enjoyed the same per capita consumption level, India alone would require 63 per cent of today's world production of this depletable natural resource.

The conclusion is clear: industrialized countries enjoy their current consumption levels precisely because the developing countries are subject to deprivation, and it is extremely doubtful from a resource point of view, whether the earth can support all nations (including the poor ones) achieving standards similar to those of the rich. Even if it were feasible for developing countries to emulate the western pattern of technology, the justice and morality of such as attempt should be considered.

The whole pattern of inequality, injustice and exploitation characterizing the current international economic order is repeated within developing countries, for almost all of them are polarized into dual societies with a society of the richest 10–15 per cent of the population separated by a vast chasm of lifestyles, incomes and aspirations from a society of the poorest 85–90 per cent consisting mainly of the rural poor. The market economy encompasses almost exclusively the richest sector which has emerged as a politically powerful, conspicuously consuming, western-oriented elite. At the same time, the poorest sector, and in particular the poorest 40–50 per cent, exist in poverty outside the market economy. The polarization is also associated with rural stagnation and impoverishment, with massive rural employment and underemployment, and with mass migration to metropolitan slums. Hopes that the benefits of industrial growth will percolate to the countryside and reduce income disparities have not been borne out by experience. In fact, it appears that the western pattern of technology upon which Indian industrialization has been based has only accentuated the evils of the dual society.

A particularly alarming result of the adoption of western modes of technology is its impact on employment. Because of the capital-intensive nature of the technology, the investment required to create jobs is extremely high, about $1,500–$15,000 per job in India. At this rate, industrialization on the basis of western technology can provide employment only to restricted numbers. The backlog of unemployed (about 20 million in India), and the new entrants to the work-force every year (about 5 million), will not be able to find employment unless astronomically large investments are made. But these are impossible in capital-starved developing countries. Hence, unemployment grows to serious proportions.

As in the situation between industrialized and developing countries, the haves can *have* only if the have-nots do *not* have, in the sense that the affluence of the elites can be preserved only at the expense of the masses. Such disparity cannot be associated with stability; it can be maintained only by force. Thus the exploitation, injustice and misery inherent in dual societies implies the immorality of the western pattern of technology. Thus not only is the western pattern not feasible; it is also immoral. But interestingly, what is immoral cannot be sustained, and what is not feasible over the long run is also immoral. Historical feasibility and morality seem to converge. An alternative pattern of technology must be implemented.

The Concept of Appropriate Technology

It is against this background that the clamour for an alternative pattern of technology — appropriate technology — has been raised. Arguments for appropriate technology have been slowly mounting for over half a century, but what is appropriate technology? Before seeking an answer, it must be realized clearly that the word appropriate has no meaning in itself, unless one specifies 'appropriate to what?'. The point is that technology is only an instrument, but like all instruments, it must be fashioned to achieve the purpose for which it is intended. So, the definition of the word appropriate must emerge from the purpose of technology in developing countries like India. Stated thus, it is obvious that since development is the objective and technology is the instrument, technology must be appropriate for development.

The experience of the past forty years shows that development must not be equated with growth. Far more important than the sheer magnitude of growth is the structure and content of growth, and the distribution of its benefits. Once a particular pattern of growth takes place, neither its structure and content, nor its benefits, can be easily altered. Growth for the benefit of the elite, for example, processed and packaged foods, expensive cloth, luxury houses, capital-intensive private hospitals, richly-endowed universities, and private cars, cannot be transformed easily into growth for the masses, that is, cheap food and cloth, low-cost housing, mass health care, education and transportation. The GNP by itself does not reveal what constitutes it, e.g., cars or buses?, or who benefits by it, i.e., the elite or the masses?

Development must be defined, not merely as growth, but as a process of socio-economic change principally directed towards:

1. satisfaction of basic human needs (food, clothing, shelter, health, education, transport/communication, etc., and employment which makes all this possible), starting from the needs of the neediest, in order to reduce inequalities;

2. social participation and control in order to strengthen a self-reliance that grows from within; and
3. ecological soundness in order to achieve harmony with the environment and make development sustainable over the long run.

This view of development is totally different from a simple GNP-maximizing approach, for whereas the former is concerned primarily with human beings, the latter is preoccupied with goods and services. The former is deliberately directed towards the neediest (who are incidentally the majority in developing countries), whereas the latter hopes that benefits will spontaneously trickle down to these underprivileged. According to the development-oriented approach, what goods and services are produced is of central importance, but this question is of little concern in the GNP-maximizing approach.

It is from this standpoint of development that appropriate technology must be defined, that is, as a technology that advances development objectives. It must promote the satisfaction of basic human needs, starting from the needs of the neediest; social participation and control; and ecological soundness. The test for the appropriateness of technology is whether it facilitates the reduction of inequalities; the strengthening of self-reliance; and harmony with the environment.

This view of appropriate technology according to which it is linked to the development process can now be compared with five other approaches which are current, namely,

1. the area approach,
2. the factor-endowment approach,
3. the resource-endowment approach,
4. the target-group approach, and
5. the market-expansion approach.

According to the area approach, appropriate technology is technology that is relevant to the area/region of interest — to the village, cluster of villages, district, region or nation. Such an area-based criterion of appropriateness ignores the fact that societies are stratified and that, while particular technologies can be appropriate to the region in which a society lives, their benefits may flow overwhelmingly to the richest and most powerful sections of that society, and thus amplify its inequalities. Since such an impact negates development, it means that what is appropriate to an area may not necessarily be appropriate for development.

The factor-endowment approach is based on the view that appropriate technology is technology that is applicable to the factors of capital and labour which the area is endowed with, so that in a capital-short, man-power rich country, appropriate technologies should be capital-saving and labour-intensive. While such an approach ensures employment-generation and manpower utilization, it ignores the product-mix question, i.e., what mix of goods and services are produced and whether such a mix satisfies the basic needs of the neediest. Since it is easy to imagine luxury goods for the elite in poor countries and/or for export to the rich countries being produced in a capital-saving labour-intensive way, it is clear that the factor-endowment approach leads to an important part of the definition consistent with development, but not to the whole definition. In other words, technologies can be appropriate to the country's endowment of capital and labour, but inappropriate for development.

The limitations of the resource-endowment approach are similar. While technologies appropriate for development must as far as possible be based on local resources, it is quite possible that technologies attuned to resources can be inconsistent with development. For instance, the manner of production may not be appropriate to the capital-labour endowments or the products that are produced may not be relevant to the neediest sections.

The emphasis on what products are produced and whose needs they satisfy is therefore of fundamental importance. This emphasis is safeguarded in the target-group approach according to which appropriate technology is technology that is suited to the needs of the underprivileged sections of society. This target-group approach comes closest to the development-oriented definition of appropriate technology proposed above, but unfortunately it facilitates a narrow and short-sighted view in which remote and long-term linkages to the basic needs of the target groups are ignored. For instance, exlusive concern with the immediate needs of the weakest sections may lead to absence or insufficiency of attention on basic goods and infrastructure, for example steel transport and power. Thus, the weakness of the target-group-based definition of appropriate technology is that it may restrict the time-horizon over which appropriateness should be considered. In contrast, the development-based definition of appropriate technology not only facilitates a balanced concern over long- and short-term development objectives, but also protects the interests of the target group by its emphasis on 'starting from the needs of the neediest'.

Finally, there is an insidious definition of appropriate technology which equates it with technologies that integrate rural areas with the urban market. This definition implicity assumes that such an expansion of the urban market into new areas is always beneficial to the new areas irrespective of the 'terms of trade' between them. This definition ensures growth, but it may not promote development.

A number of different interpretations regarding appropriate technology are in circulation — some are honest differences in understanding, others reflect alternative perceptions, still others are misconceptions, and many counterfeit versions may even have been deliberately propagated. These differing interpretations merit discussion and refutation.

Sometimes, the impression is conveyed that the case for appropriate technologies is built upon a rejection of industry and industrialization. Nothing can be further from the truth. In fact, it is considered self-evident that industrialization is essential for meeting the basic needs for growing populations, but the case is for those products, patterns and forms of industrialization that will advance the type of development described above. It is implicit in such a view that a great deal will have to be learnt from the process of industrialization of the industrialized countries, but, this process — it must be noted — includes both successes and failures and the associated lessons. Hence, development does not have to consist of a slavish imitation of the type of industrialization followed by the industrialized countries.

Similarly, it has often been assumed that the proponents of appropriate technology demand a total rejection of the so-called modern technology of the industrialized countries. In fact, what is demanded is a careful scrutiny of the economic, social and environmental implications of modern technology from the standpoint of the objectives of development, and an acceptance of those technologies (in original or adapted form) that advance the basic objectives.

Thus, what is rejected is the blind faith that all the technologies of the industrialized countries are universally appropriate, despite the specificity of the historical circumstances that spawned them. Also discarded is the naive belief that these technologies are always an unmitigated blessing, equally satisfying the interests of those who sponsor, hawk and vend them, as well as the development objectives of recipient countries.

Since the technologies of the industrialized countries have scant relevance to the basic needs of the rural poor in developing countries, and the educational, scientific and technological institutions of most developing countries pay little heed to their problems, this deprived section of humanity has no choice but to rely on traditional technologies. Unfortunately, these technologies have been completely ignored by development planners even as a starting point for innovation — not on the basis of any rigorous study and assessment, but in an a priori manner blinded by euphoria over western technologies. The fact that traditional technologies have ancient origins, and are very often embedded in apparent mumbo-jumbo, has been sufficient to exclude them from any serious attention.

The historical origins of this contempt of traditional technologies can be traced to the early period of colonial subjugation where colonial rulers found themselves ranged against the traditional technologies of ancient civilizations. If these rulers were to succeed in 'selling' their alien technologies, they had first to break down 'consumer resistance'. And thus began the powerful process of undermining the subjects' faith in their own technologies. The ethos of 'all that is rural is bad, all that is urban is better, and all that is western is best' had to be broadcast.

Those traditional technologies that have survived are the evolutionary product of a long process of selection, often stretching over several centuries. Whether in the case of agriculture, crafts, food, clothing, shelter, health or transport, there is mounting evidence that traditional technologies to meet basic needs are solutions which approach optimality within their frame of reference.

Given the conditions, materials, equipment and resources available to those ancient peoples, and the magnitude of the problems facing them, the solutions which they developed were highly rational. In many cases, traditional technologies have indeed been ingenious — for example, the navigational techniques of the South Sea islanders, or the inter-cropping practices of the ancient civilizations of Latin America, or the weather prediction skills of tribals, or the house-building techniques of Africans, or the design of the Indian bullock-cart, or . . . the list is long and incomplete.

Traditional technologies constitute a veritable treasure-house of experience, insight and methodology. Their value derives from the following inherent characteristics: environmental soundness (many ancient civilizations were wiped out because of environmentally unsound technologies, but that does not mean that they are environmentally sound for the conditions of today); extremely low capital cost and high labour-intensiveness; dependence on locally available materials; obvious orientation towards minimum needs; use of local skills and, therefore, a firm basis in endogenous self-reliance; and incorporation into the fabric of social life.

Since traditional technologies were appropriate within their frame of reference, they can serve as an excellent starting point and basis for generating the appropriate technologies essential for the development of developing countries.

Such a transformation of traditional technologies into appropriate technologies would require the following steps:

1. study of traditional technologies with a view to understanding the functions they were expected to fulfil and the conditions within which they had to operate;
2. precise definition of the limitations and drawbacks of traditional technologies from the standpoint of present-day needs and conditions;
3. analysis of the scientific content of traditional technologies in terms of the language and idiom of modern science and engineering;
4. use of this scientific insight to generate qualitative improvements with minor changes and alterations.

There is no guarantee that all traditional knowledge can be fitted easily into the framework of modern scientific knowledge. It may turn out that, in order to incorporate traditional approaches, the frontiers of modern knowledge may have to be extended — if not in principles, at least in methodology of application. There are three grounds for such an expectation. Firstly, traditional technologies are founded very much more on sense-based observation than on the instrument-based data of western technology. Secondly, they also make much greater use of an integrated, holistic approach rather than of the isolated partial view. Lastly, they are invariably based on a deep and intimate view of the environment which was so vital for the survival of ancient peoples.

The implication of the above view is that the transformation of traditional technologies into appropriate technologies for development is not a trivial matter. It requires sophisticated scientific and technological inputs. It constitutes advanced research and development work. But, the pay-offs are likely to be large for they would be much more environmentally sound, inexpensive, need-oriented, conducive to local self-reliance and in tune with local culture than the cheapened and crude versions of advanced technology usually proposed for rural needs. The point is that the transformation of traditional technologies represents a resumption of the evolutionary development of indigenous technologies, and therefore has inherent advantages with regard to social acceptability, economic viability, and rapid and widespread diffusion.

The cultural, economic and social blindness of science and technology planners in ignoring traditional technologies would not have been so serious had it not been for the fact that these technologies are fast disappearing. There is a real urgency to be attached to the task of collecting, codifying and storing information on traditional technologies, otherwise developing countries will soon lose access to a crucial source of technology which is native to their soils.

The principal aim of studying traditional technologies must be to use them as a starting point of technological advance, and not as a basis for revivalism or a retreat into the past. In some quarters, however, the argument for appropriate technologies has been misunderstood as a romantic plea for a total return to, and dependence on, the traditional technologies of ancient peoples. Such a naive return to the past must be unequivocally ruled out on the following grounds.

Firstly, the magnitude of the problems facing developing countries today is far greater than that faced by their ancestors, and population growth is perhaps the most important reason for this change. For example, the carrying capacity of land corresponding to traditional agricultural technologies has been exceeded by the growth of population.

Secondly, traditional technologies have undergone a selection process over centuries of empirical testing; hence, they are very likely to represent optimal solutions, but only optimal for the particular conditions, constraints, materials, and needs in response to which they were developed. With the emergence of new conditions, constraints, materials and needs, that optimality is likely to have eroded and rendered invalid today. The Indian bullock-cart, for instance, may have been an optimum solution in an age of transport and travel over unprepared cross-country terrains, but it is now being used over tarmac roads, which may require quite a different cart design from the traditional one. Further, materials such as teak wood which were plentiful in the heyday of this vehicle, are becoming increasingly scarce, thus necessitating new solutions.

Thirdly, the development of communication and transport has increased the range of awareness, materials, equipment and resources accessible to the rural areas. Thus, many of the rigid constraints within which traditional technologies were developed have either ceased to exist or have been transformed. This means that traditionally optimal solutions are sub-optimal today, essentially because the productivity of the capital and labour which they utilize is inadequate for the new population levels.

Still another issue which needs clarification is the scope of appropriate technology. Too often the advocates of appropriate technology have restricted their concerns to production technology. However, if technology is to be an instrument of development, it must be understood in a much broader sense as encompassing both product and production, and both software and hardware. That is, the technologies under discussion must include what type of goods and services are produced, in addition to how they are produced. They must include software or technologies concerned with ways of, and instructions for, utilizing people, machines, devices and materials. Thus, all types of technologies, and not merely production technologies, must be scrutinized for appropriateness.

By the same token, the concept of appropriate technology can be extended to all sectors — industry, agriculture, power, health, education, human settlements, transport, communication. It must not be restricted to industry alone.

Then, there is the question of the advanced character of technologies. This character should derive not from the trivial criterion of scale of prc luction, but from the extent to which the technologies embody scientific and engineering thinking.

From this standpoint, it is possible that appropriate technologies and alternative technologies need not be primitive; they can turn out to be as 'advanced' — and 'modern' in the literal sense of the word — as the technologies of the industrialized countries. In fact, this is highly probable because, unlike the technologies of the industrialized countries, there is no crowded and beaten path for the generation of appropriate technologies, and therefore, the dependence on fundamental science and engineering must be even more firm.

For a similar reason, it is unfortunate that the technologies of the industrialized countries are invariably described as 'high' technologies, in contrast with appropriate technologies which are pejoratively referred to as 'low' technologies. But, the terms high and low should depend on whether there is a high or low science and engineering input, and not upon whether the technology originates from the industrialized countries or not. Invariably, however, it is this geographical origin of a technology which determines the terms of common parlance: advanced/primitive and high/low. The underlying subconscious belief or conscious policy is the equating of all that is good with what emanates from the industrialized countries.

Since a technology is to be adjudged appropriate or inappropriate depending upon whether it advances or retards development, it may turn out that, in the case of some products, small-scale production is more appropriate, and in others, large-scale production is more appropriate. Thus, there is no universally valid inverse correlation between appropriateness and scale, so that the smaller the scale, the greater the appropriateness. In other words, small may not always be beautiful. In fact, what is likely is that, for every product or service, there is an optimum scale or size, below which appropriateness increases with scale or size and above which it decreases with scale or size. This question of optimum becomes particularly significant when one considers factors such as environmental impact or social participation, in addition to simple economic parameters such as the unit cost of production.

Very often the term intermediate technology has been used as a synonym for appropriate technology, but it has never been clear whether the word intermediate refers to a stage (neither the most recent nor the most ancient) in the industrialization of the West, or to an extent of modernity (neither the most primitive nor the most modern), or to a scale (neither the largest nor the smallest). The vagueness of the concept, coupled with opposition to the attempt of the industrialized countries to equate intermediate technology with the now-obsolete machines used by them in an earlier stage of their growth, has resulted in the term being largely abandoned today.

Interest in appropriate technology received a spurt as a result of the attention devoted to the book *Small is Beautiful* and the unfortunate demise of its author, E.F. Schumacher. This has created an impression that the concept is of western origin, but this is far from true. A large number of Indian thinkers — Gandhi, Lohia, Kumarappa, Gadgil, Raj and others — have contributed to the formulation of the concept. In fact, it was Gadgil who coined the term 'appropriate'. Other developing countries must also have contributed to the concept. That being the case, the attention paid to Schumacher and the book is one more example of an idea acquiring legitimacy and respectability if it is routed through the West. It is another instance of the cultural dependence which is a residue of the colonial past.

A widespread view among the elites of developing countries is that appropriate technology is one more manoeuvre of industrialized countries to keep developing countries backward and thus maintain domination over them (the assumption being that appropriate technologies are backward technologies). This view has gained ground because of the moves which were made, particularly in the late seventies, to 'sell' appropriate technologies to the developing countries. For example, the aid agencies of the industrialized countries actively promoted them. (It may be noted, however, that these same elites did not object to western technologies — central-station power plants, fertilizer complexes, mineral-processing conglomerates, oil-drilling rigs — despite the fact that these technologies were even more vociferously peddled by these agencies. Could it be that the elites perceived more 'benefits' in these western technologies than in appropriate technologies?)

The situation, however, is not so straight-forward for at the same time several industrialized countries are strongly opposing appropriate technology. They argue, in effect, that the only possible model of industrialization is that which was followed by them, and that developing countries must rely heavily on technology transfer to implement this model of industrialization.

These apparently contradictory positions are not without an underlying logic. The current pattern of western technologies is umbilically linked to the present exploitative international economic order, which is characterized by wide (and widening) disparities in resource-use and living standards between developed and developing countries. Hence, one way of delaying the establishment of a New International Economic Order is to perpetuate the pattern of western technology in its entirety. For, as long as developing countries are locked into this pattern, domination over them can be maintained through control over technology.

Opposition by industrialized countries is directed particularly against the type of definition of appropriate technology used here, according to which the concept includes the urban, large-scale capital-intensive sector, and insists that the products and services produced by this sector, and the ways in which these products and services are produced, must be consistent with development goals. It is in the interest of industrialized countries to restrict the meaning of appropriate technology so that it excludes the urban, large-scale, capital-intensive sector. If this restriction is accepted by developing countries, they may end up concentrating exclusively on appropriate small-scale, labour-intensive technology for the rural sector, thus ensuring the continued domination of industrialized countries over the capital-intensive technologies.

At the same time, however, the continuation of dual societies in developing countries is seriously threatened by escalating unemployment, particularly in rural areas, which is aggravated by capital-intensive western technologies. But the perpetuation of dual societies is in the vested interests of industrialized countries, because, very often, it is the ruling elites of these societies who are politically linked with the commercial interests of industrialized countries and multinational corporations. Hence, these interests advocate employment-generating appropriate technologies for *rural* development as an insurance against cataclysmic social changes. There may even be some pecuniary benefits to be derived from this advice. The economic and political compulsions which are making appropriate technology an inevitable policy in the developing countries are bound to generate an extensive and growing demand for appropriate industrial plants and devices, and industrialized countries sense in this demand a large market which can be captured. The advent of appropriate technology consultancy organizations located in the industrialized countries is evidence of this market-sense.

The suspicion that industrialized countries are selling their version of appropriate technologies as a way of capturing new markets is strengthened by the fact that many of their institutions and agencies are wholly involved in the generation of appropriate technologies for the developing countries instead of tackling the major technological challenges in the industrialized countries. These efforts are sustained, with few exceptions, by technologists who quickly acquire a standing in the appropriate technology community which they would almost certainly not have achieved in the scientific and technological establishments of their own countries. And unfortunately, these efforts are also assisted by expatriates from the developing countries who combine the material benefits of working in rich countries with the comforting thought that they are in the vanguard of the war on Third World poverty.

Furthermore, these efforts curtail and undermine the opportunity which the educational, scientific and technological institutions of the developing countries have to use the generation of appropriate technologies as a means of building technological capability and institutions and thereby acquiring technological self-reliance. In effect, these alien efforts deny the importance of self-reliance as a crucial developmental objective — hence, they are anti-development.

The dissemination of appropriate technology involves the developing countries in one or more of the following activities: field testing, evaluation, monitoring, hospitality, logistics, local language, demonstration, liaison with local institutions. This necessity of interacting with the natives is often the basis for efforts to collaborate with indigenous groups and institutions. (In some cases, however, the efforts may be motivated by a genuine desire to catalyse the building of indigenous capability and institutions.) Unfortunately, these collaborations have not been great successes, and most of them have petered out. The reasons for these disasters are many, but perhaps the most important is that it is extremely difficult for these collaborations to be between equals — the developing country group invariably winds up technologically subordinate and/or without due recognition because the credit goes to the institution of the industrialized country. Even the technological successes may turn out to be developmental failures because an impression is consolidated in the minds of the local population that successes require the intervention of foreign experts from industrialized countries.

All this is not to deny a role for the institutions of industrialized countries in the generation of appropriate technologies, but this role has to be focused on activities for which these institutions have the requisite infrastructures and special strengths, namely freely published basic research and design theory that is vital to technology development; technology assessment based on the enormous information resources available in the industrialized countries; and information dissemination on appropriate technologies to developing country groups/institutions.

The tendency of the institutions of education, science and technology to cater preferentially to the demands of elites in stratified dual societies has often led to the view that these institutions are so moribund that only non-institutional voluntary agencies can generate appropriate technologies. Such voluntary agencies may have a part in technology generation (as distinct from technology dissemination) but it can be only a marginal role in view of the inevitable limits to their multi-disciplinary expertise and their facilities for research and development. This judgement rests, however, on the valid assumption that appropriate technology is not second-class technology and that its generation is not a trivial exercise.

In this context, it is interesting to note that voluntary agencies in developing countries often have stronger links with appropriate technology groups of industrialized countries than with the institutions of education, science and technology in their own countries. Of course, the anti-appropriate-technology bias of the indigenous institutions of education, science and technology may well be responsible for provoking this preference for the foreign linkage. And, quite often, the foreign groups also prefer this linkage, perhaps because it is easier for them to deal with the less technically advanced voluntary agencies than with the technically advanced institutions of education, science and technology.

Just as the indigenous institutions of education, science and technology have an advantage in the matter of technology generation, so are the voluntary agencies usually better at technology dissemination by virtue of their direct contacts with the target groups, their dedication and commitment, and their sustained work in the field. Thus, when both appropriate technology generation and dissemination are considered, it is clear that both indigenous institutions of education, science and technology as well as voluntary agencies have clear-cut roles to play.

The role of the people in the two processes of technology generation and dissemination is quite different. An R & D institution can generate technology without the active participation of the people in the designs, calculations, experiments, equipment manufacture, etc. In other words, their participation in the purely technical stages of technology generation *per se* is not necessary. But, close consultation with the people is vital for obtaining better insights into felt needs, traditional solutions, local conditions, materials and skills, and these insights are quite essential for ensuring the appropriateness of technology. Further, the actual users and operators of appropriate technology should have a crucial role in invention, particularly in the continuous testing, refinement, and adaptation of new technologies. Indeed, it is hoped that a constant interplay between institutional and popular inventors will enhance the appropriateness of technologies.

In the matter of technology dissemination, the issue is far less subtle — the active participation of the people is a necessary condition for technology dissemination.

Some advocates of appropriate technology have themselves been responsible for creating the impression that technology alone can remove poverty, redress injustice, solve development problems, and prove a universal panacea (provided it is the right brand!). But, technology is only a sub-system of society, and the development of society hinges not only on technology but also on the other crucial sub-systems — the political, economic and social sub-system — as well as on the physical environment of society. In other words, technology is only an instrument for the development of society. Like all instruments, it must be specifically chosen and/or designed to fulfil its intended function. But the will to use the instrument, and the skill to wield it effectively, do not depend so much on the instrument as upon the users of this instrument.

Thus, the right type of technology (an appropriate technology) is a necessary condition for development, but not a sufficient condition. It is also essential that the political structure and the socio-economic framework are both committed to development goals. Appropriate technology, therefore, has both power and limits — it is an essential requisite for development, but it cannot be a substitute for economic, social and political change. Since it is not a substitute for social change, it is often concluded that attempts to generate and disseminate appropriate technologies are not only pointless, but indeed an obstacle to social change. There are a number of defects with this conclusion.

Firstly, society does not change mechanically according to some pre-ordained timetable such as: first, a political change and then, technological change. Society is a whole; its economics, politics, culture, laws, arts, technology have to hang together. Which component of society will trigger basic changes is not easy to predict; hence efforts have to be made on all fronts. In this sense, the fight for appropriate technology is an integral part of the fight for genuine development.

Secondly, if appropriate technology is incompatible with the continuation of a dual society, then it is clear that it cannot become a flourishing and widespread feature of such societies. This does not mean that attempts to disseminate appropriate technologies will not succeed in particular sectors, in many locations and in special circumstances; and that these limited and partial successes will not catalyse social change.

Thirdly, appropriate technologies will certainly be necessary some time and definitely after the desired social change has taken place. When then would they come from? They cannot be produced overnight. In fact, experience shows that the generation of any appropriate technology is a time-consuming affair involving a significant gestation period. Hence, the sooner this task is taken up, the better the chance of social change fructifying after it takes place.

Fourthly, since appropriate technologies cannot be generated and or disseminated without direct and intimate contact with the principal target groups, the urban and rural poor, those involved in these processes of generation and dissemination are likely to undergo asensitization and de-alienation. Thus, appropriate technology work is a powerful method of conscientizing scientists, engineers, social scientists and development workers.

There is a view that appropriate technology is a matter of expediency, a transitory measure and a tactical device to cope with the current predicament of developing countries. But the definition of appropriateness in terms of genuine development means that appropriate technology is an instrument for the achievement of a new society, in which social participation and control, and harmony with the environment are as important as satisfaction of basic material and non-material needs. Thus appropriate technology is a component of a new strategy, not a mere tactic — it is part of a new vision, and not a short-cut to old delusions!

20

Traditional is Appropriate
Ecologically Balanced Agriculture in Sri Lanka

G.K. Upawansa

Modern agricultural techniques have played havoc with Third World societies. Even the most staunch defenders of high technology agricultural methods must acknowledge that ecologically beneficial systems of agriculture, based upon traditional experience, have been disrupted; that methods of maximizing production in the short-run are showing up long-term adverse effects; and that values and attitudes of farmers are changing. The new technology is altering the farmer from a sturdy and responsible individualist into a mere cog in a purely mechanical production and marketing process. Initial disturbance of the farming system set up a chain reaction of consequences. The end result is that modern agriculture created a dependent agriculture, and a system of exploitation of farmers who compose a major part of the population of developing countries.

Sri Lanka, with over twenty-five hundred years of recorded history, has suffered greatly. Its civilization flourished around an indigenous farming system — with irrigation systems that still excite admiration — up to the time of western domination. It had many distinctive features which constituted the basic principles of the system.

The most important feature was timely planting. Knowledge gleaned out of long experience of climatic rhythm was used to obtain the maximum benefit of seasonal rains, and minimize crop damage and failure. There is a proverb enshrining this wisdom *Kal yal bala govithan karanna*, simply meaning stick to the time and season for planting. Every agroclimatic region developed its own rule of thumb for time of planting. Dry zone sowing, for example, in the north, was calculated with reference to the lunar moon, beginning with the new moon approximately between 15 September and 15 October. The timing varies from year to year. The formula used is, *Wapmulata biju isanna*, sow with the beginning of the new moon. In the dry zone in the south, seed was sown towards the latter part of September, the timing again depending on the position of the sun and the moon. Elaborate calculations were made in order to avert pest damage and disease. In the wet zone sowing was also done according to solar and lunar calculations. This traditional practice reduced damage from pests and diseases. Even today we can avoid brown plant hopper damage and grassy stunt virus disease if we adopt timely sowing of paddy.

The second feature was minimal tillage. Paddy fields in low-lying areas were accordingly trodden three times, at intervals of about a fortnight, by groups of buffaloes or cattle, then levelled and sown. In fields with hard soil or less water, two buffaloes together drawing a country plough were used to break up the soil. Here too, the main operation involved treading. Using this method, the depth of land preparation was limited to about three centimetres. With deep preparatory tillage, using tractor-drawn implements, the chemical nature of rice soil is affected by the turning over of the soil. Tillage operations on high land were also better with shallow scraping. Minimum tillage not only saved energy, but crops too grew better.

The third principle was mixed cropping. Several crops — grains like kurakkan, maize, pulses and vegetables — were grown together. Sometimes short-age fruits, like papaw and sweet melon, were interspersed. The mixed crop establishes itself fast with the first rains, and covers the soil, so that soil erosion during the torrential monsoon rains is minimized. Mixed crops promote better photosynthesis and reduce competition for nutrients because of differential plant preferences. The pulses fix surplus nitrogen which is made available to other crops. Crops are sometimes mixed in such a manner that one crop runs for one full year, through to the next season. An example is kurakkan and chillies. Once the kurakkan is harvested the chillies spread and cover the soil. The kurakkan stubble acts as a mulch and conserves soil and moisture, controls weeds and finally adds organic matter to the soil.

The benefits from mixed cropping to the farm family were a steady food supply, notwithstanding weather changes and damage to some crops; extended use of available family labour throughout the season and year; no pest buildup, as with monocropping, and no pest control measures needed.

A further feature was the adoption of different cropping patterns for irrigated paddy and highland. The cropping pattern can be defined as a sequential planting of different crops during different periods of time in the season, or year, depending upon the climatic rhythm, diurnal variations, and expected weather conditions. The benefits of this practice were similar to those of mixed cropping, the two practices complemented each other.

Land preparation, broadcast sowing and irrigation water greatly reduced the weeds in paddy fields. Only a random hand picking of weeds was needed. In Chenas too very little weeding was required for the first four seasons or so, before the land was abandoned. Severe outbreaks of diseases were apparently not known. This is not because there was an absence of disease but because they were not a significant problem. However, pest damage was recognized very early on and control measures were incorporated in religious festivals and in traditional ceremonies known as *Kem*, which involved a kind of magic.

To give some idea of pest control measures involving biological control, I shall describe three traditional measures. The first practice involved leaving a portion of the paddy field adjoining a thicket for birds. This portion, identified as *kurulu paluwa*, means bird damage. Though details are not available, it is possible that this portion may have been cultivated with a paddy variety preferred by birds, so that bird damage to the main crop was minimized. The underlying purpose was to attract birds so that they would pick off the insects. The second practice entailed allowing birds into the community orchards belonging to all the farmers, during off seasons. With a good bird population pest damage was minimized. Thirdly, out of the many religious functions and *kems* one or two are explained to indicate the theory behind these traditional practices. When leaf-eating caterpillar damage was observed, the following *kem* was performed. At sunset a round section of a young plantain stem was fixed to a stake driven into mud, forming a small receptacle or platform on which cooked rice, pulses, flowers and a lighted wick dipped in coconut oil were placed. This was done in a few spots distributed over the affected and adjoining area. Birds would be attracted from afar by light. When they perched on the unstable platform, it would fall on to the paddy field. When the birds picked at the fallen food, they would see the leaf-eating caterpillars which are a delicacy to them. Within two days the pest damage would have been brought under control.

Farmers used certain plants for pest control in grain stores as well as in growing crops. The harvested well-dried rice and pulses were protected by a little ash and lime leaves. The thrips attack which appears in paddy during the early seedling stage was controlled by placing a few chopped pieces of euphorbia coated with latex in the water. There are many other localized practices.

For many centuries Sri Lanka flourished on a basic farming system with a variety of modifications to suit each agronomic region. This consisted of an irrigated paddy plot, a homestead and a *chena*. The homestead had fruits and coconut intercropped with annuals like vegetables, and pulses, and crops like chillies. The village tank was the source of water for animals and for the irrigation of the paddy and also kept the ground water level reasonably stable. The tank itself was a source of food providing fresh water fish, lotus seeds and suckers, *olu* seeds and *kekatiya*, for example. The tank bed, paddy fields and abandoned *chenas* were the grazing ground for cattle and buffaloes belonging to farmers. The *chenas* were cultivated with grains like kurakkan, pulses and vegetables such as pumpkins and other gourds. The farmer had every thing he needed except for commodities like salt and clothes; the people enjoyed a balanced diet. This farming system still exists in many parts of the island: Hambantota district is one such area; Meemure, close to Hunnasgiriya in the Kandy District, is another such example. Here a stable ecological system is being consciously maintained.

The destruction of this farming system was spurred on by the introduction of coffee to the central hills in the early nineteenth century and by the introduction of the tractor which replaced the animals. With intensive preparatory tillage soil proper becomes a mass of minerals. Crops do not thrive well on minerals, therefore fertilizers have to be added. As the water-holding capacity of minerals is less, irrigation is required, even during a short spell of drought. Run off during heavy rains is inevitable, and this causes erosion. Because of abnormal conditions and artifical fertilizers the crops are not strong and healthy and are easily damaged by pests and diseases. This necessitates the spraying of pesticides and fungicides. For all these operations Sri Lanka has to import tractors, equipment, agrochemicals, all of which keep the farming community dependent on industrialized nations. In addition, the adverse effects of pesticides have resulted in a number of people dying from poisoning, and in an alarming number being hospitalized.

As the memory of traditional methods is still in the minds of the people and as some still practise them it would be easy and logical to introduce well tested traditional agricultural practices incorporating some useful new findings. We need to return to an ecologically balanced farming system with a higher productive capacity to cater for an increasing population, the higher energy demand of present day farmers, and to produce food items without harmful chemicals.

The starting point for reviving the farming system could commence with practices already inherited by the farmers with the subsequent or simultaneous introduction of appropriate low cost production methods, such as those indicated below.

1. The non-burning of organic matter.
2. Minimal tillage.
3. Timely cultivation.
4. Selection of suitable crops and crop combinations, preferably those with high photosynthetic efficiency and high nitrogen fixing capacity.
5. Optimum planting distances and densities.
6. Choice of a planting system to cover the ground completely.
7. Mixed cropping, including perennial trees where multi-tier poly culture can easily be adopted.
8. Rotation of crops, particularly the annuals.
9. Keeping the ground always covered with some kind of plant so that solar energy, water and plant nutrients in the soil would be trapped and produce organic matter.
10. Planting trees, preferably leguminosae, to be used as fodder during drought periods and also as a green manure for crops.
11. Rear animals to produce valuable manure, and provide a raw material for bio-gas production.
12. Depend primarily on renewable sources of energy for the operation of the farm and the farm family. Bio-gas is an extremely useful source which

enriches the environment and supplies a versatile, clean, safe fuel. Other renewable sources are wind, solar energy, producer gas and hydro power.

Such practices would produce many beneficial changes: an increase in the organic content of soil; the establishment of a balanced eco-system; minimize pest and disease; increase productivity; steady the supply of all food items of high quality; save energy and control pollution; and most important of all, create an independent farming system based on self-reliance.

The crop-livestock-energy integrated farming system would fulfil the present and future needs of the developing world. Undoubtedly the future of farming will rest on such a system. The characteristics of the system are that:

1. It is possible to incorporate innovations in crop or animal husbandry production.
2. It produces its own inputs and lessens purchased inputs.
3. Capital requirement is considerably reduced.
4. It is an intensive system.
5. It has a high and continuous labour requirement with remarkably high labour productivity.
6. It is not limited by the size of the farm.
7. Almost all waste is recycled.
8. It increases productivity and profits.
9. It improves soil fertility and conserves soil and moisture.
10. It controls pollution and maintains an ecological balance.
11. It conserves and regenerates energy from renewable sources.

The integration of the three components (crop-livestock-energy) is brought about by the utilization of waste; this can be described as the linking force. The strength of integration depends on the quantity of wastes used, the greater the proportion of waste used the stronger is the integration. Efficiency depends on the utilization multiplicity of wastes. This multiplicity of usage can be achieved either by repeated use of a particular waste, or by using a waste for different purposes, as for example, in the use of fresh straw for mushroom production, partly decayed straw for bio-gas generation and the sludge as manure for crops. Using bio-gas sludge as manure, animal feed and as a medium for culture of algae is an example of a waste used for different purposes. Efficiency can also be improved by increasing the crop livestock and energy combinations. A variety of crops, different kinds of animals and several sources of energy (e.g. bio-gas, solar and firewood), facilitates more efficient recycling. Reusing animal residue as animal feed in particular, is important and also brings about a steady state to the system.

The integrated system causes an upward spiral of production and reaches an equilibrium point when all wastes are utilized. Although this is possible, it is difficult to reach the saturation point. Saturation of the system occurs when all natural resources like rain and solar radiation are combined with all available appropriate techniques of agriculture, animal husbandry and energy. At a given point of time there should exist a gap between what is practised and what is available. This gap provides incentive and opportunity for the continued improvement of the system.

I should now like to look at a simple working unit at Galaha in the Kandy District. The extent of the land is approximately one-eighth of an acre. A small portion of the plot is set apart for fodder crop which is rotated with vegetables. The bunds of the terraces too are planted with fodder crop. The farmer has a small poultry pen. The entire area of land has thinly scattered fruit and minor export crop trees allowing enough sunlight for the vegetables. The bio-gas unit has been constructed next to the cattle shed so that the washings of the cowshed are channelled to it. As the fodder cultivated is not sufficient, an additional quantity of grass is cut from the road sides and waste lands. In addition, tree fodder such as Glyricidia is fed to the animals. Fodder, grass and vegetable refuse are fed to the animals. The dung and urine together with other residues are washed down to the bio-gas unit. The manure that comes out is used for the fodder, vegetable and other crops. Because of a complete and high nutrient and organic content, including hormones with growth promoting substances, this bio-gas manure (which is free of pathogens, parasitic organisms and weed seeds) greatly increases the production per unit area of land. This high productivity is further enhanced by such agricultural practices as mixed cropping, multi-tier polyculture and appropriate planting systems and plant densities. These practices create a natural but controlled eco-system that has inbuilt biological controls. There is no requirement for pesticides and fungicides as there are no pests and diseases. The complete cover provided by the vegetation, together with the organic matter that is added, conserves both the soil and moisture. While maintaining a natural ecological balance in such a rich environment the farmer gets milk, fruit, vegetables, manure feed and bio-gas fuel for his farm and his family. A monthly income would depend on the intensity of the farming, the types of animals reared on the farm and on the capital and management skills used.

The dung and urine cattle-shed wastes of two animals can generate bio-gas for the entire cooking and lighting of a family of five or six. The manure production is sufficient for the cultivation of an acre of land. Besides these, milk and the animal power for tillage would also be available to the farmer.

Farm sizes normally vary from the size of this unit to those hundreds of acres in extent. The number and the kinds of animals that can be reared on a farm vary with the size of the farm, the climate and the experiences of the operator of the farm. In a very small farm a few birds or ducks can be reared. When the farm is slightly bigger and some herbage is available, a goat or a sheep or a few pigs — or in larger farms, cattle and buffaloes — can be reared.

Despite all the benefits of an integrated farming system there are nevertheless some constraints to be considered as well:

1. Lack of research and flow of information to extension personnel on farming systems. In Sri Lanka the research is still concentrated on separate studies of varieties, techniques and agro-chemicals but not on farm management or integrated farming systems. Now, with the appearance of the computer and advances in statistics, models and statistical tools which are available for complicated analysis, it is possible to study the effect of all these factors in an integrated system.
2. The persistence of a western European influence where the agriculture is based on completely different climatic conditions.
3. Policy-makers are guided by those who lack practical experience. They often offer academic models and untested theoretical concepts.
4. Powerful propaganda by agro-chemical industries.
5. A lack of recognition of the valuable grassroots experiences of the farmers which should also be incorporated in developing farming practices.

Attempts to increase farm family-income and agricultural production have to be viewed from the perspective of the whole farm rather than that of a particular crop or a specific technique. In increasing agricultural production and improving the quality of life of farmers energy plays a key-role and therefore production of energy should be given its due place in a farming system. The only farming system that incorporates all the factors mentioned is a crop-livestock-energy integrated farming system, a system that must be revived to face the crisis caused to the Third World by modern science.

21

Traditional is Appropriate
Lessons from Traditional Irrigation and Eco-systems

D.L.O Mendis

The remains of a vast system of irrigation works from ancient Sri Lanka are still very much in evidence today. Indeed, some of these ancient works have been functioning for centuries without interruption, in spite of sporadic disruptions to the general organization of society on account of internecine strife, foreign invasions, or the advent of malaria. The Dutch, who occupied the maritime provinces for roughly 150 years from about 1658, were interested in restoring some of the larger works such as the Giant's Tank in the Mannar District in the north-east of the island, but believed that some of these stupendous systems had not in fact been completed, or could not have functioned as intended, on account of technical errors such as wrong levels. It has subsequently been shown that it was the Dutch who erred in their levelling, and not the old Sri Lankan engineers.

The British, who took over the maritime provinces from the Dutch and conquered the whole island in 1815, set about restoring some of the larger reservoirs such as Kantalai in 1873, Kalawewa in 1887, and Parakrama Samudra in 1943. They too could not comprehend the full scope of the achievement of the ancients in spite of the insight of some individual engineers such as Colonel Woodward, and Henry Parker, who had a proper appreciation of the ancient irrigation works. In particular, British engineers did not understand why a very large number of small village tanks (reservoirs) had been built in the dry zone of the island, even though many of these supported impoverished villages, thus earning the description of 'tank villages'. The old irrigation works of Sri Lanka can be described as consisting of river diversion systems, including diversion structures and diversion channels, large storage reservoirs, medium-scale storage reservoirs, small village tanks or storage reservoirs.

R.L. Brohier has shown how the large storage reservoirs and diversion channels in the Rajarata were all interconnected. More recently, it has been surmised that the medium-scale reservoirs were located near urban settlements. The numerous small village tanks in the dry zone, as stated above, have always been something of an enigma to irrigation engineers. The one-mile to an inch topographical maps of the island show nearly 15,000 of these, of which over 8,000 are in working condition today. Tradition has it that some 30,000 of these small tanks had been constructed down the ages, and there is a reference in the chronicles to 20,000 in the ancient province of Ruhuna alone in the twelfth century. In 1923, after heavy monsoon rains in the dry zone, a failure of one of these small tanks in the Mannar District triggered off a series of failures of further small tanks where the spill waters of one lead into another down the chain. The resultant flood damaged the railway line near Medawachchiya, with much loss of life and damage to property. The Irrigation Department thereafter started a study of the ancient village tanks, and ten years later, J.S. Kennedy, then deputy director and later Director of Irrigation, presented a landmark paper to the Engineering Association of Ceylon, entitled 'Scientifc Evolution and Development of Village Irrigation Works in Ceylon'. This comprehensive study was published as a handbook, and became the basis for the restoration by the Irrigation Department of selected village works lying abandoned, or in need of improvement, in various parts of the dry zone of the island. However, it also led to a misconception regarding the usefulness of these village tanks, which was to become the basis of a misconceived dogma in the Irrigation Department in the years to come.

Kennedy stated in his paper that because there were so many small village tanks in the dry zone, it was unlikely that all of them had functioned at one and the same time. This could have meant that all these tanks were not meant to be used for irrigated agriculture continuously, but that each tank could be rested with the fields dependent on it lying fallow, from time to time. This is indeed quite a likely hypothesis. Unfortunately, however, irrigation engineers later mis-interpreted Kennedy as meaning that the small tanks were intended for irrigated agriculture only until such time as they were replaced by some large reservoir or reservoir cum diversion system. The small tank would thus have been a stage in the development of irrigation systems in ancient Sir Lanka. This unfortunate misconception became the unofficial dogma of the Irrigation Department. In 1956 it was given legitimacy by Brohier, one of the acknowledged authorities in the field, who presented a four-stage theory for the evolution and development of the ancient irrigation systems at the fiftieth jubilee year of the Engineering Association of Ceylon (when it became the Institution of Engineers by Royal Charter). Later, in 1971, Joseph Needham republished this theory in his monumental work, *Science and Civilization in China*. This theory justified the replacement of small village tanks by large storage reservoirs.

Meanwhile the restoration of the large ancient reservoirs, started in colonial times, was continued after independence in 1948. Some of the ancient channel systems were also restored, but the interconnection of large reservoirs and channels was never completely recognized, despite Brohier's famous paper to the Ceylon Branch of the Royal Asiatic Society in 1935. As a result, restoration of the large reservoirs and channels was actually done on a more or less piece-meal basis, and not with a recognition of the grand leitmotiv behind their layout.

Immediately after independence the government launched the Galoya project modelled on the TVA project in the USA. The most important feature of the Galoya project was the construction of the Senanayake Samudra, by far the largest storage reservoir ever constructed in this country, seven times larger than the biggest ancient reservoir. Fortunately, the existing Pattipola-ara irrigation scheme in the Galoya valley, consisting of river diversion and storage reservoirs, had an extent of 35,000 acres, making it the largest existing irrigation scheme in the whole island. This existing scheme benefited directly from the construction of the Senanayake Samudra, because the drought resisting capability of the new large reservoir combined with the existing irrigation scheme to give immediate benefits. The extension of the existing scheme to benefit a total of 120,000 acres of irrigated land, with new settlers, was actually therefore more readily justified than if the whole project had been an entirely new one. Nevertheless the Galoya project has also run into many difficulties, not the least of which is a shortfall in the expected returns. In this respect, Galoya is no different from many other major irrigation settlement projects based on restoration of major ancient irrigation schemes.

The next big investment in the agricultural sector after Galoya, was the Walawe project. The Uda Walawe reservoir headworks were virtually the last major irrigation headworks constructed without massive foreign aid in this country. Although there was an input of technical assistance from Czecheslovakia, the project was controlled and implemented by local engineers at all stages, and the cost at less than Rs50 million, in 1968, was very low compared with costs of later headworks constructed with foreign aid.

The Walawe project could be used to illustrate the Brohier theory that large reservoirs must replace small village tanks. The Uda Walawe reservoir submerged a system of small tanks, one leading into the other along the tributary branches of the Walawe ganga. Ever since its construction, however, the Walawe project has been plagued with what are called 'water management problems'. Predictably, foreign aid has been sought and obtained to study and solve this problem, so far without much success. Usage of water is very high, and farmers have been blamed for wasting irrigation water. Since they are the only people actually dependent on the water for their livelihood, it is not at all clear why they should waste it. Nevertheless excessive consumption of water from storage is held to be the farmers' responsibility. As a result, a new approach to finding a solution for the problem has emerged recently, described as a method to achieve farmer participation in the management of the scheme. Whilst this is a step in the correct direction, so long as it is tied up with foreign aid for 'rehabilitation' of the project, it is still not likely to yield a final complete solution.

After Walawe, another project was mooted in the south of Sri Lanka, called the Lunugamvehera or Kirindi oya reservoir. Conceived about two decades ago by technocrats in the Irrigation Department, the project was opposed by others in the Ministry of Planning. With a change of government in 1977, the new minister in charge of irrigation made his first big decision about the location of the proposed reservoir. At a recent meeting, he recalled how he had to make this major decision just two weeks after assuming office. He had followed a strategy attributed to Napoleon that it was acceptable to lose a battle or even to lose a war, but never time. Since investigating a new site some twenty miles upstream of the site that had already been investigated would delay the project, he had given the go-ahead for construction of the Lunugamvehera reservoir. The Minister conceded that only time will tell whether he made the correct decision or not. An engineer in the audience declared that history would absolve the Minister for his decision which had been given in good faith, but it would never absolve the technocrats who had selected the wrong site for detailed investigations, ignoring the alternate site until too late.

The Mahaweli development project was started in the mid-sixties as a United Nations study project. An impressive team of foreign engineers and their local counterparts was given a mandate to study the possibility of diverting the waters of the Mahaweli ganga, the largest river in the island, to adjacent river basins in the dry zone. This study was first announced as a thirty-year plan for the total development of land and water resources in the Mahaweli ganga and the adjacent river basins, and the date of completion of the project was reckoned to be the year 2,000.

The Mahaweli project may broadly be described as consisting of two major components: the headworks or storage reservoirs and major diversion structures on the one hand, and the downstream development and settlement areas on the other. New large storage reservoirs were to be used to generate hydropower, and in every case the investment could virtually be justified in terms of the energy benefits alone, especially after the oil price hike in 1973. The water available in the storage reservoirs could therefore be made available for irrigated agriculture virtually free as indeed irrigation water has been available to cultivators in this country from time immemorial; but timing of water issues would then have to depend on electricity generation priorities. So far, however, there has been no direct conflict between the needs for power generation and the needs of irrigated agriculture, although the whole scheme is still at an early stage of implementation, at least in so far as the settlement programme is concerned.

When the downstream development and settlement area was designed at the beginning of the Mahaweli development programme, the existing village tanks shown on the topographical sheets were not considered to be necessary. As a general rule they were to be levelled off, and the land recovered for cultivation. This was of course in keeping with the notion that small tanks were inefficient compared with large storage reservoirs.

However, the first area to be settled or actually resettled (because it had been settled under the restoration of the Kalawewa in colonial times) was System H. When the proposal to level off the existing small tanks in this area was made known, there were protests from the villagers in the area, who alone knew the true significance of these small tanks in the total system. The then Chairman of the Mahaweli Development Board, Mr H.de S. Manamperi, decided to live in some of these villages and see for himself why the farmers valued these small tanks so much. After this experience he was convinced that the small tanks should be left, and gave instructions that the plans should be revised to accommodate many of the existing small tanks, and this was done.

This important episode illustrates the fact that there were some fundamental errors in the assumptions previously made by irrigation engineers that the small tanks would become redundant when large storage reservoirs and diversion systems were constructed. It was apparent that it was necessary to re-think basic concepts about irrigation systems.

Subsequently a seven-stage theory for the evolution and development of ancient irrigation systems was proposed. These seven-stages included:

1. Rain-fed agriculture.
2. Seasonal or temporary river diversion irrigation, using sticks and stones to build temporary diversion structures on poor foundations and practising flood or inundation irrigation on river banks.
3. Permanent river diversion irrigation, using permanent river diversion structures made of stone blocks or of brick masonry on good (rock) foundations.
4. Development of river diversion channels following a falling contour with a single earth embankment on one side, in which stone masonry or brick masonry structures were incorporated, namely weirs and spillways.
5. Invention of the sluice (*horowwa*) with access tower (*bisokotuwa*) based on the experience of operation of weirs and spillways on diversion contour channels.
6. Construction of storage reservoirs incorporating the sluice for controlled issue of irrigation water.
7. Damming perennial rivers either by using the sluice for temporary river diversion, or by the twin tank method.

Although these stages followed each other in a time sequence, any new stage did not replace a preceding stage or stages, but supplemented it or them. Thus for example, seasonal or temporary river diversion came after rain-fed agriculture had been practised for a long time. Thereafter both rain-fed agriculture and temporary or seasonal river diversion irrigation were practised for another lengthy period of time until the next stage, permanent river diversion, was thought of. Rain-fed agriculture as well as all the other stages remain in use right up to the present day.

This theory is also consistent with the well-known fact that river diversion irrigation is known to have been practised from a very much earlier date than irrigation from storage in all the earliest river valley civilizations such as those in the Tigris-Euphrates, the Indus, the Nile, and the rivers in China. River diversion irrigation represents a mastery of water management in the dimension of space, whereas storage represents a mastery of water management in the dimension of time. The former would have occurred at an earlier point in time than the latter in any natural sequence of evolutionary development.

However, any theory for the stage-by-stage evolution and development of irrigation systems or irrigation eco-systems, must also fit into and accommodate theories for the stage-by-stage evolution and development of social formations, or the evolution of cultural complexity.

Morton Fried has identified four types of society, namely non-ranked non-stratified societies, ranked societies, stratified societies, and states. Elman Service has divided societies into four classes described as bands, tribes, chiefdoms, and states. How does the seven-stage theory for the evolution and development of irrigation systems fit into these concepts?

We may assume that hunting and gathering that preceded rain-fed agriculture corresponded to Fried's non-ranked non-stratified society, and Service's bands. Rain-fed agriculture and domestication, called the neolithic revolution, would correspond to the beginning of ranked society or the beginning of tribes, that came after bands. Temporary river diversion irrigation would have been well within the ability of tribes in ranked societies, but permanent river diversion irrigation systems would have required more organized labour and administrative skills so that at that stage Fried's stratified societies and Service's chiefdoms would have emerged. During several centuries of this stage in the evolution and development of irrigation systems and social formations in ancient Sri Lanka, diversion contour channels with their weirs and spillways would have been developed. The stage was then set for the next great technological quantum jump, the invention of the sluice with its access tower. After the invention of the sluice it was possible to build storage reservoirs with control arrangements to issue stored irrigation water for irrigated agriculture. Small, medium and large irrigation storage reservoirs were built thereafter. This would have coincided with the evolution of the state in ancient Sri Lanka, and this is the beginning of the historic period from about the fourth century BC.

This outline also accommodates all the formative social forces such as population growth and environmental circumscription, irrigation, and class conflict, which are known to have contributed to breaking down kinship relations and establishing societies based on occupational, social and economic stratifications.

The physical features of the ancient irrigation systems of Sri Lanka also fit into Kent Flannery's analysis of cultural complexity based on general systems theory. Flannery said that segregation and centralization were two key processes responsible for social formations and social change. The small village tanks correspond to segregation, while the interconnected large storage reservoirs and diversion channels correspond to centralization.

The small village tank represents an irrigation eco-system as described in E.P. Odum's comment: 'It is the whole drainage basin, not just the body of water, that must be considered as the minimum eco-system unit when it comes to man's interests.' As mentioned previously, there was a reference to the existence of 20,000 small tank villages in the province of Ruhunu alone in the first half of the twelfth century. The present extimate of 30,000 small tanks in the whole dry zone is not excessive, even though this represents a density of approximately two small tanks per square mile in the whole dry zone. If on average we take two small tanks to make up a single village settlement, these 30,000 small tanks could support a total of 15,000 village settlements, even today.

In ancient times, each of these segregated villages would have had its own economic, religious, social and perhaps even its own political sub-systems, that fitted into the larger complex of systems in the whole national social matrix. The interconnected system of large storage reservoirs and channels would have supported that larger matrix, and represented the element of centralization in the analysis of society in ancient times according to Flannery's theory. If this analysis is accepted, the small village tank becomes the natural unit for the segregated or decentralized subsystems. Each tank village would have its own ecological or natural subsystem.

In the settlement plan of the Mahaweli Authority of Sri Lanka today, there are now settlement centres described as villages, hamlets and townships. This layout is not based on the layout of the ancient irrigation system, but on the irrigation system which does not accommodate small tanks to anything but the barest minimum density.

However, if the analysis presented in this article is correct, the ancient tank villages should have been restored and used as the basis for settlements. It is argued that if this had been done, a more ecologically stable system would have been made possible, despite the destabilizing effects of intensive inputs that are currently in vogue. This is turn could have led to new attitudes of self-reliance and non-dependence, not only among the settlers themselves, but also among the bureaucrats and technocrats responsible for designing and implementing this vast project. In sum it would have been a unique example of learning from the traditional irrigation eco-systems of ancient Sri Lanka.

Is it too late, even now, to adopt such a policy?

Appendix
The Penang Declaration on Science and Technology

Modern science and technology is in a state of acute crisis. This crisis manifests itself in several forms. The most obvious are in the end products which are often directed towards destruction, waste and alienation. From disasters at Chernobyl and Bhopal to the depletion of the ozone layer, pollution and devastating environmental degradation, depletion of natural resources, science and technology manifests itself in the lives of the majority of inhabitants of our planet as a poison. From sociobiology to bio-engineering, from robotics to fifth generation computers, science and technology are generating unemployment and under-employment. There are numerous examples to show that science and technology are not improving the material and spiritual conditions of the people of the Third World. More sophisticated science entails greater problems and fewer solutions; reliance on capital-intensive technologies breeds dependence on non-indigenous resources and the gross misuse of available human and material resources.

There is a growing awareness that there is something intrinsically wrong with the very nature of contemporary science and technology. Even in the most fundamental of the scientific enterprises, namely mathematics and physics, there has been a period of epistemological uncertainties for over two decades. In biology, the techniques of recombinant-DNA and the possibility of creating and unleashing new and deadly forms of mutant species and even the cloning of human beings, has brought the nightmare of Frankenstein very close to reality.

Reductionism — the dominant method of modern science — is leading, on the one hand, towards meaninglessness in physics, and on the other, towards Social Darwinism and eugenics in biology. There is something in the very metaphysics of modern science and technology, the way of knowing and doing of this dominant mode of thought and enquiry, that is leading us towards destruction.

The romantic notion of science as the pursuit of pure, unadulterated 'objective' truth, with the scientist working in isolation from mundane social reality — like a hermit, trying against impossible odds to understand some sort of objective reality — has now become dangerously untenable. In fact, after a decade of world-wide research, it is now clear that science is mediated through a social process. Indeed, some philosophers and historians of science argue that science *is* social process. Social forces, operating at the level of global economy and structures, and the national political economy, as well as in the scientific groups themselves, shape the character, content and style of modern science and technology. Scientists are strongly committed to beliefs and certain cultural ethos which compels them to flatten diversity and complexity into a uniformity. In addition to this belief system and cultural ethos — which manifest themselves in the propositions that scientists embrace — science has its own power structure, reward systems, peer groups, all of which combine to ensure that science is closely correlated with the existing, dominant and unjust, political economic and social order of the world.

Domination and control are inherent and integral parts of the current scientific and technological enterprise. These concepts are at the very heart of scientific methodology and the present process of creation and generation of science. From the inception of modern science, at the beginning of the European Renaissance, the goals of science were articulated in terms of domination and control. The *philosophes* of the seventeenth-century Enlightenment movement, in complete contrast to the intellectual heritage of Islam and other civilizations on which they were constructing a new world, divorced reason from values, and elevated it to an arch-value. Reason thus became the dominant mode of knowing to the exclusion of all other alternatives. The notion of the domination and control of nature gradually changed into the domination and control of non-European people by means of the use of the scientific method and the linear rationality incorporated within it. The present modern scientific and technological enterprise has evolved within the imperial experience of Europe and North America and the colonial experience of the Third World.

The crisis of science manifests itself in the Third World as well. Modern science and technology has dislocated Third World societies, destroyed traditional cultures and played havoc with the environment of Third World nations. It has also replaced a way of knowing, which is multidimensional and based on synthesis, in Third World societies with a linear, clinical, inhuman and rationalist mode of thought. From the Green Revolution to massive incorporation of modern medicine to the waste of valuable and scarce resources on research into fashionable areas such as cancer research and nuclear power, western science and technology has systematically underdeveloped Third World societies in the name of scientific rationality. No figures or indicators can convey the loss of lives and resources that these social engineering experiments — induced by

Western 'experts' and 'consultants' — have brought to our countries.

In most developing countries the transplant of science and technology has not taken root. Neither the science nor the scientists have an organic relationship with local problems, resources and the pressing needs of the society. It is often irrelevant, wasteful, unproductive, imitative and bears the hallmark of a second-rate and second-hand product.

Moreover, the system of modern science and technology in the Third World has grown at the expense of the pre-European scientific cultures. Before the domination by the European imperial powers, Third World civilizations and their own elaborate and sophisticated systems of knowledge and craftsmanship flourished.

In North Africa and the Middle East, Islamic science was the main problem-solving paradigm. Both in terms of quality and quantity the output of Islamic science, from the seventh to fourteenth century, remains unmatched. Muslim scientists laid the foundation of algebra and trigonometry, measured the circumference of the earth, studied the properties of light and motion, examined the human body and discovered the circulation of blood, and obtained results at whose accuracy one can only marvel. Yet they did all this in a framework of thought and enquiry which integrated facts and values in a metaphysical framework. Muslim scientists never accepted the tyranny of method but sought to develop and evolve methods in conformity with the nature of enquiry and within a clearly defined ethical matrix.

In Chinese civilization too, science and technology flourished. Much of our arithmetic originates from the work of Chang Ts'ang (d.152 BCE); and our geometry from the classical treatise of Wang Hsiato-t'ung (d.727 BCE). Chang Ch'iu-chien's (d.650 CE) work on medicine is still a source of wonder. Here again a system of science and medicine evolved in a metaphysical framework that emphasizes synthesis and promotion of certain norms and values.

From ancient times the Indian subcontinent has been known for its technologies of agriculture, metallurgy, textiles, ship-building, architecture and medicine. These technologies were widely practised in the whole of the South and Southeast Asian region till as late as the eighteenth century as documented in the various European accounts of the period. These technologies were characterized by their simplicity as well as by their sophistication. Each of them had their own separate *shastra* which outlined the fundamental scientific principles. These various *shastras* were themselves founded on the philosophical basis provided by the various *darshanas*, the schools of philosophy which define the logical, epistemological and methodological structure of Indian thought. The guiding principle of all these sciences and philosophies was that the world in itself was the repository of truth; the purpose of science and technology was merely to enable people to live happily and healthily in this world rather than changing and manipulating the world.

Southeast Asian traditions are also replete with countless examples of systems of indigenous science and technologies which have been destroyed by western systems. This destruction manifests itself in the negation of the history of science and technologies of traditional civilizations and cultures as well as in the suppression and destruction of indigenous medicine, local architecture and building techniques, and ecologically sound farming and irrigation practices.

Given the destructive nature of contemporary science and technology, and the fact that it is controlled and directed by industrialized states and multinational corporations, it is essential for Third World countries to create their own indigenous bases for the generation, utilization and diffusion of scientific and technological knowledge. Third World countries should co-operate with each other in this endeavour. Moreoever, the whole notion of transfer of technology and importing science should now be abandoned.

Evolving indigenous scientific culture requires Third World scientists, technologists, decision-makers and activists to appreciate the true value of traditional science and technologies. Traditional technologies and medical systems should be upgraded, developed and promoted. They should form the basis for the evolution of indigenous, but thoroughly contemporary, systems of alternative technologies and health care. Similarly, national problems should be solved within the framework of an indigenous mode of thought and enquiry and with locally available resources.

Only when science and technology evolve from the ethos and cultural milieu of Third World societies, will they become meaningful for our needs and requirements, and express our true creativity and genius. Third World science and technology can evolve only through a reliance on indigenous categories, idioms and traditions in all spheres of thought and action.

Science, Technology and Natural Resources

Before the colonial conquests, the natural resources of the Third World were utilized through technologies based on local expertise and knowledge and which were small in the scale of exploitation. With colonization came the immense demand for the natural resources of the colonies, be they forests, minerals or agricultural products, to fuel the increasing requirements of the industrial revolution. This was the first stage in the destabilization of functioning local technologies; it undermined their resource base and market. This process of direct transfer of natural resources was not possible in the colonial era. However, as the process of 'development' was entrenched as a national goal, to be achieved with the aid of international finance, the process of resource transfer continued to pay for the imported inputs required for the process of development. The latter included expertise, technologies, equipment as well as luxury consumer items.

This process of development merely serves the purpose of the easy marketing of obsolete technologies and of extending access to the industrially advanced countries in the North to the remotest natural resources of the countries in the South. In the late sixties, the process of development touched the agriculture of the Third World by introducing the technologies of the Green Revolution. In the decades that followed, this process manifested itself in more undemocratic control over the land use in the Third World as international technologies and finance entered the management of land and water use in a big way — including the area of forestry. On the other hand, industrial growth in the south concentrated on polluting industries transferred from the north where environmental consciousness did not allow such industries to function anymore. These included textiles, dyeing, tanning, hazardous chemicals and the nuclear industry.

It was thus possible for the industrially advanced countries to have the first industrial revolution by transferring the natural resources of the colonies, and then, in the 1970s, the second revolution, leading to the clean service society in the North with the transfer of obnoxious industries to the South.

Until about the 1980s the transfer of obsolete technologies was the practice, but thereafter, with the advent of biotechnologies, the latest technologies from the North began to touch the remotest villages of the South in a decentralized manner, proving that the small need not always be beautiful. It is in this era of a rapidly changing relationship between technology and natural resources that we have to locate our role.

We see our intervention as:

1. Creation of a civilizational response from Third World countries for the development of resource-prudent technologies and as enhancing the control of local bodies on the decisions related to natural resource use.
2. Use of land and water to be guided on a sustainable basis as to satisfy the local needs first, starting from the needs of the neediest.
3. Defending our crop and plant genetic resources from destruction.
4. Actively opposing the dumping of obsolete or polluting technologies in the name of economic development and actively encouraging options for economic development with resource-prudent, non-polluting indigenous technologies.
5. To ensure equitable access to resources and information on all technologies to be used in a region, including all possible environmental impacts.
6. To increase people's participation in the choice of technologies and management of natural resources with the objective of choosing ecologically sensitive technologies of resource use.

Science, Inequality and Inability to Meet Basic Needs

The failure of modern science is manifested by the irony, on one hand, of the development of technological capacity powerful enough to meet the basic needs of every human given an appropriate arrangement of social and production systems; and, on the other hand, of the fact that more than half the world's population (and something like two-thirds the Third World's people) live in conditions where their basic and human needs are not met. This tragedy is rendered even more catastrophic by the evidence that the same technological capacity that has facilitated the irrational composition of products, is also so powerful that it has enabled the destruction or depletion of a very high proportion of non-renewable resources in the world. Day by day, this gigantic technological capacity uses up more energy, extracts more minerals, chops down more forests, results in more loss of topsoil, and pollutes more water, more land mass, more air and even the stratosphere. At current rates of production, many critical resources will run out within a few decades.

There is a finite stock of world resources available, and in the process of production a portion of that stock is used up every year. The Gross National Product about which all nation states are so obsessed is only an annual *flow* which is very much dependent on the available stock of natural resources. Most resources are non-renewable. The more they are depleted, the less are they available for use in production *in future*. In other words, the higher the GNP at present, the lower it will be in future, when the effects of resource depletion are felt. GNP is only a flow dependent on availability of stock; when the stock runs out, the flow will dry up. This most simple and elementary of facts is almost completely omitted in economics textbooks. It is barely in the consciousness of the planners and politicians who plan our future and rule our lives, or in the consciousness of the scientists and technologists who have made possible the rapid depletion of resources through the development of technological capacity.

Moreover, the rapid extraction and utilization of resources is very unequally carried out in terms of control and benefits; 80 per cent of world resources are used up in the developed world and only 20 per cent in the Third World. This unequal distribution also determines the nature of goods to be produced. To produce for the elite market, high-tech technologies are created to produce high-tech products, such as video recorders, compact discs, computers, motorcars, and services like medicine, tourism abroad and even tax-evasion legal programmes. A large portion of developed world GNP is spent on such consumer goods and on producing capital goods or technologies to make these consumer goods. Since national incomes in the Third World are also unequally distributed, a large portion of these resources are depleted for producing the same high-tech consumer products as are enjoyed in the developed world, and in importing capital-intensive technologies to make these elite consumer goods. Thus, only a small portion of world resources flows towards the processing of basic goods required by the poor majority in the Third World for their survival, and the making of simple capital goods which are the technologies used by poor farmers and small industrial craftsmen or enterprises.

In this on-going process of resources depletion and irrational use of resources, the main impetus and dynamics are located in political economy, the socio-economic systems, which give rise to competition for growth between companies and between nations. But the role of science and technology is crucial. If the level of technology is low, then we may still have the same inequality, but the degree at which resources are depleted would be less. In reality, however, technological levels are increasingly coming under pressure of competition between firms and countries (not only in the economic but also military spheres), and so the depletion of resources also increases rapidly. Moreover the ever increasing technological capacity of the developed world leaves the Third World even further behind, thus in itself widening the inequality gap between nations.

In 1980, nations of the North, with only a quarter of the world population, earned 80 per cent of the Gross Global Product (GGP). In the South, three-quarters the world population claimed only 20 per cent of the world income. Since 1980 the world has become even more unequal. Due to the colonial experience, the Third World remains dependent on the developed world for trade, loans, investments and technology. In the past few years, increasing amounts of funds have flowed from the South to the North. In 1985 alone, US$74 billion left the Third World on its debt account alone: it obtained only $41 billion in new loans but had to pay $114 billion in debt servicing. If one includes outflow of profits by transnational companies in the Third World, capital flight from the Third World and the capital deficit of Middle East exporters, the 1985 outflow of capital from the Third World in 1985 alone would be US$230–240 billion. If we also include the US$65 billion lost due to the fall in commodity prices (an *Economist* estimate), the Third World's loss would be US$300 billion in one year. In 1986 the situation would have been even worse with the collapse of oil prices and the fall in prices of other commodities. Total loss could be US$300–350 billion. Given this gigantic flow from the South to North, it is clearly ludicrous to say that the North is giving aid to the South. Whatever aid is given is a mere drop in the ocean of what flows from South to North, and even this drop is tied to conditions.

The North's grip over modern science and technology has contributed to the exploitation of the Third World's economic weakness. The rich countries use their industrial and agricultural technologies to produce surplus goods which they are unable to use themselves (part of the problem of over-development, or 'over-accumulation'). So they dump the surplus cereals or other crops or materials on the world market, causing prices of Third World commodities to collapse, and thus reduce incomes and living standards of the poor. Modern technology and information systems have also enabled transnational banks and companies to expand into developing countries and draw them into the world market further. After being drawn deep into the world market, the Third World finds the rich countries putting up protective tariff barriers to block the entry of their industrial goods. They find that the rich countries have developed new technology for their own advantage. For instance they have reduced their usage of the Third World's raw materials by finding substitutes and by using less materials per unit of product. As a result, export prices and earnings in the Third World have fallen drastically at a time when they have to shell out more funds to service foreign debts.

Within Third World countries, the same structure of inequality exists at national, regional and local level. Thus, the national composition of goods also follows the same pattern; luxuries for the upper group, middle-class goods for the middle level and basic goods or less than that for the bottom 70 per cent.

In the commercialized sector, firms compete with one another for higher market shares so that they can maintain or increase profits. Firms with insufficient profits may have to close down or be taken over by a stronger rival. Expansion and growth is thus built into the system of inter-firm competition. Modern technology plays a vital role in expansion, both in seeking more 'productive' ways of producing and in developing new products or new models. Thus, modern science and technology is used in the service of the firms' desperate need to expand.

In the socialist countries there is a strong competition to keep up with the rich capitalist countries, both economically and militarily. There is thus also a strong tendency for developing capital-intensive technology and aiming for maximum growth. Thus, the ethos of industrialism is built into both capitalism and socialism.

In the Third World, the nature of development follows that of the North, except that ours is a dependent form of development. Growth takes the form of depletion of resources for export to the North, and the use of surplus from exports and from foreign loans to build expensive infrastructure and to invest in capital-intensive technology which mainly benefits big firms or big farmers. The commercialized sectors, with superior financial and physical resources and technology, penetrate, invade and take over the traditional, viable sectors, thus dislocating a large portion of people from their livelihood and homes. For instance, small fishermen using ecologically-sound production systems are displaced by big commercialized trawler boats which destroy the marine ecology by overfishing and the use of destructive gear. Or else food-crop farmers have their lands taken back by landowners or bought by either governments or private companies to be converted into middle-class housing estates, or free trade zones for industries, or for highways, etc.

In terms of many criteria, such as provision of employment, community or producer control over technology and production process, equity and ecological soundness, the indigenous technologies of the Third World are superior to the types of modern technology which have invaded the Third World. Yet these indigenous technologies are being wiped out under the impact of the commercialized sectors and under the threat of the consumer culture which entices tastes away from local to western culture, fashion and products. Thus, being sucked in a dependent manner into the modern world system has been disastrous for Third World nations whose futures in terms of sustainable development would have demanded the rational use of their resources for the genuine development of their people. It is time therefore for a re-orientation of the concepts of science, technology and development.

Proposals

1. There must be a radical reshaping of the international economic and financial order so that economic power, wealth and income is more equitably distributed, so that the developed world will be forced to cut down on its irrationally high consumption levels. If this is done, the level of industrial technology will also be scaled down. There will be no need for the tremendous wastage of energy, raw materials and resources which now go towards production of superfluous goods simply to keep 'effective demand' pumping and the monstrous economic machine going. If appropriate technology is appropriate for the Third World, it is even more essential as a substitute for the environmentally and socially obsolete high-technology in the developed world. But it is almost impossible to hope that the developed world will do this voluntarily. It will have to be forced to do so, either by a new unity of the Third World in the spirit of OPEC in the 1970s and early 1980s, or by an economic or physical collapse of the system.

2. It is in the Third World that the new ecologically sound future of the world can be born. In many parts of the Third World there are still large areas of ecologically sound economic and living systems, which have been lost in the developed world. We need to recognize and identify these areas and rediscover the technological and cultural wisdoms of our indigenous systems of agriculture, industry, shelter, water and sanitation, medicine and culture. We do not mean here the unquestioning acceptance of everything traditional in the over-romantic belief of a past golden age which has to be returned to in all aspects. For instance, exploitative feudal or slave social systems also made life more difficult in the past. But many indigenous technologies, skills and processes which are appropriate for sustainable development and harmonious with nature and the community are still integral to life in the Third World. These indigenous scientific systems have to be accorded their proper recognition, encouraged and upgraded if necessary.

3. Third World governments and peoples have first to reject their obsession with modern technologies which absorb a bigger and bigger share of surplus and investment funds, in projects like giant hydrodams, nuclear plants and heavy industries which serve luxury needs. We must turn away from the obsession with modern gadgets and products which were created from the need of the developed world to mop up their excess capacity and their need to fill up effective demand.

4. We need to devise and fight for the adoption of appropriate, ecologically sound and socially equitable policies for the fulfilment of needs such as water, health, food, education and information. We need appropriate tecnologies for agriculture and industry, and even more important we need the correct prioritizing of what types of consumer products to produce. We can't accept appropriate technology producing inappropriate products. We need technologies and products which are safe to handle and use, durable, fulfil basic and human

needs, and which do not degrade or deplete the natural environment and resources. And perhaps the most difficult aspect of the fight is the need to de-brainwash the people in the Third World from the cultural penetration of our societies, so that lifestyles, personal motivations and status structures can be delinked from the system of industrialism, its advertising industry and creation of culture.

5. Finally, whilst a new science for the masses cannot succeed unless there is an accompanying or preceding change in social structures, it is also true that a change in socio-economic structure alone is insufficient for developing a new sustainable order. Control and distribution of resources is a crucial determinant of social order but a change in this aspect alone is insufficient and could lead to similar problems without there being an understanding of the limits of resources and the environmental, health, ethical and cultural aspects of science and technology. Therefore there can be no meaningful reform of science without a change in society at large. There can also be no meaningful reform in social structures unless there is a change in the understanding of science and its proper application to serve the people and to be in harmony with nature.

Linkages between Economics and Modern Science

The linkages and interrelationships between economic factors and modern science work in both directions — economic factors have an impact on science and technology direction and policy, and science and technology aid in fulfilling the economic designs of dominant powers. It is only by conceptualizing this bi-directional loop that the full implications can be spelt out.

To put the discussion in perspective, reference is made to an incident narrated by Dr Richaria. During his time at the Madhya Pradesh Rice Research Institute (MPRRI), the institute was close to making an important breakthrough on developing High Yielding Varieties (HYV) of rice from local genetic resources. These varieties, while having the potential of dramatically increasing the output of rice, also required little artificial fertilizing. The International Rice Research Institute (IRRI) based in the Philippines, together with western interests, immediately referred to the World Bank. The bank characteristically offered a US$4 million loan to the Madhya Pradesh government for rice research on 'condition' that the MPRRI was scrapped! The MPRRI was indeed put on ice and Dr Richaria made redundant.

This example illustrates the general point that economic and political domination by the West has infused a love of western science and technology into the minds of Third World policy-makers, which makes them blind to any other equally viable alternatives. At the same time the prohibitive cost of converting science into technology, leaves only the option of importing technology from the West. Expensive, high-tech in particular, normally benefits only a small sector of the population mostly in the urban areas. Governments, in turn, are cajoled into providing expensive infrastructure to make this technology operational. The country's resources are spent on such projects and additional funds are borrowed to make good the foreign exchange shortfall, thus adding to the already crippling debt burden.

More insidious is the effect on the equitable distribution of wealth resulting from the anti-democratic nature of this technology. As has been demonstrated by events like The Green Revolution and Operation Flood, the impact of technology is pronounced in shifting resources from rural areas to urban areas, thereby reinforcing the disastrous process of the impoverishment of rural peoples. Such projects also require foreign inputs like fertilizers, insecticides, pesticides and irrigation machinery. These, in turn, add to the debt burden year after year, without providing any credible level of import substitution.

The impact of the economic domination on science and technology, particularly in the development of military technology is well documented. Indeed, western states have been referred to as military industrial complexes. Now, since more than half of the expenditure in the leading western countries is devoted to such efforts, the implications for western science and technology and their outposts in our countries are too frightening to contemplate. Further doses of such technology can only increase the Third World's subservience and dependence on the West.

If survival with dignity be the major objective, then all these linkages need to be understood and strategies for their unravelling put forward. A major plank of any such strategy should be the delinking of the Third World from the secular dynamic which institutionalizes the hegemony of the West. Also present should be a plan to cultivate confidence in indigenous values so that local creativity can flourish and evolve into a more relevant way of knowing and utilizing nature without the alienating, dehumanizing and inequitable characteristics of western science and technology.

Recommendations

1. Increasing awareness of the linkages between economic forces and the operation of science and technology. This should encompass:
   (i) Awareness of military research budgets and their intended recipients. The extension of such projects/grants into the Third World can be rejected by mobilizing popular opinion.
   (ii) Identification of prestigious high-tech projects which have little benefit for the masses, but which entrench western technology in Third World countries and increase the crippling debt burden at the same time.
2. Identification of major 'externalities' in science and technology programmes and projects. Thus, those that require excessive borrowing or result in perpetuating inequity can be made the targets of popular protest.
3. Science and technology policy-making institutions be infused with credible political scientists to help in working out the total impact of such policies.
4. Studies be initiated to chart out more equitable ways of knowing and

utilizing nature. A critique of western science and technology is a necessary pre-requisite for this task.

5. Ways to graft the critical concept of 'sustainable' use of resources on any emerging paradigm for utilizing and knowing nature.

Science and Hazards in Technology, Work Process, Products and High-Tech Holocausts

With the startling rise in the occurrence of major high-tech disasters all over the globe, the credibility of modern science has been seriously questioned. These disasters, which have taken place in capitalist and communist countries, and which are associated in the public mind with modern science and technology, reflect the crisis in this system of knowledge.

The Bhopal gas disaster, the Three Mile Island nuclear accident, the Chernobyl nuclear explosion, the Challenger tragedy, the Minamata tragedy of Japan, and now, the Sandoz-Ciba Geigy-BASF pollution of the Rhine river, have become the major turning points on the road of popular disenchantment with modern science.

The association of science and violence operates, however, at more conscious levels than may be seen in such industrial disasters. In the Vietnam war, for instance, science collaborated closely in an obscene military programme based on the use of chemical and biological weapons that eventually ended in ruining and destroying the entire living environment of the Vietnamese people.

Yet, none of these developments, horrendous as they are in their consequences, can match those arising out of the close and active alliance of modern science with the nuclear weapons industry and the nuclear arms race. The intensity of the planned violence inherent in this arsenal is not reduced just because it has not yet been used. With the expansion of modern science and technology, this madness has also passed on to Third World governments and the science they support.

Thus it can be seen that modern science has become the major source of active violence against human beings and all other living organisms in our times. This association of science and violence, as exemplified in what we have stated above, cannot be dismissed as being of an accidental nature; industrial disasters are not accidents, the barbarities of war cannot condone scientific activities in peacetime. Third World and other citizens have come to know that there is a fundamental irreconcilability between modern science and the stability and maintenance of all living systems, between modern science and democracy. In our view the very idea that science should be free of democratic control has been responsible for much of the violence associated with the scientific system in the past.

Recommendations

1. High-tech and large scale industrial units must henceforth be subject to democratic consensus in all countries. Opposition to undemocratically imposed science should be supported everywhere, across the globe.
2. Governments in developed countries should introduce legislation to forbid the export of hazardous products and industries which are banned in their own countries.
3. Third World countries should strengthen their capacity to monitor, regulate and control the import of hazardous industries and products. Legislation should be enacted to ensure:
• occuptional health standards equivalent to standards in the industrial countries;
• adequate controls over the various types of pollution and environmental degradation;
• that hazardous products are not allowed entry into the country.
4. Third World governments should provide political support to ideas, processes and institutions that utilize non-western science and technology.
5. Third World governments should pool resources and action to deal firmly with the perpetrators of all high-tech disasters.
6. A $1 billion fund should be raised by Third World governments to take care of victims of modern science, high-tech disasters, and this fund should also be used for the care of victims of the First and Second Worlds.

Science and Racism

Racial discrimination involves far more than racial prejudice or even inequalities of opportunity; it is an integral part of science, supposedly the most objective and neutral of human enterprises. The racial connotation of the IQ debate, generated by the work of Eysenck and Jensen, is now well known. The recent work of sociobiologist Edward Wilson is not as well known. Sociobiology seeks to promote a notion of humanity that sees intrinsic inferiority in the genes of certain racial and social groups. Eugenics, a 'science' which was abandoned in the early 1930s, is set to make a comeback. Sperm banks containing sperm of intellectually superior people have been established. Those with the means and desire to produce offspring of certain racial and intellectual purity will have access to appropriate sperms in the near future. Recombinant-DNA techniques are on the verge of cloning human beings; science has a habit of turning today's faction into tomorrow's fact.

In science education, racism starts early during secondary schooling. The dominant western model of schooling, by now adopted in most countries, is geared to classifying students for their future roles in the capitalist labour market. Although all students have the capacity to learn, their ability to learn depends partly upon their different cultural backgrounds. Today's competitive individualist structure of schooling widens the differences that students bring to school; it stratifies them along the lines of culture (race), as well as class and gender. In addition, teachers' (often unconscious) prejudices influence students' choice of school subjects and levels.

Thus, classroom practice and formal testing procedures turn out to define many students as being of low ability because they either cannot or will not submit to the criteria that the schools set for achievement. Expectations of these children are correspondingly lowered. Occupational hierarchies then appear as the natural result or as 'racial' differences in ability, diligence, etc.

At many workplaces, especially in the West, it is the cultural groups labelled as having lower ability who are relegated to the worst jobs or unemployment. Modern science and technology, as servants of capitalist power, have helped intensify that exploitation.

To take just one example, the micro-electronics industry promises us all the benefits of labour saving devices, yet it creates drudgery for some and unemployment for others. In the West is generates a two-tier labour force, in which Third World immigrants and their children are stuck in the bottom tier. In the USA, the mass production of integrated circuits on which the entire industry depends reserves the lowest-paid and most dangerous jobs for Asian and Hispanic workers.

Western multinationals have also exported that model, along with its chemical processes so hazardous that they would not satisfy western safeguards for health and safety. At their service, Southeast Asian governnments have eagerly competed to offer these multinationals the most attractive terms for exploiting their countries. The Free Trade Zone (FTZ) in Penang, for example, employs not the local unemployed but mainly rural Malay women, whose cultural traditions are manipulated in deference to the authority figures who enforce the rigid discipline of the factories. The deference is reinforced by legal restrictions on the workers' right to organize. In the absence of any effective requirements for protecting the workers' health and safety, many become unemployable after a few years. Thus, in the name of providing employment, Third World governments squander their countries' resources in collaboration with the multinationals.

Recommendations

Schools should oppose racist stereotyping in several ways:

1. Course content: science courses should not repeat the claims by some scientists that there are important racial differences that have a genetic basis but emphasize the scientific evidence against the existence of distinct races. History courses should teach race as a concept that was devised by imperialist countries to justify exterminating or exploiting so-called inferior races.
2. Structure of education: to avoid 'ability' labelling, the pedagogy should de-emphasize exercises based on memorization and rapid note-taking, in

favour of ones based on group discussion and co-operative problem-solving. Assessments of students should de-emphasize competitive timed tests, in favour of group assessments of how well the students have helped each other to learn.
3. Cultural differences: schools should maintain a respect for the cultural backgrounds of all students. At the same time, they should make students aware of how cultures have been shaped historically.

In the workplace efforts should be made to overcome the employers' power to manipulate cultural differences for dividing workers and for super-exploiting those considered inferior.

1. Wherever low-status (low paid and/or dangerous) jobs are associated with one cultural group, there should be efforts to upgrade those jobs and remove the stigma.
2. To facilitate solidarity, all restrictions on workers' rights to organize should be lifted, as well as all restrictions on their rights to express themselves in print or public assembly.
3. Employers should be forced to divulge any information relevant to workers' health and safety.

Science and Sexism

Modern science is based on and continually reproduces unequal relations which give white, western, middle-class men power. Behind the facade of the male, scientific and technological elite there is an invisible support network of women: assembling the equipment, cleaning the offices and the laboratories (often hazardous in itself), and providing material and emotional support.

Women's relatively disadvantaged position in this sexual division of labour within science and technology means that they have little say in the decisions within those worlds. Yet, in the 1980s, science and technology are playing ever more powerful roles in shaping their lives. For rural women, this can involve coping with the latest piece of machinery which may render her labour obsolete, ineffective or more difficult, or with pesticides which endanger her (and hence her unborn children), or her family; or she may cook by burning biomass, the smoke of which is a serious health hazard. In many cases, technological innovations either push women off the land or draw them into the factories of the cities. In others, the introduction of new technology has led to environmental pollution and deterioration and resource depletion. This strikes directly at their subsistence through pollution of water sources and denudation of forests and topsoil. Although the quality of life of the whole family is affected, it is women who bear the brunt of this suffering.

In urban settings, women's lives are often even more pre-structured by scientific experts. The designs of their homes and the range of domestic products at their disposal emerge from technologists who are largely unaware of and unconcerned with women's needs and interests. Nevertheless, in both rural and urban environments it is women who cope with the problems and disasters created by technology: dealing with droughts, shortages and famine, and caring for the victims of illnesses and accidents.

Gender-biased designs in the area of fashion, for example, have reduced women to the status of commodities. Most designers of women's apparel and footwear are men in the First World who deliberately accentuate the female body for their own pleasure and titillation. Through the mass media these fashions have been imitated and adopted by women in the Third World. Often this has been achieved at the expense of their health and nutrition, for example, factory women in Malaysia have been known to work overtime, forego their meals to earn and save money to pay for new clothes, shoes and cosmetics.

But science and technology also invade the privacy of women's lives. They find themselves used as guinea pigs in technological trials in contraception and birth control. For instance, women in Bangladesh and longhouse communities in Sarawak were given the injectable contraceptive Depo-Provera without being informed of its ill effects. Innovations in this realm, as in many others, are designed mainly by men, for profit, to be used on women. Their rights to control their own bodies have been usurped by medical expertise and the technocratic power of transnationals. High-technology monitoring of pregnancy and regulation of birth through ultrasound monitoring, mechanical foetal heart monitoring, amniocentesis are further features of the medicalization and dehumanization of women. In the case of the ultrasound monitor, it can be used to determine the sex of the foetus and has very often led to the aborting of female foetuses in patriarchal societies. This bias has been further enhanced with the advancement made in biotechnology today. Apart from this, dangerous drugs (e.g. psychoactive drugs), cigarettes and alcohol are promoted to women as a panacea for their problems which often arise as a result of their socially defined roles and through expectations of them in modern society. Food technology has also led to a decline in breastfeeding and infant nutrition with the introduction of infant formulae and weaning foods, which are both costly and hazardous. This craze for profits has led to thousands of infant deaths and untold suffering for millions of children and their mothers in the Third World.

In the field of mass communications, the advertising and movie industry has advanced the commoditization of women to an extent which is unparalleled in human history. Women have been reduced to objects of sex and violence in many insidious forms.

In addition to these immediate dimensions of women's encounter with science and technology, they also witness the sapping of resources for military technology. It must be understood that militarization and the buildup in modern nuclear armaments in the First World has been achieved at the expense of Third World resources, the plundering of which affects women first and foremost. Once again, decision-making in military science and technology is mainly controlled by men, while most women would not support the build-up of the arsenal.

Today, many women are protesting against the insidious and growing power of scientific and technological experts. They are challenging their hold over their lives.

Proposals

1. Women workers such as those in the new micro-electronics industries, are paying the price for this technological revolution. More stringent regulation of their working conditions (including health and safety measures requirements of shift work, etc.), and rates of pay are required. They should be guaranteed full workers' rights (including the right to join trade unions) and the support of other trade unionists. Elsewhere women should demand appropriately designed machinery and gadgets to suit their needs and laws must be enacted to ensure that this is carried out.

2. Governments should enact laws to safeguard the dignity of women. Advertisers and the mass media industry should not be allowed to promote or depict women as sexual objects or as neurotic patients in drug advertisements for psychotropic drugs.

3. In the era of reproductive rights, women must have more control of their own bodies. This necessitates, in the first instance, the banning of dangerous contraceptives such as Depo-Provera (many of which have already been banned in the West). It also means that resources should be devoted to non-chemical, and accessible forms of birth control. Women must no longer be subjected to the experiments of manufacturers keen on profit in this sector.

4. Related to this, women should have more freedom to choose the conditions of pregnancy and childbirth. This requires a firm control of medical and technological interventions in these processes and the encouragement of breast, rather than infant formula feeding. Laws should be enacted to protect women from these dangerous technologies.

5. Governments should ban infant formula and ony allow its use in exceptional cases. It should be made available on prescription like pharmaceutical products.

6. The sale and marketing of tobacco and alcohol should be restricted. High taxes should be levied on these products to discourage consumption.

7. There is a need to develop relevant unhazardous technology to lighten the burden of women, especially in the rural areas of the Third World. Such use of science and technology would make greater sense than the present allocation of resources on military technology.

8. There needs to be a strong representation of women in decision-making about science and technology. This should include all groups implicated in the struggles to transform the worlds of science and technology. This is to guarantee that they have a say in the construction of new forms of knowledge and

expertise appropriate to their lives; that products will be designed with their needs and priorities in mind.

9. The re-allocation of resources away from military technology towards immediate nutritional, housing and other social needs would effectively provide more support for women. As the primary care-takers most women are literally carrying the burden of this over-investment in militarized technology.

Science and Militarization

Increasing numbers of scientists and technologists from many countries are using their scientific knowledge for destruction whereas it is the responsibility of scientists to use their knowledge for the betterment of humankind. Such use of science has led to the invention of ever more effective weapons of destruction.

It is clear that the social, economic and political problems of the world cannot be solved by military means. All countries must refrain from using the threat of force or force itself to settle disputes, including disputes within national boundaries. No country should allow the military forces of another country on its soil, as in Afghanistan, nor should any country use its mercenary forces against another as is happening in Namibia, Angola, Mozambique and Central America. It is imperative that scientists refrain from participating in the militarization of science and that ordinary citizens do their utmost to reverse the militarization of our common world.

World militarization is accelerating at an exponential rate. Fifty million people are engaged in military activities worldwide, including 500,000 scientists and engineers in military research and development. This would account for approximately 20 per cent of the world's scientists and engineers engaged in military work during the 1970s. In the USA and UK 40 per cent of the scientists are engaged in military research.

Global expenditure on military research and development in 1980 was US$35 billion in 1985. Expenditures for all military facilities and activities reached US$900 billion in 1985. The US military budget alone is about $300 billion a year.

In the Third World military spending has increased by five times since 1960 and since then the number of countries ruled by the military has increased from twenty-two to fifty-seven. In 1980 the Third World spent $116,872 million on the military and only $40,827 million on health. For every physician there are twenty-five soldiers in developing countries. The worst use of modern science and technology in the Third World is the utilization by despotic regimes and local ruling elites of weapons developed by modern western science and technology to pacify their own people and preserve existing exploitative structures.

The biggest scandal in Third World development is that these countries utilize a large share of their nations resources and current budgets to purchase military weapons. Nuclear arsenals have the capacity to destroy every person on the planet many times over, and still nuclear weapons continue to be built every day. The superpowers have 50,000 nuclear weapons between them with a destructive power one million times that of the atomic bomb dropped on Hiroshima. Yet both superpowers have acknowledged officially that a nuclear war cannot be won because civilization as we know it would be destroyed. The most dangerous feature of the arms race is its increasingly destructive capacity which, when coupled with regional military confrontations, could lead to total nuclear annihilation.

Military advancement has been costly in terms of damage to the environment. Nuclear testing by both the USA and France has caused many inhabitants of the Pacific Islands to suffer from cancer, leukemia, and radiation sickness. Some of these islands have become totally uninhabitable through damage done by this testing.

We face the threat of not only nuclear but also biological and chemical devastation. The US Congress has just voted to start production of binary nerve gas weapons although this action is a violation of existing treaties. All nations, including the superpowers, should abide by existing treaties such as the Geneva Protocol of 1925 and the Biological Weapons Convention of 1972. These treaties have been signed by almost all the nations of the world and ban the use of any such weapons. Despite this ban, Iraq has used certain of these weapons.

Sponsored research programmes in academic areas such as political science, anthropology, biology, chemistry and physics have been used against the interests of Third World people. For example, the US Army sponsored the Pacific Bird Project in Bangkok to record the migratory pathways of many species of birds from their breeding grounds to their wintering grounds. This research was done by Asian civilian bird lovers who had no idea their work was for the US Army. This data could later be used to introduce diseases into certain areas, as research has also been done to see if birds can be carriers of disease. Insects and animals can be used in biological warfare to make people sick or to kill them. Toxic chemicals can also serve as weapons to attack people, livestock and crops. The US is actively developing new types of chemical weapons and is preparing to mass produce them. The US Congress is also being asked to fund dangerous testing of biological agents.

The US Army used bombs in the Vietnam war which destroyed all life in an area of more than 3,000 square metres by absorbing all the oxygen in the area. Bodies of hundreds of people who had died for no apparent cause, without wounds on their bodies, were found. The weapon (code-named Fuel Air Explosive) kills by making use of a chemical to deprive humans and animals of oxygen. Other chemicals such as Agent Orange were used by the USA to contaminate water systems and destroy forests and crops in Vietnam. These chemicals have been found to cause cancer and deformity of the foetus in those exposed.

Science has also been used for the torture of human beings in many countries of the world. This particularly obnoxious abuse of science must be condemned.

Military research has also employed cruel experiments on live animals to test the effects of these inventions.

Proposals

1. The world's political, economic and social problems are not solvable by military methods. There should be no military intervention by any country in any other country. This includes the placement of military bases.

2. Money and resources spent on military research and development should be diverted to finding means to fulfil basic needs.

3. Developing countries should not fall into the trap of purchasing armaments or accepting so-called military aid to suppress their own populations, to set up military bases, or to prepare for military conflicts with neighbouring states.

4. Efforts to establish nuclear-free zones around the world should be commended and supported. In particular the stand of the people and government of New Zealand in preventing the harbouring of nuclear warships should be commended.

5. The militarization of outer space, to which the present Reagan administration is resolutely committed, must be halted. We encourage the scientific community to joing the 6,500 scientists and engineers in the USA who have signed a pledge refusing to engage in Star Wars research.

6. There should be an immediate halt to all nuclear weapons testing in the Pacific.

7. There should be an immediate halt to torture and cruel experiments on animals for military purposes.

8. The scientific community should sign a pledge against research into chemical and biological warfare.

9. The USA and the Soviet Union should sign a treaty to abolish research and development of nuclear weapons.

10. India and Pakistan should refrain from devoting their limited scientific, technological and economic resources to their ongoing nuclear programmes. Let South Asia become a nuclear-free zone in every sense of the word.

11. The use of chemical weapons in the Iran–Iraq war should be condemned and stopped immediately.

12. Scientists of the world must stand united in pre-empting any attempt by South Africa and Israel to further their political ends through the superiority of their military technology, including nuclear power.

13. No military research should be done under cover of civilian institutions, either government or private. This applies to all countries without exception. Third World countries should be wary of any US government sponsored programmes in areas involving political science, anthropology, biology, chemistry and physics. The results of the research could be used for the purposes of developing military capacity, especially chemical and biological warfare.

14. Torture of any form must be condemned. Scientists must not offer their services in the creation or implementation of tools used in the torture of people.

Energy

Energy planning and consumption in the Third World is skewed to meet the needs of the elites and minority, while the energy needs of the majority is unfulfilled. For example, in Sri Lanka less than one per cent of the total population accounts for almost 45 per cent of the total domestic consumption of electricity in the country.

The energy consumption between the First and Third World is also very unequal. An average Indian for instance, consumes less than one thousandth of the energy consumed by an American.

Modern science and technology, which is based on the unscrupulous exploitation of the earth's limited resources, is also very inefficient. On the other hand traditional societies, which were based on agriculture and small scale industries, are efficient and less destructive. For traditional agriculture the energy efficiency is much higher than that of modern agricultural practices because they use freely available solar energy. In many cases, they are 50 to 250 times more efficient than the new technologies.

The western approach to nutrition which emphasizes animal protein is also highly inefficient in terms of energy. Animals consume ten times the food energy that they deliver.

Modern technology that consumes more energy than it produces makes up the deficit by exploiting Third World countries. This is done by extracting energy and resources from the Third World at very low prices. Third World countries that imitate the western models of industrialization are aiding the West in their own exploitation. Dangerous and polluting forms of energy like nuclear power should not be used. The use of 'peaceful' nuclear power has now led to a large number of countries having the capacity to build their own atomic bombs. Nuclear power plants are also potential military targets for hostile countries. With these plants spread all over the world, a conventional war can now easily be turned into a nuclear holocaust. The nuclear waste from the nuclear plants also creates radioactive contamination which can last for hundreds of years.

Recommendations

While looking for alternative development paths the Third World countries must develop a sound energy policy which is efficient and serves the needs of the majority. The first condition is that agriculture must remain an energy producing and not an energy consuming sector. Third World countries should embark on a project of developing and utilizing the renewable natural sources of energy, such as solar, wind, wave and biomass energy, in such a way that they are not amenable to monopolistic control.

The planning and development of energy resources must be based on an understanding and evaluation of the needs of the majority with top priority accorded to the basic need for dignity for all. In the first phase it should be done with low energy intensity growth so that there is sufficient time to develop the benign energy sources. Planning and forcasting energy requirements for a society should be done by identifying the basic human needs and these needs should be studied explicitly to estimate the energy required for their satisfaction. It is recommended that a Third World network be formed for energy policy, consisting of people who are prepared to work on a needs-based approach to energy planning.

Countries which have nuclear power plants should gradually phase out the use of nuclear power. In other countries, governments should be urged not to use nuclear power. Non-governmental organizations and the public should join the protest against the use of nuclear power and armaments.

Agriculture

After serious examination and detailed deliberation on the status of agriculture in the Third World this conference has come to the conclusion that:

> The resources, techniques and practices of the Third World in the fields of agriculture, irrigation, forestry, animal husbandry and fisheries are in immediate danger of being wiped out under the impact of policies favouring modern western practices in all these areas. The new practices that are being introduced are inherently incapable of efficiently utilizing the resources available in the Third World countries and sustaining the Third World populations. These practices are adversely affecting the productivity of land and are destabilizing the ecological balance. They are worsening the dependence of the Third World on the industrialized countries for knowledge, techniques and inputs in areas where indigenous knowledge and resources are capable of, and were till recently, meeting the needs in a satisfactory fashion. Moreover, logging and forest degradation has also caused tremendous soil erosion in Third World countries. Agriculture and connected practices being the major way of life of the Third World societies there is an urgent need to defend the indigenous knowledge, practices and resources in these areas from the western onslaught. We therefore recommend that the Third World countries should make joint efforts for the restoration of the indigenous way of life in these countries. Meanwhile the following steps should be undertaken as emergency measures:

1. A Third World Agriculture Documentation and Research Centre should be established to document the indigenous science and practice of agriculture and

irrigation in the countries of Southeast Asia. The centre should also collect information on the available agricultural and irrigation resources in these countries, and work on finding ways and means to put these resources to the best possible use through indigenous know–how and practices. The Centre may have its headquarters in Sri Lanka and a regional office in each of the countries in South and Southeast Asia.

2. A Third World Foundation should be established for the preservation of the agricultural genetic resources of the Third World. The Foundation should:

  a. Persuade and help the governments of South and Southeast Asian countries to take steps for the collection and preservation of the agricultural genetic resources in their countries.

  b. Provide technical and financial help for the immediate establishment of long-term storage facilities for the genetic resources in these countries.

  c. Encourage research on the indigenous methods of collection and storage of genetic resources prevalent in these countries.

3. The governments of the Southeast and South Asian countries must be persuaded to use only the tree-species that are indigenous to this region in their afforestation programme. To help in this, a Tree Exchange should be established. The exchange should prepare a directory of the tree-species in the region of South and Southeast Asia, and help the various governments in obtaining the appropriate species from the region for their afforestation programme. The Tree Exchange could be established in Malaysia or Indonesia.

4. A regional livestock organization for the region of South and Southeast Asia should be established. The organization should prepare a data base of the livestock breeds of the region and provide facilities for exchange of livestock breeds within the region. The governments of the region should be encouraged to restrict cross-breeding of livestock within the breeds available in the region.

5. The Consumer Association of Penang and Third World nations should be requested to establish immediately a commission to consider the impact of the Green Revolution in South and Southeast Asia. The commission should study:

  a. The origins of the policy of introducing western technologies and exotic seeds in the countries of the region.

  b. The effect of the new technologies on the self-dependence of the countries in matters of agricultural knowledge and inputs.

  c. The effect of the new technologies on the productivity of law in these countries.

  d. The socio-economic impact of the new agriculture technology and practices.

  e. Implications of the new technologies for the genetic pool and environment of the region.

6. CAP (Consumer Association of Penang) and TWN (Third World Nations) should be requested to establish immediately a commission to investigate the

impact of the introduction of modern dairy practices in South and Southeast Asia. The commission should study:

  a. Origns of the policy of introducing western dairy technologies in these countries.

  b. Effect of the new technologies on the self-dependence of these countries in matters of dairying.

  c. Expenditure made on dairying infrastructure in these countries.

  d. Effect of the new practices on the production of milk and meat in these countries.

  e. Effect on the quality of meat and milk.

  f. Effect of the new practices on the genetic base of indigenous livestock.

  g. Socio-economic impact of the new practices.

  h. Impact of the fluctuations in the European dairy market on dairying in these countries.

7. CAP and TWN should be asked to explore the possibilities of establishing commissions to study:

  a. The impact of the introduction of modern fishery practices in the region of South and Southeast Asia.

  b. Impact of large dams and irrigation practices on soil, water, tables, water drainage and health in the countries of the region.

8. CAP and TWN should be requested to explore the possibility of a regional organization for looking after the consumer interests of the region of South and Southeast Asia in the area of agriculture, food and drugs. A watch must be put on the activities of agribusiness in this area, and should collect and convey information of all new agricultural inputs (chemical, biological and mechanical), drugs and food products entering the area. For this purpose it will be necessary to establish a network of similar organizations in all countries of the region. The regional organization will co-ordinate information gathered in these countries and convey it to other countries in the region.

9. All these steps should be undertaken by people in the region and with resources from the region.

Health

The western medical health system has infiltrated all societies in the world. Its apparent success is not due to its scientific credibility, but due to its aggressive salesmanship. In fact it has had incredible failures and perpetuated serious crimes, but all this is hidden from the public by the bribing of governments and medical personnel. The medical industry is a powerful political force in the modern world; health care is no longer concerned with the health of the people. Growth in the area of pharmaceutical products has led to the proliferation of useless and dangerous drugs in Third World countries. For example, in Bangladesh before 1982 some 1,700 worthless medicines were available. Presently in Malaysia, there are over 25,000 preparations on the market.

Similarly, health education in the Third World is merely a copy of the western model with an emphasis on urban, curative health care. These systems have an unquestioned reliance on medical technologies which are imported to great cost. At present, health research is being carried out on 'exotic diseases' which absorb massive amounts of resources and funds which have to be diverted from the more serious diseases which affect the majority of people in the Third World. For example, increasing amounts of money is poured into Aids research, while the search for a cheap and readily available form of immunization for Hepatitis B which is a serious disease afflicting Asian populations, is neglected. Research and development in tropical diseases like malaria and diarrhoea is neglected because it is not lucrative.

Health for the majority of the peoples of Asia, Africa and Latin America and for growing number of marginalized people in western nations has become a serious problem. It is adversely affected by the poverty and insecurity that has developed during the last two centuries. There is sufficient evidence to prove that western science and technology, intrinsically wedded to the capitalistic ethic, have been the main ideological and material instruments in the destruction of the self-reliance of indigenous cultures all across Asia, Africa and South America.

Highly decentralized indigenous and self-reliant health care systems based on the use of locally available plants, animals and minerals have also been suppressed. Over thousands of years they have developed a widespread folk culture that deals with primary health care problems, and forms of this culture survive even to this day in Asia, Africa and Latin America. The most significant feature in these systems is its autonomous and self-reliant nature supported by an oral tradition of knowledge. In many Asian societies there also exists a comprehensive indigenous science which has its empirical roots in folk traditions. There is evidence that the indigenous health traditions carry the potential of making communities and particularly rural ones in the Third World, entirely self-reliant in their primary health care needs.

There is an awareness today in non-western societies of the falsity of the myths about the poverty of their own cultures that has been inculcated by the western world. There is a new awakening to the strength and potential within their own roots. In the young generation the intoxication with the West is more or less coming to an end, particularly after their being witness to the terrible destructions that the modern civilization has brought in its wake. In the area of health the biggest challenge before the non-western world is to revitalize its own indigenous health care system. This is a long-term work that calls for devoted and steady efforts, because many 'current' weaknesses due to the 200 years of suppression have to be overcome.

In the immediate transitory phase an urgent task is to try and rationalize the western system of medical care that has been so exploitative of the people.

Proposals

1. Third World governments should adopt a rational drug policy based on the WHO policy on essential drugs, which recommends that a total of 200 drugs would be more than sufficient for a country's drug requirement.
2. Related to this policy, it should ban the use of dangerous and inefficacious drugs.
3. It should promote the use and manufacture of generic drugs.
4. The above should be important elements in the formulation of a national health policy.
5. This health policy should also incorporate the use of indigenous medicine.

On Revitalizing Indigenous Health Traditions

1. To document urgently the existing state of health traditions, the ailments they claim to treat, the medicinal plants, minerals or animals parts that they use, and social ethos in which they function.
2. To evaluate the indigenous traditions with the help of indigenous sciences and not western medical science. The western system has entirely different principles, concepts and categories and so is not competent to interpret indigenous traditions
3. To lobby with WHO to establish Asian institutes for the research and study of traditional medical sciences.
4. To promote at local levels thousands of herbal gardens and medicinal forests throughout the Third World.
5. To create networks amongst non-governmental organizations to facilitate scientific exchange on various theoretical and practical aspects of traditional medical sciences.

Telecommunications and Micro-Electronics

Micro-electronics, telecommunication and new information technology are seen by many as the most significant scientific development in the second half of this century. Technological developments in this sphere have led to predictions of an 'information revolution'. Some technocratic 'futurologists' claim that this new technology could democratize work and society. But all evidence suggests that the micro-electronics industry is moving towards a tremendous concentration of capital, knowledge and power in a small number of countries and co-operations.

Rather than eradicating unpleasant work, micro-processors have led to a degradation of skills, dignity and job interest for the majority of those working with new technology. Only a few people, based in the developed countries, are engaged in the creative aspects of this work, with their results affecting millions of others. Control of micro-electronics research and development, product development manufacture, and the marketing of telecommunications services is perhaps more concentrated in the hands of a small number of powerful transnational corporations than most other major industries. Research and development of microchips is concentrated in the United States and Japan, based on standards, codes and concepts which are wholly western in origin. Even European corporations are now finding themselves unable to compete with the US and Japan in the capital-intensive area of micro-electronics research and development.

Most of the companies involved with electronics and telecommunications research and development are highly dependent on military contracts for a large proportion of their work, and thus they collaborate closely with the military establishment at the developmental stage and give priority to military uses of the technology. Many of the civilian applications of micro-electronics result only indirectly from this military work, and take second place when there are conflicting interests.

Unlike the capital-intensive research and development phase, some parts of micro-electronics production and labour-intensive. Thus, although controlled by small number of corporations, the actual production has been spread around the world to take advantage of cheap, unorganized labour, and the various financial incentives offered by governments eager to attract foreign investment. The stages of electronics manufacture located in developing countries generally offers few opportunities for transfers of skills to workers or transfer of technology. These manufacturing plants are highly mobile, and, as conditions change, can be relocated or closed down without serious loss to the company. Electronics manufacture has now emerged as one of the least stable forms of foreign investment in developing nations.

Despite the clean image of the industry, workers in electronics factories are exposed to a very large number of hazards from a rapidly expanding range of highly complex chemicals, the health effects of which are only just beginning to be documented. In addition to chemicals which cause cancer, reproductive problems and poisoning, the nature of electronics manufacture can subject workers to eye-strain and stress.

In the application of electronics to telecommunications, the control and domination persists. A very small number of corporations design and deploy satellite and undersea cable systems, while another handful of corporations operate the telecommunications services on these systems. These companies are almost entirely North American and European. The cultural biases inherent in the design and operation of these international communications systems in such matters as language, conceptual codes, short-forms, clearly enforces a standardization and conformity on users which ultimately changes the contents of communication, not only the form.

The major telecommunications systems are structured to ensure that information useful to corporations banks can be extracted from developing nations and used to gain competitive advantage. The most efficient communications channels run throughout the North, and link the South with the North, but do not link the developing nations to one another. With these, corporations and banks can swiftly transfer financial, economic, political and labour information to enable them to out-manoeuvre competitors, and destroy political opposition to their plans. With this technology in place, national governments are losing their power to regulate the activities of transnational banks and corporations in their own countries. As Dieter Ernst has said, these new computer, communications and control technologies have made it possible for managements to 'synchronize, on a worldwide scale, decentralized production with a strictly centralized control over strategic assets'.

The control and structure of the world's telecommunications system also assists the penetration of western entertainment, news, and advertising to the South, thus further asserting western cultural and political values, and expanding the market for western goods.

Just as companies use the telecommunications systems to amass business information, national governments are able to abuse this information technology for social control in policing, surveillance, and the creation of highly efficient data bases on opposition groups and dissidents, thereby violating rights to privacy and laying the groundwork for repression on a massive scale.

As with the production of the telecommunication equipment, the operation of telecommunications services seeks to maximize profits, and thus the organization and management of telecommunications work takes inadequate precautions against health hazards for those workers operating the system, such as radiation dangers and eye-strain from video display screens and stressful alienating work situations.

Proposals

1. Some nations, such as Brazil, have reserved their micro-electronic production for local companies only and restrict the import of electronic goods, thereby developing an indigenous micro-electronics industry. This greatly reduces dependence on foreign goods and can create products more appropriate to local needs and conditions. Efforts by developed nations, led by the United States, to block such moves by demanding the opening of these markets as part of trade agreements should be resisted.

2. Governments should be urged to accept that far from being a clean industry, the complex and sophisticated chemicals and processes used in the electronics industry present severe health hazards, the full extent of which have yet to be realized. Measures should be taken to identify and control health hazards in high-tech industries.

3. Scientists, engineers and designers in the Third World should become highly conscious of the values implicit in micro-processor technology, and strive to expose them, and where possible, eliminate these through imaginative reprogramming and redesign.

4. The control of the micro-electronics industry by a small number of key transnationals is based on a monopoly of technology and know-how. Third World governments and non-governmental groups should look critically at the present criteria for copyright and patent laws imposed upon them by the developed nations, and see if more balanced alternatives cannot be put in place, particularly in such priority development areas as health, housing, education and food production.

5. Non-governmental organizations must make every effort to ensure that the access to the latest telecommunications systems does not remain the exclusive right of the foreign corporations, local corporations and governments, but should also be made open to non-profit, non-governmental organizations and individuals for the free exchange of ideas.

6. At the same time, Third World governments must take measures to ensure that telecommunications systems are not abused by transnational corporations to transfer funds rapidly from country to country thereby causing economic instability, or to amass and transfer information with which they can exert unfair influence or control over local economies, national governments, employees, or local companies.

7. Third World nations should implement registration and control of data bases which ensure that neither companies nor governments themselves can amass information about individuals which can subsequently be used to violate their right to privacy, curtail their civil and political rights, or to gain unfair commercial influence over them. At the same time, any such law controlling computerized information should not be phrased so as to allow governments unrestricted access to information held by non-governmental or opposition groups.

8. In order to reduce the gap between the powerful and the powerless, efforts by grassroots groups to learn about information technology, adapt it, and where possible set up small-scale decentralized and democratic alternatives should be encouraged. However, when applying this technology, all organizations should consider carefully the health and hierarchical effects of any equipment they introduce.

Appropriate Technology and Industrialization

Appropriate technology and industrialization is that which advances development, where development is looked upon as a process which leads to a life with dignity through satisfaction of basic needs, starting with the needs of the neediest; self-reliance; and non-destructive harmony with the environment.

There are three aspects of appropriate technology: its selection; generation and dissemination. Selection is from the available pool of technologies covering the whole spectrum from the technologies of the industrialized countries (which should be scrutinized with utmost care regarding their tendencies to amplify inequalities, undermine self-reliance, destroy the environment and promote violence) to traditional technologies (either in their pristine form or after transformation). In all these three aspects of selection, generation and dissemination Third World networking has a crucial role to play. Co-operative selection of appropriate technologies is essential, particularly when the Third World is confronted with the technologies of the industrialized countries.

The scope of appropriate technology should cover small-scale, decentralized and large-scale, centralized technologies. Attention should be focused both on hardware and software, and on all sectors (not only industry, but also agriculture, health, transport etc). However, there should be a specific emphasis on appropriate technologies that satisfy (in a self-reliant, ecologically sound way) the basic needs of food, shelter, clothing, health, education, particularly of the neediest sections.

Proposals

1. Developing countries should evolve their own appropriate technologies and their own techniques in various areas — agriculture, industry, health, care, housing, water management, transport, energy and so on. It is recognized that the appropriateness of a particular technology or product would also partly depend on the conditions of a particular country. Such appropriate technologies should as far as possible make use of local resources. They should be relatively simple to operate with skills which can be passed on, based on sound ecological principles and of a small scale suitable for family or community use.
2. Appropriate technology should be a policy not only for the Third World but also and especially for the industrialized countries. These rich countries should not continue to operate capital intensive technologies, producing relatively luxurious or superfluous products, which use up the world's resources. If the Third World adopts resource-efficient appropriate technologies but the rich countries continue to use resource-wasting industrial technology, then the present unequal distribution of resources and power will be perpetuated. Thus concerned individuals and groups in the industrialized countries should make efforts to convince their people and governments of the necessity to change their technologies and techniques of production and reorientate their approach to science accordingly.

Science Education

Science education in the Third World is a colonial legacy, rooted in the western system of education and has no relevance to our societies. It was designed to create a cadre of workers whose job was to carry out programmes planned and designed in the West. The result of this unfortunate legacy, which has not been completely abandoned in a majority of the Third World countries, is that the efforts of Third World scientists and technologists are merely extensions of the programmes of their western mentors.

This situation has been worsened by the exodus of a large number of scientists and technologists from the Third World. Those going to the so-called advanced western societies have returned home with reinforced western ideas, exacerbating the problem of indigenous science rather than alleviating them. A majority of these students and researchers, of course, never return at all and this contributes heavily to the depletion of the poorer countries' resources in the form of the brain-drain. Even those who do manage to come back to their home countries usually continue their scientific research along lines established in their doctoral theses carried out under the tutorship of a foreign scientist. In other words, foreign-trained scientists are the greatest germ-carriers of the western virus against which our societies are seeking immunity.

In order to remedy this situation, it is essential that the education of scientists and technologists in the Third World takes place in such a way as to ensure that they not only retain their cultural and social moorings, but that their scientific interest is maintained in solving problems pertinent to their indigenous environment. Clearly such a system needs to be based on a system of education which appreciates the value of indigenous scientific and technological culture.

The realization of this goal necessitates re-shaping the science syllabus at school and university level as well as dictating a conceptual change in the framework of science teaching itself. The goal of scientific education is to produce an imaginative and dedicated personality, of creating an individual who is both resourceful and responsible. The teaching of science therefore should never be divorced from the value-system of the indigenous civilization. The students should also develop a critical faculty so that they may judge the cultural and ieological bias of western science and technology.

Proposals

1. Science students should learn social realities — economics, politics, culture — particularly about the domination of the Third World in the world system.
2. Science students should be given a thorough exposure to the relations between science and its effects on society, including its potentially harmful effect. Ethics of science, especially the social responsibility of scientists, should be central to the education of the scientist.
3. Science students should be made aware of indigenous science and its roots — to study the elements of indigenous medicine, shelter, food, industry, transport.
4. Science should not be for the elites but directed towards farmers, fishermen, workers, etc. to help them improve the livelihood of the people. Science education should be spread to the masses.
5. The ecological and environmental aspects must be central in science education, expecially the interrelatedness of various natural elements. Students and researchers should focus attention on the destructive nature of man's

activities on the environment and the ways of avoiding these and they should seek to rehabilitate nature whenever possible.

6. Science research priorities should be given to the identification of positive indigenous technologies, scientific values and systems, knowledge and processes on various spheres like agriculture, industry, medicine, shelter, etc. These systems should be defended and improved.

Science Policy and Management

Only a handful of developing countries have explicit science policies geared towards the creation of a scientific and technological infrastructure in the country. Often this infrastructure has been developed at the expense of national independence with the new science and technology institutions in the Third World becoming an extension of scientific establishment of the industrialized countries. Moreover, countries with explicit science policies have tended to focus on prestige areas of science and high-technology projects. Much of the established infrastructure has thus tended to be irrelevant to the needs and requirements of the country.

The majority of the Third World countries do not have a declared science policy. However, an implicit science policy is in operation everywhere. The overall emphasis is on the transfer of technology, establishment of fashionable research centres through technical and financial aid, and the use of foreign consultants in solving local problems. Within this framework of explicit and implicit science policies, a management structure that relies on hierarchy and one directional, top to bottom, communication, and a suffocating bureaucracy has been adopted. This management structure has isolated the decision-makers from the rank and file scientists and technologists as well as from the local working conditions and work environment.

Proposals

1. Governments and scientists in the Third World should review their present bias towards modern, capital-intensive technology, mainly imported or imitated from the West. A profound understanding of the inappropriateness and destructive nature of these technologies should be fostered in Third World national science policies.

2. Science policies in the Third World should focus on establishing an indigenous base for the generation, utilization and diffusion of science and technology. They should not promote the transfer of technology or a reliance on scientific research done in the industrialized countries, or on technical assistance or on foreign consultants. They should aim to promote and upgrade traditional and modern indigenous sources of knowledge and know-how. Our cultural environment, from the tribal to the civilizational level, is suffused with potential technologies, scientific insights and methodologies that could and

should be used to provide a necessary organic linkage with our own roots. New management structures that take into account that science is a political and social process should be encouraged, initiated and institutionalized.

3. Today's science and technology are closely interlinked and the new technology is almost directly science based. In order that Third World countries break out of their dependent condition they should seek out and establish cross-linkages with each other. Such cross-linked groups would in partnership generate a science and technology which is self-reliant, basic needs oriented and ecologically sound.

4. The new search for science and technology should be pursued by encouraging viable groups which are both socially conscious as well as aware of their own discipline. Mechanisms should also be evolved to generate critical masses around such groups so that a creative and society-oriented science and technology is produced. Mechanisms should also be evolved to insulate such socially aware, creative groups from the pressures of various vested groups, while they are encouraged to develop linkages with the mass of their people.

Index